T0136475

Re-Creating Nature

Re-Creating Nature

SCIENCE, TECHNOLOGY, AND HUMAN VALUES IN THE TWENTY-FIRST CENTURY

JAMES T. BRADLEY

The University of Alabama Press
Tuscaloosa

The University of Alabama Press
Tuscaloosa, Alabama 35487-0380
uapress.ua.edu

Typeface: Minion and Optima

Cover image: Gold printed circuit-like bonsai, abstract electronic scheme;
courtesy Martin Janeček
Cover design: David Nees

Library of Congress Cataloging-in-Publication Data
Names: Bradley, James T., 1948 - author.
Title: Re-creating nature : science, technology, and human values in the
twenty-first century / James T. Bradley.
Description: Tuscaloosa : The University of Alabama Press, [2019] | Includes
bibliographical references and index.
Identifiers: LCCN 2019003451| ISBN 9780817320294 (cloth) |
ISBN 9780817392437 (ebook)
Subjects: LCSH: Biotechnology --Moral and ethical aspects.
Classification: LCC TP248.2 .B73 2019 | DDC 660.6 --dc23 LC record available
at https://lccn.loc.gov/2019003451

For my grandchildren and their descendants

Contents

Illustrations

TABLE

Preface

Humans are at the cusp of taking control of their own future evolution and possibly that of the entire biosphere. Available now are biotechnologies capable of facilitating the designing of future generations of *Homo sapiens*. Along with forever changing humans as we know ourselves today, scientists and entrepreneurs are close to creating, re-creating, or selectively exterminating specific species, both wild and domestic. Ongoing national and international brain research projects may finally provide an explanation for human consciousness itself and then offer the means for cognitive enhancements and other alterations that go far beyond current pharmaceutical adjustments to brain function. This new knowledge about the human brain coupled with advancements in nanotechnology, computer science, artificial intelligence, and robotics has some thinkers already suggesting that by the middle of this century the line between fully human and machine will have blurred to the point that some autonomous robots will seek personhood and some humans will commit themselves rationally and emotionally to robots as lifelong partners. How should we view ourselves and the entire living world as we engage with this brave new world of biotechnology? Recognizing the deep antiquity and interconnectedness of all forms of life through thousands of millennia is a good starting place.

Earth's present biosphere, the planet's thin crust containing living organisms, evolved over the past 3.8 billion years. Present human nature has been evolving through the past three million years, the approximate age of our genus *Homo*. Evolution produced a web of interdependence between all living things, from microbes to mussels, conifers to chickadees, and between life and the planet's physical and cyclical attributes. Stunning examples of this interdependence are all around us on grand scales, regionally localized scales, and underfoot. Green plants, including photosynthetic algae, use sunlight, water, and carbon dioxide to build their cells and tissues. The by-product created by that activity is the oxygen needed to sustain animal life—from amoebae to human beings.

Recently, I witnessed the astounding convergence of nearly half a million migrating sandhill cranes in and around the Platte River near Kearney, Nebraska. Their annual spring migration from Texas and Mexico to Canada, Alaska, and Siberia funnels them through just one or two counties in America's heartland. Here they spend two to three weeks foraging in the fields by day and roosting on the river's shallow sandbars to avoid predation by coyotes and foxes by night. Two-year-olds perform mating dances to establish pair bonds that endure for their twenty-year lifespan. How long have these cranes been migrating and dancing? Nobody knows exactly, but Nebraskan rock contains ten-million-year-old crane fossils. Closer to home, your abscessed tooth, lunchtime yogurt, and petrol for your car would not exist were it not for the action of anaerobic bacteria whose ancestors dominated the earth billions of years ago. More than half a billion years ago, single-celled microorganisms at the bottom of seas located in what is now the western United States and the Middle East decomposed the dead bodies of plant and animal life that sank into their deep abode. Products of that decomposition gave rise to oil and natural gas. In fact, every year humans harvest, by drilling and pumping, the product of about one million years' worth of recycled plant and animal bodies.

The point is that all life and the physical aspects of our planet are interconnected in vastly complicated, ancient, interlocking webs of dependency that humans may now be equipped to manipulate at will. The future of the biosphere and of human nature itself, however we choose to describe it, depends largely on how we develop and use modern biological technologies. As eloquently stated by Renaissance philosopher Pico della Mirandola in his *Oration on the Dignity of Man*, humans are distinguished from other animals by being choice makers.[1] In his *Oration*, writing mainly for a fifteenth-century Jewish and Christian audience, Pico had God telling the first humans how they differ from other creatures: "The nature of all other beings is limited and constrained with the bounds of laws prescribed by Us. Thou, constrained by no limits, in accordance with thine own free will, in whose hand We have placed thee, shalt ordain for thyself the limits of thy nature . . . with freedom of choice and with honor, as though the maker and molder of thyself, thou mayest fashion thyself in whatever shape thou shalt prefer. Thou shalt have the power to degenerate into the lower forms of life, which are brutish. Thou shalt have the power, out of the soul's judgment, to be reborn into the higher forms, which are divine" (Pico della Mirandola [1486] 2005, 287).

Choices we make now will determine not only the kind of world our children, grandchildren, and their children will live in but also the kind of humans they will be. The choices encompass a broad range of decisions that modern biotechnologies press upon scientists, nonscientists, and policy makers. They range from choices about research funding and projects to decisions related to learning, teaching, marketing, buying and using, voting, and personal life styles and goals. This book's primary objective is to stimulate informed thought and conversation about current and future scientific endeavors, their present and potential applications, and the many ethical and social issues emerging from them.

Difficulties of and opportunities for choice making for specific technologies appear within six chapters focused on specific topics, including human stem cells and embryo research, embryo selection, human genetic enhancement, synthetic biology, genetically modified organisms (GMOs), genomics and gene editing, nanotechnology, human brain research, and robotics and artificial intelligence. Each of these technologies is already the object of many books, university courses and programs, and specialized professional journals. Coverage of these topics here is sparse by comparison due to space limitations and the book's objective of provoking thought.

Bracketing the six biotechnology chapters are an introductory chapter about genes, molecules, cells, and humanity's place in nature (chapter 1) and an examination of responsibilities that scientists and nonscientists must fulfill for humankind to thrive in the age of biotechnology (chapter 8). The book concludes with a discussion on the importance for humankind to identify long-range goals for itself and how best to prepare for the process of goal selection (chapter 9). In chapters 8 and 9, I offer my personal views on how humans can best nurture good living in the era of biotechnology. In all other chapters, I refrain from giving personal opinions on ethical issues in favor of presenting a variety of positions on the issues and encouraging the reader to develop her/his own informed views. Each chapter concludes with a set of questions to stimulate personal thought and group dialogue. The questions have no correct answers but aim to fuel the public discussion needed to shape wise policies and decisions at personal, state, national, and international levels.

Acronyms like DNA, RNA, and CRISPR are defined when they first appear in the text. For easy reference when they appear later, a list of all acronyms used in the book with their definitions is included. Notes and sources for additional information accompany each chapter, and appendices supplement chapter information on DNA function, neurotechnologies, and sources of scientific information for nonscientists. A comprehensive reference list and detailed index complete the book.

The book's audience is all who wonder at nature and contemplate humanity's place in it. As I wrote, I held in mind those who marvel at the beauty and complexity of life itself—from the smallest microbe to the maze of intersecting ecosystems that comprise Earth's biosphere—while at the same time viewing humankind's burgeoning abilities to manipulate and redesign life with a tangled mixture of hope, fear, optimism, resignation, and courage. I have written with the goal of making science enjoyably understandable for inquisitive, self-informing nonscientists. Students and colleagues in the natural sciences, humanities, social sciences, and engineering should also find plenty to chew on in these pages.

Note

1. For a discussion of the relevance of Pico's ideas for our modern age of biotechnology, see the companion volume on this subject, Bradley, *Brutes or Angels: Human Possibility in the Age of Biotechnology* (Tuscaloosa: University of Alabama Press, 2013), 1–2.

Acknowledgments

Many persons contributed to this project with stimulating conversation, suggestions, encouragement, and other tangible help. Foremost among these are my musician wife, Sue, who helped to ensure that the information I strive to share and the questions I pose are compatible with the questions and insights that nonscientists may have. Other family members contributing valuable ideas and encouragement to write about biotechnologies and ethics include my late parents, Donald L. Bradley and Marjorie Bradley Scholl, Marilynn Bradley, Sara and John Compaglia, and Laura and John Bodt. I benefited from illuminating conversations with numerous friends and colleagues including Raymond Allar, Kenneth Cadenhead, Gerard Elfstrom , Michael Hein, Clark Lundell, Gary Mullen, Randolph Pipes, Michelle Sidler, Andrea Vazsoniyi, Frank Werner, members of the adult study group known as the Library Class at First Presbyterian Church in Auburn, Alabama, members of the two adult classes at Auburn First United Methodist Church and the Auburn OLLI program that studied *Brutes or Angels*, my previous book on biotechnology and ethics, members of Accademia della Crusca in Auburn, members of the Thinking Man's Book Club in Auburn, my genethics course students at Auburn University, and participants in Sunday morning forums at the Blue Hills Unitarian Universalist Fellowship in Rice Lake, Wisconsin. I am further indebted to Kenneth Cadenhead, Christine Schnittka, and Lana Saffert, who gave their time to read and comment on specific chapters, and to Sue Bradley and Susan Harris who read and edited the entire manuscript. A special thank you goes to Elizabeth Motherwell, senior acquisitions editor for natural history and the environment at the University of Alabama Press for her encouragement, patience, wisdom, and work with this project. Unless credited otherwise, the illustrations were created by designer and photographer Janna Sidwell (www.facebook.com/jannaclairesidwellphotography/). Her patience, professionalism, dedication to the project and artistic talent are largely responsible for making the scientific concepts accessible to nonscientists. Finally, my deepest gratitude goes to my companion in life, Jackie Sue, for encouragement, countless conversations about science and ethics, and her patience and endurance during my geographical and mental absences while writing.

Abbreviations

AAAS	American Association for the Advancement of Science
AI	artificial intelligence
ALS	amyotrophic lateral sclerosis
AMA	artificial moral agent
ATP	adenosine triphosphate
BRAIN	Brain Research through Advancing Innovative Neurotechnologies
BRCA1 and BRCA2	breast cancer proteins 1 and 2
BRCA1 and *BRCA2*	breast cancer genes 1 and 2
BTBI	brain-to-brain interfacing
Cas9	CRISPR-associated nuclease number 9
CBP	China Brain Project
CFRB	Coordinated Framework for the Regulation of Biotechnology
CRISPR	clustered regularly interspaced short palindromic repeats
CRISPR/Cas9	designates the gene-editing technology employing elements of the bacterial CRISPR system
DARPA	Defense Advanced Research Projects Agency
DNA	2-deoxyribonucleic acid
EEG	electroencephalogram
ELSI	ethical, legal, and social issues
EPA	Environmental Protection Agency
ESC	embryonic stem cell
FDA	Federal Drug Administration
fMRI	functional magnetic resonance imaging
GEO	genetically engineered organism

GMO	genetically modified organism
GRIN	genetic, robotic, information, and nanotechnology
HBP	Human Brain Project
HGP	Human Genome Project
HEP	Human Epigenome Project
hESC	human embryonic stem cell
HSCs	hemopoietic stem cells
Hz	Hertz
ICM	inner cell mass
IHGSC	International Human Genome Sequencing Consortium
IVF	*in vitro* fertilization
iPS	induced pluripotent stem
MAD	mutual assured destruction
MEG	magnetoencephalography
MEMS	microelectromechanical system
MIT	Massachusetts Institute of Technology
MRI	magnetic resonance imaging
NAS	National Academy of Sciences
NASEM	National Academies of Science, Engineering, and Medicine
NIBC	nanotechnology, biotechnology, information technology, and cognitive science
NIH	National Institutes of Health
NNI	National Nanotechnology Initiative
NSABB	National Science Advisory Board for Biosecurity
NSF	National Science Foundation
OMIM	Online Mendelian Inheritance in Man
OSTP	Office of Science and Technology Policy
PET	positron emission tomography
PGD	preimplantation genetic diagnosis
PMI	Precision Medicine Initiative
RNA	ribonucleic acid
SBP	standard biological part
SCNT	somatic cell nuclear transfer
SDI	Strategic Defense Initiative
STEM	science, technology, engineering, and mathematics

TMS	transcranial magnetic stimulation
TPS	Technology Protection System
UCD	urea cycle disorder
UCS	Union of Concerned Scientists
UGV	unmanned ground vehicle
USDA	US Department of Agriculture
WHO	World Health Organization

Re-Creating
Nature

1

Cells, Molecules, Genes, and Nature

A cell is regarded as the true biological atom.
    —George Henry Lewes, *The Physiology of Common Life* (1860), 297

Twenty-first-century humans live at a remarkable time in the history of the 13.8 billion-year-old universe. Earth coalesced from a gaseous cloud of elements and other interstellar debris just 4.6 billion years ago, and living cells adorned the planet about 3.8 billion years ago. *Homo sapiens* is a mere 200,000 years old. If the length of one's entire arm represents the time life has ranged over the Earth, all of recorded human history is as airborne fingernail dust from a single swipe of a nail file. Our species is very young compared to many others. For example, dinosaurs ranged the planet for 135 million years. What is remarkable about being alive now is that during our moment in time, the universe is coming to understand itself, perhaps for the first time. What I mean by that becomes clear through the first of the following questions this chapter addresses:

1. Why can we state that humans are literally star dust?
2. What are atoms, molecules, cells, and genes?
3. What is the relationship between DNA and protein, and why does it matter?
4. What is a genome?
5. Why is it important to think now about how humans will use bio-technologies many generations hence?
6. What should every school girl and boy know about life on Earth?
7. How does research with cells and genes intersect with human values?

Life, Nature, and Their Future

In this section, we examine questions 1–5. We will gain some fundamental information about the matter of which we and other living things are composed. This information is valuable when we later consider biotechnologies including brain scanning, monitoring, and manipulation; genetically modified organisms; gene editing; artificially intelligent robots; and more. Hopefully, we will also acquire a viewpoint about life that aids in making decisions about how best to develop and use powerful biotechnologies. We begin with our very beginning, the origin of the atoms in our bodies.

Humans Are Star Dust

It is a fact that we are literally star dust. Atoms comprising the cells of all living things were forged inside one or more stars more massive than the sun. The smallest and most plentiful atoms in the universe are hydrogen (H) and helium (He). They comprise the bulk of burning stars. Stars are gigantic fusion reactors in which gravity forces H atoms together to form He, releasing huge amounts of atomic fusion energy in the process.[1] The most powerful nuclear weapons are atomic fusion bombs in which H nuclei fuse to produce He, just as happens inside the sun. Energy emanating from the sun is directly or indirectly responsible for all of life's activities, our weather patterns, and geological phenomena such as weathering and erosion. Sunshine powers the water cycle: precipitation, evaporation of surface water to form clouds, condensation, and back to precipitation. This water cycle formed the Grand Canyon, the Badlands of South Dakota, and Niagara Falls.

But what about us and star dust? When the H fuel of a star runs low, the outward radiation of energy and matter is insufficient to fully counter the inward collapse of the star due to gravity. The collapsing star creates unimaginably enormous forces in the star's interior, resulting in a cascade of atomic nuclear fusions that creates all 115 known elements in the chemist's periodic table.[2] Finally, the huge imploding star explodes as a super nova and spews its forged elements into space as a vast cloud of star dust from which planets ultimately coalesce. Life arose on at least one such planet about four billion years ago. Perhaps it has on others as well.

Some of that star dust from an ancient super nova is self-conscious, learning about its origin and about the chemistry that made it conscious. We are truly star dust, and our particular configuration of star dust realizes this fact. This self-conscious bit of the universe now has the technological acumen to take charge of its own future evolution and perhaps even to entertain itself by animating inanimate star dust and designing entirely new beings in the visage of synthetic organisms and autonomous, thinking robots. That is a remarkable thing. We are the universe contemplating itself and recreating nature.

What does this cosmic conversation have to do with the subject of this book, biotechnologies and their ethical and societal implications? It puts contentious issues into perspective by emphasizing our cosmic commonality. The fact that we each are bits of the universe attempting to understand ourselves and each other, our place in nature, and the larger whole of which we are integral parts can be an equalizer and a unifier. I believe that this knowledge nurtures respect, tolerance, and empathy between us and for the rest of nature, and virtues like these can facilitate wise decision making about developing and using technologies able to reform human nature and nature itself.

Humans are sentient, reasoning, components of the universe with the capacity of foresight to gauge the long-term effects of their actions on other components of nature. Reason and foresight set us apart from the rest of the universe as we now know it and obligate us to protect our planet from potential harmful effects of some of the very technologies discussed here. In 1949, ecologist and conservationist Aldo Leopold argued that an action is right if it promotes the integrity, stability, and beauty of the land and that it is wrong if it does otherwise. The overarching mission of this book is to provide information about twenty-first-century technologies with transformative powers for both right and wrong, according to Leopold's definitions, and to show a path forward that promotes right action and minimizes wrong action.

This book is for persons with an urge to know, think, and converse about how modern technologies, primarily biotechnologies, are changing individual lives and society. Its reading requires no formal background in biology or chemistry. Only an elementary understanding of cells, molecules, and genes, which is provided by this chapter and at various other points along the way, is needed. Biotechnologies are technologies based on biology. They examine, manipulate, mimic, or redesign nature's biology so it befits us to know a bit about biology, the study of life from atoms on up.

Molecules, Cells, and Genes

Atoms comprise molecules and molecules comprise cells. Genes are portions of very large molecules called DNA, an acronym for its formal biochemical name deoxyribonucleic acid. Atoms like carbon (C), oxygen (O), hydrogen (H), nitrogen (N), sulfur (S), and phosphorus (P) comprise the molecules of living things. The most abundant molecule in living things is water (H_2O) whose formula indicates that one water molecule contains two atoms of H and one of O. Another small molecule, carbon dioxide (CO_2), is produced by animals as a by-product of metabolism and used by plants during photosynthesis.[3] Plants growing hundreds of millions of years ago incorporated the carbon in CO_2 into large molecules like cellulose that comprise plant tissues. When plants in ancient forests died and were covered with soil, they decayed and were compressed into coal and oil. Burning those large, primeval carbon-containing molecules as fossil fuels releases their C atoms into the air as CO_2, one of the greenhouse gases driving our current period of planetary warming. Other often-mentioned molecules of life include epinephrine, serotonin, dopamine, glucose, fructose, vitamins, amino acids, estrogen, and testosterone. Finally, all living things contain some very large classes of molecules referred to as the "big four": proteins, nucleic acids (DNA and RNA), lipids (e.g., cholesterol, saturated and unsaturated fats), and polysaccharides (e.g., cellulose and starch).[4] We return shortly to two of these macromolecules of life, DNA and protein, but first consider some principles of cell biology.

Cells are the structural and functional units of all life on Earth, much as buildings are the building blocks of cities. As do buildings, cells come in many different types, more than two hundred in the human body. But unlike buildings, all cells come from preexisting cells. These two statements, that all living things are comprised of cells and that all cells are begotten by cells, constitute the cell theory. Observations recorded by three German scientists between 1838 and 1855 led to formalization of the cell theory and its elevation to a level of certainty enjoyed by other scientific theories such as the heliocentric theory for the solar system, the germ theory for disease, the atomic theory of matter, and the theory of evolution for biodiversity.

Known cells belong to one of three domains based on genetic analyses: Bacteria, Archaea, or Eukarya (fig. 1.1). Single-celled, microscopic organisms (microbes) comprise the Bacteria and Archaea. Eukarya too contains single-celled organisms like yeast and amoebae, but it also includes all multicelled plants, animals, and fungi. Bacteria is by far the most diverse of the three domains. In fact, a 2016 genetic study of microbes found in California meadows and deep sea vents turned up numerous new phyla belonging to a previously unknown evolutionary limb of the Bacteria domain (Hug et al. 2016). Bacteria's other limb also contains thousands of species including familiar beneficial and pathogenic microbes that live in our gut, on our skin, in our nostrils and other orifices, and in the soil. Bacteria give us infections

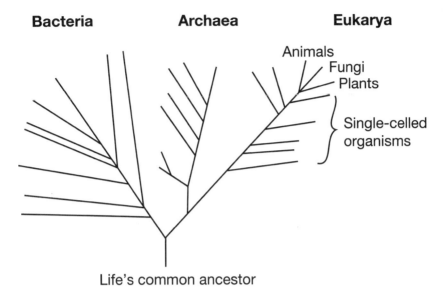

Bacteria **Archaea** **Eukarya**

Animals
Fungi
Plants

Single-celled
organisms

Life's common ancestor

Figure 1.1. The three domains of life. (Courtesy of the author)

and food poisoning, cause Lyme disease, and produce rotten apples. They also create yogurt and cheese, enrich the soil with nitrogen, and promote good health via our gut microbiome.[5]

Archaea are Bacteria-like in appearance, small singled-celled microbes without the internal structures described below for Eukarya. Surprisingly, in the late 1970s, Archaea were discovered to be genetically more closely related to the Eukarya (which includes humans) than to the Bacteria (Fox et al. 1977), an evolutionary relationship confirmed many times over since then. Archaea live in extreme environments like the hot springs in Yellowstone National Park, high salinity spots like the Great Salt Lake, and places with high methane levels. Before Bacteria and Archaea were recognized as two distinct evolutionary groups of cells, all the bacteria-like cells and certain algae were lumped together under the name *prokaryote*.[6]

The Eukarya domain includes organisms from amoebae and yeast to aardvarks and yellow jackets, from mushrooms and hyenas to maple trees and humans. Eukaryotic cells differ structurally from cells in the other two domains by partitioning their genetic material (DNA) away from the rest of the cell inside a membrane-bound nucleus. The word *eukaryote* derives from the Greek *eu-* (good, true) and *karyon* (nut or kernel). Thus, cells of eukaryotes possess a true nucleus, a kernel-like object near the center of the cell visible with a conventional light microscope and very prominent in cells

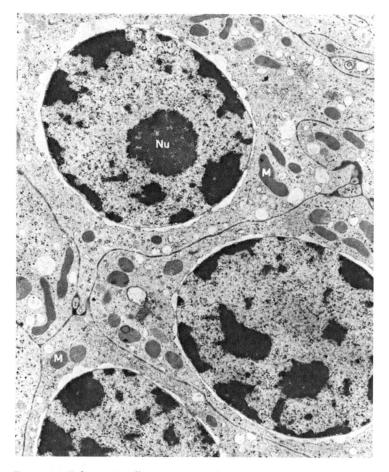

Figure 1.2. Eukaryotic cell structure seen by transmission electron microscopy. The photograph is of a gland near the brain in crickets that produces a hormone controlling insect growth and development. Portions of three cells and their nuclei are shown. Each nucleus contains the genetic information necessary for development of a complete, living cricket. A specialized part of the nucleus, the nucleolus (NU), is visible in one cell, and numerous mitochondria (M) are in the cell's cytoplasm surrounding each nucleus. (Courtesy of the author)

viewed by electron microscopy[7] (fig. 1.2). In chapter 2 we see how cloning and mitochondrial therapy involving three-parent babies make use of the fact that eukaryotic cells compartmentalize their DNA inside a nucleus.

In summary, atoms from a supernova comprise the molecules that comprise cells, which emerged on Earth four billion years ago. The chain of life remains unbroken since its beginning. Genes provide molecular continuity between generations of living things. What is a gene?

The term *gene* was introduced by Danish plant geneticist and physiologist, Wilhelm Johansen, in 1909. At the time, *gene* was coined in opposition to *pangene*, a term referring to invisible gemmules of hereditary particles in Charles Darwin's 1868 theory of pangenesis for heredity. Darwin did not know about genes or DNA, but he did note how particular traits moved from parents to offspring and on down the generations. To explain this, Darwin hypothesized that different cells and tissues in an organism produced their own tiny, specific, unseen particles of hereditary information that collected in the reproductive organs and eventually entered the developing embryo to direct tissue and organ formation. In 1865, Austrian monk Gregor Mendel reported experimental evidence for invisible "factors" of genetic information and laws governing their passage between generations, but his work was not appreciated or its significance understood until 1900, eighteen years after Darwin's death.

Working with the fruit fly *Drosophila melanogaster* between 1908 and 1915, American geneticist and evolutionary biologist Thomas Hunt Morgan gathered experimental evidence linking the concepts of Mendel's genetic "factors" and Johansen's genes with a recognition that hereditary factors reside in chromosomes visible with conventional light microscopes. From then through the 1940s, several biologists, including one of my biggest heroes, Harvard biologist Ernst Mayr (1904–2005), put evolutionary biology on a mathematical and genetic footing and established what came to be called the modern evolutionary or neo-Darwinian synthesis.[8] Soon the genetics of entire populations of organisms were described mathematically, and the evolutionary roles of genetic diversity within and between populations became analyzable.

The discovery of DNA's chemical and geometric structure by Erwin Chargaff, Rosalind Franklin, James Watson, Francis Crick, and others in the early 1950s ushered in the present era of molecular biology and biotechnology. Pulitzer Prize–winning physician-scientist Siddhartha Mukgerjee (2016) wrote a magnificent natural history of the gene for the enjoyment of nonscientists, so I need not expand on the gene's history here. Mukgerjee guides readers from the sixth-century BCE ideas of Pythagoras that hereditary information is carried in semen through stepwise refinements of our understanding of heredity into our current era of genomics and burgeoning ability to edit virtually any gene in any organism on Earth.

So just what is a gene? In its natural form, a gene is a portion of a very long DNA molecule. In the cell, each of these long DNA molecules comprises one chromosome, and each chromosome contains hundreds to thousands of genes. Genes can be isolated from the context of their larger DNA molecules or simply synthesized in the laboratory without the additional DNA that normally surrounds them. In this case, the shorter segments of DNA are still called genes.

Most genes belong to one of two types: structural or regulatory. Structural genes contain information needed by the cell to produce other types of molecules, RNA, and/or proteins. Regulatory genes function to control the activity of other genes in the same cell. If a gene encodes some RNA or protein that in turn regulates the activity of other genes, we speak of that gene as a structural gene that codes for a regulatory gene product. Soon, we consider the protein products of genes further, but first we will briefly reflect on DNA itself.

All three domains of life encode genetic information within molecules of DNA. Some viruses encode their genetic information in a similar but different molecule called RNA, but biologists do not consider viruses to be alive, so here we focus just on genes made of DNA. Recall that a single molecule of DNA may contain many units of genetic information called genes.

DNA is a long, linear molecule comprised of four different structurally related subunits linked together. It is like stringing together beads of four different colors to make a chain thousands of beads long. If one assigned meaning (words or letters) to the sequence of bead colors, information could be encoded within and transmitted by such a chain of beads. It is the same for DNA molecules comprised of specific sequences of the four molecular subunits of DNA, abbreviated as A, T, G, and C. We need not concern ourselves with what these letters stand for or what their particular chemical structures are. It is enough to know that a gene consists of a particular sequence of these four subunits within a single strand of DNA and that the subunits are called bases. The length of the sequence of bases in a gene varies from fewer than one hundred to several thousand depending upon the gene's function. Most of the genes of interest to us here code for proteins. Appendix 1 describes how cells use information in DNA's base sequence to create proteins via the so-called central dogma of biology. Proteins are basic to life. In fact, virtually every structure in our cells, tissues, and organs and their proper functioning is directly or indirectly due to the properties of proteins. Knowing a few things about them will help us appreciate how genetic biotechnologies work.

Like DNA, proteins are large molecules comprised of a linear array of smaller subunits. The subunits of proteins are amino acids. The cell uses

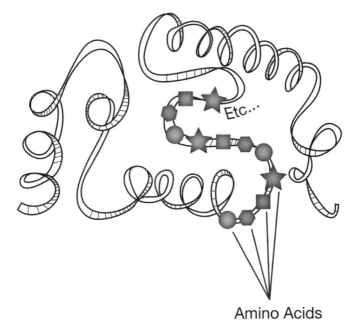

Amino Acids

Figure 1.3. Proteins may contain right-handed coils, sheet-like structures, random-appearing coils, and other amino acid arrangements that contribute to their overall three-dimensional shapes, which in turn determine their functions inside cells. Courtesy of the author)

twenty different amino acids to build its thousands of different proteins. A single file of amino acids joined together creates a protein, and sometimes two protein chains wind around each other to create a multichained protein. Proteins bend and fold to acquire complicated three-dimensional shapes determined by interactions between the amino acids along their length (fig. 1.3). For example, if one amino acid is positively charged and another one negatively charged, they will be attracted to each other causing a bend in the region of the protein molecule lying between them. Similarly, two negatively charged amino acids repel each other causing a bend in the chain of amino acids, taking the two like-charged subunits further apart. Thus, the sequence of amino acids determines the three-dimensional shape of a protein, and the three-dimensional shape in turn gives each protein the properties crucial for its particular function in the cell.

Earlier, we categorized genes as either structural or regulatory. All proteins are encoded by structural genes via the central dogma of biology (appendix 1). We can also group proteins into categories based on what they do in the cell. Nearly every protein falls into one of four categories based on what it does: structural, catalytic, regulatory, or nutritional. The structural proteins build things like cell membranes, hair, fingernails, silk, spider web material, muscles, and thousands of other structures inside cells or produced and secreted by cells. Catalytic proteins are the same as enzymes. Enzymes facilitate biochemical reactions. If our catalytic proteins suddenly become inactive, our cells die. In fact, this is what happens if our body temperature gets too high, either from a fever or heat stroke. There is an optimum temperature (98.6°F, 37°C) at which our enzymes maintain their active, three-dimensional structure. Above that temperature, enzyme structure and function are compromised causing the biochemical reactions that enzymes catalyze, which are necessary for life, to slow down or cease operating at all. Examples of the thousands of biochemical processes catalyzed by enzymes include the digestion and metabolism of food molecules to release and capture energy, synthesis of life's necessary molecules like DNA, RNA, and proteins, muscle contraction, and maintenance of proper pH and electrolyte levels in the blood.

Regulatory proteins include insulin and growth hormones and neuropeptides like oxytocin, endorphins, and encephalin that regulate a range of brain functions including those associated with eating, social behaviors, reproduction, reward responses, memory, and learning. Nutritional proteins include egg yolk proteins in birds, reptiles, fish, insects, sea urchins, and the platypus as well as mammalian mothers' milk proteins. A few proteins are multifunctional. For example, one protein called topoisomerase is both structural and catalytic since it helps to give structure to chromosomes and also catalyzes a reaction necessary for DNA replication. Speaking of chromosomes, let us take a closer look at them.

Even if many years have passed since high school biology class, most people know more than they realize about chromosomes. Most persons are familiar with the fact that humans have twenty-three pairs of different chromosomes. Some readers may even have had their genes analyzed by National Geographic's Genographic Project or a DNA testing company such as 23andMe, MyHeritage, or LivingDNA. For about $200, you can have the chromosomal DNA from a saliva sample tested to reveal the roots of your ancestry, your carrier status for over thirty different genetic disorders, and personalized genetic information about what makes you unique. Most readers also know that chromosomal trisomies, having three instead of two copies of certain chromosomes per cell, can cause serious genetic conditions like Down syndrome (chromosome 21 trisomy) and Edwards syndrome (chromosome 18 trisomy).

In the late 1800s, biologists using traditional light microscopes discovered dark-staining bodies near the center of dividing cells of plants and animals. The objects partitioned themselves equally between the two daughter cells in a process called mitosis and were named chromosomes from the Greek words *chroma* (color) and *soma* (body). We now know that each chromosome is made of roughly equal masses of DNA and protein. The DNA in each chromosome is a single, very long, double-stranded molecule in which each strand contains a linear array of the four bases (A, T, G, and C) mentioned earlier. The base sequence of the DNA is different for each of the different chromosomes, and each chromosome contains a different set of genes within its DNA. In humans, each chromosome carries from a few hundred up to two thousand genes for a total of roughly twenty-two thousand genes. An exception is the tiny Y chromosome in male cells that carries only about forty-five genes. The many different proteins associated with chromosomes of humans and other eukaryotic organisms mainly serve one of two functions: either to package the long DNA molecule into a manageable form or to regulate gene activity. All of the DNA from each of the different chromosomes in a cell of an organism is called the genome for that species (and for that individual).

Genomics is the examination and analysis of the base sequences in genomes. Much about the evolution of life is learned by comparing the DNA base sequences in the entire genomes of different species. In general, the more similar their genomic base sequences, the more closely related species are. For example, humans share 98.8 percent of their DNA base sequences with chimpanzees and bonobos, 93 percent with monkeys, and 84 percent with dogs. To summarize, genes reside in large DNA molecules that comprise chromosomes, which all together contain the genome for a particular species or organism, which in turn is contained within virtually every cell of that organism (fig. 1.4).

The Human Genome Project (HGP) of the 1990s successfully determined the exact sequence of all of the 3.1 billion A, T, G, and Cs in the DNA of human chromosomes. One surprising finding of the HGP is that less than 1.5 percent of the human genome is devoted to coding for proteins. Ethical issues emerging from the HGP include personal genetic privacy, the risk of genetic discrimination, and the patenting of genetic information.

A new procedure called CRISPR for easily, rapidly, and accurately ed-

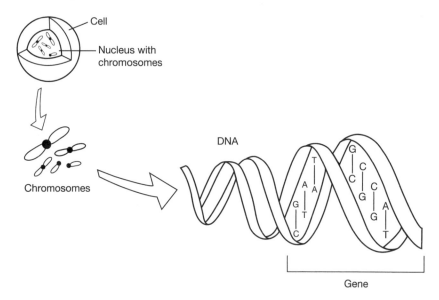

Figure 1.4. Relationships between DNA, genes, and chromosomes. Each chromosome in the cell nucleus is a single, long, double-stranded DNA molecule containing the sequences of bases (A, G, C, and T) for thousands of genes. The DNA comprising all types of chromosomes inside the nucleus of a single cell is the organism's nuclear genome. (Courtesy of the author)

iting the genetic information in virtually any chromosome of any species on Earth gained worldwide attention in 2015. In chapter 4, we consider CRISPR's implications for the future of life on Earth and the ethical issues raised.

Living Nature and the Future

By *nature,* I mean the world of things and phenomena not directly created by humans. Admittedly, a clean line does not separate natural from synthetic. For example, varieties of agricultural corn are derived from naturally occurring species of maize in Mexico that underwent generations of selection, hybridization, and more recently genetic engineering by humans. The natural and synthetic coexist in all row crop plants, domestic animals, and many ecosystems. Still, there are germ plasms not intentionally altered by humans all around us from fireflies and snow leopards to native wild flowers and soil bacteria. In this section, our focus is on *living nature,* natural things that are alive or that pertain directly to life. These include individual organisms, species, communities of organisms and ecosystems, the relationships of organisms within their communities, and human beings as components of an interlinked, interdependent living world. The beautifully balanced flow of energy and nutrients through a healthy, unpolluted, and otherwise unspoiled coral reef ecosystem, alpine meadow, and wilderness desert are phenomena of living nature.

In separating the living from nonliving components of the natural world with the term *living nature*, I intend no slight to the inextricable links between the living and nonliving. Soil, water, air chemistry and quality, climate, and many other geological, meteorological, and other physical features of nature both set limits upon and are shaped by the life forms associated with them. Even Earth's mineral diversity is influenced by life (Grosch and Hazen 2015). My reason for focusing on living nature here is that the biotechnologies we discuss directly target (or mimic) living things rather than things like climate, geological fault lines, or the height of the water table.

Of importance when considering technologies that could permanently change human nature and the nature of the nonhuman living world is knowledge about what science says about the origins and nature of life. This knowledge fosters an appreciation for who we are and for our interdependent relationship with the rest of nature. Such an appreciation is a prerequisite for wise use of life-changing biotechnologies. What follows is a list of what I believe every school girl and boy ought to know about themselves and the living nature around them. I take responsibility for each included and excluded item from this list. A philosopher, clergyperson, politician, physicist, or even another biologist might have a different understanding of life and therefore produce a somewhat different list.

1. Atoms that comprise all living and nonliving things are products of natural processes happening at the cosmic level.
2. All life is a product of natural processes, and therefore the components of living systems, from molecules and cells to tissues and organ systems, obey the laws of chemistry and physics.
3. The first cellular life on Earth appeared about four billion years ago, about six hundred million years after the formation of Earth itself. Nobody knows the details of the processes that led to the first living cells, but prebiotic chemists and synthetic biologists are making rapid progress in re-creating the steps by which organic molecules formed on the primitive Earth and ultimately polymerized

and assembled into organized structures capable of metabolism, reproduction, and genetic mutation, characteristics we associate with life.

4. All life on Earth shares a common ancestry. Life's diversification occurred over its nearly four-billion-year-long history on Earth via natural, evolutionary processes. The appearances and diversification of multicelled organisms, vertebrates, mammals, primates, the Great Apes, and bipedal hominins including humans are all products of biological evolution.

5. Current evidence is that modern humans arose in Africa two hundred thousand to five hundred thousand years ago.[9] Subsequent migrations out of Africa into Europe and Asia resulted in adaptations to local environments and emergence of what some persons call "races." All humans belong to the same species, and no living subspecies of humans exist.[10]

6. Humans distinguish themselves from other animals by using foresight to imagine likely distant future consequences of potential acts in the present. By comparing these consequences in the mind, humans may choose from an array of possible present acts that are most likely to produce the desired outcomes. Such foresight and the ability to hold several possible outcomes in one's mind at once facilitate moral choice making and long-range planning. Prolonged sacrifice to achieve long-term goals is another product of human foresight.

7. In addition to foresight, human nature includes an insatiable curiosity about the environment, a need to identify meaning in life, tribalism, hope, creation and use of symbols, and a tension between altruism and self-promotion.[11] The human characteristics I have listed are mental attributes. There are also physical attributes that work in concert with the mental ones, such as a larynx positioned to allow complex speech, bipedalism that allows hands to be used in symbol formation (e.g., signing, creating art, writing, dancing), and an expanded cerebral cortex facilitating foresight, imagination, investigation, and moral choice making.

8. All life forms are inextricably linked together by complex webs of interdependency. This is especially true for the members of specific biological communities and ecosystems but extends to incorporate all ecosystems into a planetary network of living nature's

interdependency. Humankind is a component of the planetary network of interdependency and absolutely relies on countless other organisms for its existence. Interdependency of the components of Earth's biosphere is a product of life's four billion years of evolution, and ecologists are just beginning to understand the links between its multitudinous components.

9. No mysterious "vital force" is necessary for matter to be alive. Principles of chemistry and physics apply equally to inanimate and living matter, including the most complex structure known in the universe, the human brain. This suggests that we may ultimately understand and be able to manipulate life at the molecular level.

10. Humankind now has an understanding of living things that enables it to radically change life on Earth, including human nature itself. Human-designed changes to life occur so rapidly as to be utterly foreign to the evolutionary processes that created Earth's current biosphere. The outcome of our changes to living nature will reflect our foresight and choices.

11. Despite the fact that more than half of all Americans and many other people throughout the world believe that humans and other living things are planned creations of a supernatural being, there is no scientific evidence that this is true.[12] Neither are there aspects of the living world that necessitate presuming it is true. Unguided evolutionary processes, including random, heritable variation and natural selection appear sufficient to explain the marvels of living nature.

12. The fact that modern science validates Charles Darwin's great idea about the origin of species does not mean that science disproves the existence of supernatural beings or realms of reality inaccessible to our senses or technological extensions of them. Neither does it mean that one cannot accept science and also hold such beliefs.[13] Beliefs in supernatural beings, realms, or acts are simply not within science's purview to confirm or invalidate. What the reality of naturalistic evolution does mean is that, so far as we can know, any plan for the future of the biosphere, its individual components, or human nature must originate with the one animal possessing foresight and an ethical sense, *Homo sapiens*. It is up to us to accept this responsibility and apply our best knowledge, wisdom, and courage to it. Having this responsibility is part of human nature.

In this book, I do not suggest specific long-range goals for humanity itself or humanity's actions upon the biosphere; rather, I describe technologies that make setting such goals an urgent matter and suggest how we can prepare ourselves to decide on long-range goals for our species and the planet. A public informed about the capabilities and ethical issues raised by these technologies is a necessary prerequisite for the setting of long-range goals, and my objective here and in an earlier work (Bradley, 2013a) is to contribute to that prerequisite. We consider how to prepare ourselves for choosing long-range goals in chapter 9. Ethical issues associated with modern cell and molecular biology are not all directly related to the long-term future of humankind. Many involve privacy, ownership, justice, fairness, and simply treating human beings decently. We consider two such examples in the next section of this chapter.

Cells, Genes, and Human Values

What do cell and molecular biology have to do with human values? Actually, quite a lot. How and from where do we get cells for research? What are the goals of cell research? What types of cells do researchers use in the laboratory and in clinical settings? What is actually done with those cells? How do researchers obtain and use information about genes? Whose genes do they examine and under what conditions? These and other questions define major ethical issues associated with modern cell and molecular biology.

Many ethical questions raised by modern biotechnologies are directly linked to research on cells and genes. For example, what moral status should we assign to cells of very early human embryos, any one of which could theoretically give rise to a fetus? How should we deal with the eugenic potential of preimplantation genetic diagnosis, the technology of single-cell biopsy for embryos created by *in vitro* fertilization (IVF)? Is there really anything morally wrong with cloning humans from the DNA in a single cell? Will the analysis of subtle genetic differences between individuals with differing geographical ancestries lead to better medical treatments or to discrimination? Should we invest heavily in cell and genetic research aimed at greatly retarding human aging? What constitutes wise uses for synthetic biology and the creation of new genes and life forms that have never existed during life's 3.8 billion-year history on Earth?

In the next chapter, we review some dramatic advances in these and related areas of research and consider how these developments relate to human values. In subsequent chapters, we look closely at how knowledge about cells and genes creates ethical dilemmas for agriculture, ecology, the future of the biosphere, the future evolution of *Homo sapiens*, medicine, brain biology, and robotics.

In the remainder of this chapter, we consider two examples of how research on some specific cells and genes takes us quickly into the world of moral decision making and human values. The cells are HeLa cells derived from the tumor of an African American woman, Henrietta Lacks, who died from cervical cancer in 1951 at the age of thirty-one. The genes are *BRCA* (breast cancer) genes that normally protect persons from malignant tumor formation but that, when mutated in certain ways, create an increased risk of cancer.

HeLa Cells

Shortly before Henrietta Lacks died from an aggressive adenocarcinoma of the cervix, in 1951, doctors at Johns Hopkins Hospital took healthy and cancerous tissue samples from the patient without her or her family's permission or knowledge. Dr. George Otto Gey, a physician and cancer researcher at the hospital, received the samples and began experimenting with them. Soon, he discovered that the cancerous cells grew very well under laboratory conditions, much better than any other human cells previously cultured (fig. 1.5). Gey took a single one of Henrietta Lacks's cancer cells, allowed it to reproduce thousands of new cells, divided this culture into subcultures, and allowed the cells in these subcultures to reproduce and repeated this many times. The cell cultures remained vigorous after many divisions, and in fact, the cells were dubbed "immortal" to distinguish them from other cells that lived for only a few days in laboratory cultures.

To this day, researchers use HeLa cells worldwide to investigate a wide range of problems in medicine and cell biology. HeLa cells are invaluable for learning how various toxins, drugs, cosmetics, bacterial and viral pathogens, and radiation affect human cells. They also contribute immeasurably to cancer and AIDS research, gene mapping, and to understanding the biochemical basis for the phenomenon of life itself.

HeLa cells appeared on the scene at a fortuitous time for the United States, which was in the grip of a polio epidemic in the early 1950s. Jonas Salk at the University of Pittsburgh used HeLa cells to grow the polio virus in his laboratory and develop his famous vaccine against the virus in early 1952. In 1954, the National Foundation for Infantile Paralysis (the March

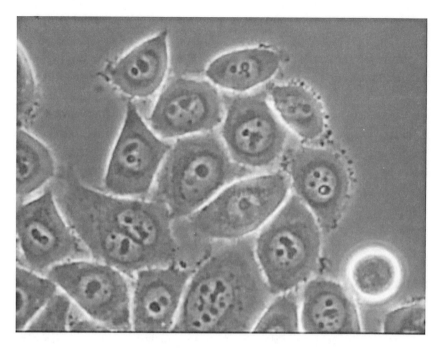

Figure 1.5. HeLa cells in laboratory culture. "Immortal," cervical cancer cells from Henrietta Lacks have been laboratory grown since 1951 and are now used in medical and basic scientific research throughout the world. Unlike other types of human cells, which stop growing after a set number of cell divisions in culture, HeLa cells lack senescence under laboratory culture conditions. (Microscope Imaging Station, Exploratorium, San Francisco. www.exploratorium.edu)

of Dimes) organized a massive field test of Salk's vaccine. The test involved nearly two million school children. More than half of the volunteers received the vaccine, while the rest received a placebo. Since HeLa cells were susceptible to the polio virus, were the first human cells successfully immortalized in laboratory cultures, and could be efficaciously shipped to laboratories all over the country, they were chosen for use in a neutralization test for the vaccine. For this test, blood samples from children inoculated with the vaccine were mixed with active polio virus and then added to vials containing healthy HeLa cells. If a blood sample contained a sufficient quantity of antibodies (a class of proteins produced by our immune system) against the virus, the active viruses would be incapacitated and the HeLa cells would remain healthy. If a blood sample did not contain enough antibody to neutralize the active viruses, the HeLa cells would become infected and die.

A HeLa cell factory was established at Tuskegee Institute (now University) in the small town of Tuskegee in rural southeast Alabama. Two well-known cell culturists worked there, and their mission was to produce and distribute tubes of HeLa cells growing suspended in culture medium to twenty-three vaccine testing laboratories around the country. Eventually, the Tuskegee laboratory supplied HeLa cells to any scientist interested in experimenting with them. The purloined, immortal cells of Henrietta Lacks are directly responsible for the defeat of polio in the United States and worldwide and for the gain of invaluable information about the basic biology of human cells. Nearly eighty thousand scientific articles and eleven thousand patents related to HeLa cells bear witness to the cells' scientific and economic importance.

Over the years, different strains of HeLa cells representing spontaneous mutations in the DNA of the cells were developed. Several of these HeLa cell strains, all derived from the original strain created by Dr. Gey, were commercialized, providing profit for private companies or institutions. Remarkably, it was not until the early 1970s that the Lacks family first learned of the existence of HeLa cells. The information came to them in a roundabout way through phoned requests from researchers for Lacks family members' blood samples. The researchers suspected that other cultured cells in their laboratories had become contaminated with the extremely resilient and prolific HeLa cells, thereby compromising the reliability of years of research. With blood samples from relatives of Henrietta, the researchers would be able to compare Lacks family DNA with DNA in their cell cultures to determine whether their laboratory had a HeLa contamination problem. Other researchers requested medical information from Lacks family members. Understandably, the family was stunned and hurt to learn that the family matriarch's cells had been used so extensively for so many years without their knowledge. Although no laws had been broken, ethical guidelines and common courtesy had certainly been breached. The many slights to the Lacks family and the entire HeLa cell story were brought to public light by the impeccably thorough research of science writer, Rebecca Skloot in her book *The Immortal Life of Henrietta Lacks*.

Astoundingly, genomic researchers published genome information for a strain of HeLa cells in 2013 without permission from or formal notification of any Lacks family member. To their credit though, the researchers later withdrew their publication of the HeLa genome and apologized to the family. Meanwhile, a more detailed genomic study funded by the National Institutes of Health (NIH) in the United States was about to be published. In this case, neither authors nor reviewers of the paper raised the genomic privacy issue in relation to the Lacks family. Later in 2013 but before publication of the NIH-funded paper, the family received some redress due largely to the initiative and sensitivity of NIH director Francis Collins.

Collins met with Henrietta's children and grandchildren three times in 2013 to hear grievances about the way they and Henrietta had been treated over the years. In the end, a unique genome privacy pact between NIH and the Lacks family was struck (Callaway 2013). The family agreed that researchers can apply to obtain HeLa cell genomic information under a restricted-access system that links the genomic data to information about traits and diseases associated with that genome. Their requests will be reviewed by a NIH committee, two members of which are Lacks family members. Researchers must agree to use the information for biomedical research only and not to contact Lacks family members for medical records, blood samples, or other reasons. In addition, publications that use the HeLa genomic data must acknowledge the contribution of Lacks and her family to the research.

The HeLa cell story highlights current controversies and weaknesses in federal law relating to privacy in genomics. For example, the US Supreme Court ruled in 2013 that law enforcement personnel may gather DNA samples from unconvicted suspects arrested for a crime. Genomic data from such samples may end up in criminal justice DNA databases supported by the Federal Bureau of Investigation. No clear policies dictate whose DNA data ought to be stored in such databases or when an individual's genomic data should be removed from the system.

Surreptitious DNA testing is another DNA privacy concern. Anybody could sample your DNA from a spot of blood, mucus, feces, spittle, or even a licked envelope. They could then have health-related, personal trait, or parentage tests performed on it without your knowledge by any of several companies that analyze DNA without requiring the consent of those whose DNA is being tested. No federal laws restrict surreptitious DNA testing, and although several states have laws governing genomic privacy, their contents and enforcement vary between states. On the NIH's website is a concise summary of the legal state of genomic privacy in the United States (NIH 2015). Of particular interest to readers may be a description of how the US government attempts to provide medical benefits for its citizenry via research databases linking genomic information to health-related traits while also protecting patient privacy.

BRCA *Genes*

BRCA1 and BRCA2 are proteins whose normal function is to protect persons against cancer. By convention, the names of genes themselves are italicized. Thus, the BRCA1 and BRCA2 proteins are encoded by the *BRCA1* and *BRCA2* genes, respectively. *BRCA1* is on chromosome 13, and *BRCA2* is on chromosome 17. Both BRCA1 and BRCA2 are so-called tumor suppressor proteins that mend breaks in DNA molecules. Such breaks can occur naturally prior to cell division or may result from exposure to radiation in medical or other settings. When unrepaired breaks accumulate in critical areas of the genome, uncontrolled cell division leading to cancerous tumors can result.

Researchers know of more than 1,600 mutations for each *BRCA* gene, and certain of these mutations give an increased risk of breast cancer in both women and men. Some of the same *BRCA* mutations that increase one's risk for breast cancer also increase the risk for ovarian and prostate cancer. In fact, 35 to 60 percent of women with certain *BRCA1* mutations get ovarian cancer at some point in their lives compared to just 1.6 percent of women without the mutations. For women with certain *BRCA2* mutations, the risk of ovarian cancer is 12 to 25 percent versus 1.6 percent for women without those mutations. It is important to note that not all of the known mutations in the *BRCA* genes cause an increased risk of cancer.

In the 1990s came discovery of specific mutations in *BRCA1* and *BRCA2* that increase cancer risk and founding of the molecular diagnostic company Myriad Genetics, Inc., in Utah. Myriad obtained patents on the use of *BRCA* genes and developed the BRCA*Analysis* test to assess risk for hereditary breast and ovarian cancer based on base sequences of patients' *BRCA* genes. Myriad's patents on the *BRCA* genes gave the company a monopoly on testing patients' DNA for the mutations in *BRCA* genes that cause increased cancer risk. For nearly fifteen years, women with a family history of breast cancer had no choice but to pay high fees for gene testing at Myriad. Depending on insurance coverage and whether other family members had been tested previously, the cost of the BRCA*Analysis* test ranged from hundreds to thousands of dollars.

According to Karuna Jaggar, executive director of Breast Cancer Action, a California-based grassroots organization for education and advocacy for breast cancer patients, Myriad's monopoly on the breast cancer genes compromised women's health in ways beyond limited access of genetic diagnosis due to high cost. Myriad's gene patents effectively blocked competitors from developing alternative tests or targeted therapies for cancers linked to *BRCA* mutations. This in turn prevented patients from getting second opinions and led to inadequate information, diagnosis, and treatment for underserved populations.

In June 2013, after a series of federal circuit court rulings and appeals, the US Supreme Court ruled in *Association for Molecular Pathology v. Myriad Genetics*, a case testing the legitimacy of Myriad's patents on the *BRCA* genes. In its unanimous decision, the Court held "that a naturally occurring DNA segment is a product of nature and not patent eligible merely because it has been isolated" (US Supreme Court 2013). The ruling broke the monopoly on testing the *BRCA* genes that Myriad had enjoyed for so many years, but this did not suddenly make things better for women in need of the tests because Myriad retained its proprietary database built from data on its thousands of patients' and their families' medical histories.

Myriad's proprietary database for thousands of mutations in the two *BRCA* genes link specific mutations to the past and future health status of patients and their family members. Recall that more than 1,600 mutations are known for each of the two *BRCA* genes. Only some of these are widely recognized as causing increased cancer risk. When some other mutation in *BRCA1* or *BRCA2* turns up in a patient's DNA, Myriad is likely to have medical history data on the mutation and can determine whether it is harmful or harmless to the patient's health. Other gene-testing companies are unlikely to have this information since they lack the medical history database that Myriad developed during its period of *BRCA* gene monopoly.

If there are no data on a mutation, the mutation is called a *variant of unknown significance* (VUS). Over the years, Myriad collected so much data from so many patients that its tests reveal far fewer VUSs than do other companies' tests. In fact, Myriad claims that the fraction of its tests resulting in a VUS is below 3 percent compared to 20 percent for the company's European competitors (Conley et al. 2014). So there is an obvious health advantage for a patient to use Myriad's BRCA*Analysis* test even if it costs much more than a similar test offered by another company because the chances are better than 97 percent that Myriad will have information on whatever *BRCA* mutations show up. That would bring peace of mind if the mutation is innocuous and possibly a life-saving heads-up if it is not. Other companies may submit their patients' test results to Myriad for interpretation, but the cost for this is passed on to the patient. Myriad stopped contributing to public databases in 2004. All genetic and health data collected since then is unavailable to other companies; yet, Myriad has access to public databases to which other companies contribute. In the end, patients pay the price in money, worry, or undiagnosed health problems for this asymmetric sharing of and access to data.[14] Whether Myriad should be required to make public its proprietary database on *BRCA* mutations is a difficult legal and ethical question. Some argue that a testing laboratory's accreditation and/or FDA approval of its tests for marketing ought to be tied to data sharing. Others argue that Myriad deserves to keep its database private, especially since it contains personal lifestyle information about its past clients. Courts of law may eventually need to balance the right of privacy against the common good in the case of Myriad's database.

Chapter Summary

Humans and all other creatures, plants, and microbes are literally offspring of the universe itself, made of molecules with atoms forged inside stars. Cells in today's organisms have an ancient heritage stretching back 3.8 billion years. Two large biological molecules with special significance in many biotechnologies are DNA and proteins. Both are long linear polymers of smaller subunits. In DNA, each subunit contains a base (A, T, C, or G), and the sequence of bases in DNA segments called genes determine the structure of the proteins a cell makes. Subunits of proteins are amino acids. The cell uses twenty kinds of amino acids to build proteins. The sequence of amino acids in a protein determines its three-dimensional shape, which in turn specifies the protein's function(s). Chromosomes each contain one long DNA molecule carrying many genes, and in eukaryotic cells (those with a nucleus), many proteins associate with the chromosomal DNA. Virtually every cell in an organism has copies of every chromosome characterizing that species. All of the DNA molecules from all the different chromosomes comprise the genome for an individual or for the species. Genomics is the analysis of the base sequence in an organism's (or species') DNA, its function, and how it compares to the genomes of other organisms. Knowledge about and manipulation of cells and genes raise many ethical issues.

Examples come from HeLa cells and the *BRCA* genes. HeLa cells, a line of human cancer cells used around the world since the 1950s, raise issues of patient privacy and control over biopsied tissue. *BRCA* genes raise issues of gene ownership, commercial monopoly of health-related genetic testing, and the proper use of patients' genomic information and health records.

Questions for Reflection and Discussion

1. What is your reaction to the knowledge that you are literally star dust? Had you heard of this before now? If most people in the world learned and accepted this information, would it make any difference in the realms of politics, religion, or society in general?
2. Do you believe that Henrietta Lacks's family has now been adequately compensated for the scientific community's decades of use of HeLa cells without its knowledge?
3. Myriad Genetics, Inc., built their proprietary database relating *BRCA* mutations and health prognosis at its own expense over many years; yet, without its monopoly on *BRCA* testing, it could not have done so. Now that Myriad's patents on the *BRCA* genes are invalidated by a Supreme Court ruling, do you believe that the database should be made public? Give reasons for your answer.
4. If your genome were sequenced, what kind of information would you like to know about yourself? What, if any, type of information would you like to remain unaware of?
5. What is wrong with this argument: All humans are part of nature so all human actions are natural and therefore ethically all right.

Notes

1. The nucleus of one H atom has one proton. When two H nuclei are forced close enough together, they fuse to produce a new atomic nucleus with two protons, the He nucleus.

2. It might be more accurate to say 113 elements since H and He were present before the star's collapse. Yet, there surely are more elements yet to be discovered. Four of the 115 known elements were discovered since 1995.

3. Photosynthesis is the biochemical process whereby green plants use water, CO_2, and the energy in sunlight to produce sugar, which the cell then polymerizes into large plant molecules including cellulose and starch.

4. Ribonucleic acid (RNA) has a structure very similar to DNA.

5. The term *microbiome* refers to a community of many species of bacteria. The gut microbiome varies between persons and is important for good health including a strong immune system. It may take up to two years for a person's gut microbiome to recover after treatment with a round of antibiotic.

6. *Prokaryote* derives from the Greek *pro-* (before) and *karyon* (kernel or nucleus), so the term means "before a nucleus," referring to the fact that Eukarya have a nucleus and the prokaryotes appeared on the Earth before the Eukarya. The nucleus is a specific compartment inside cells of Eukarya that contains the genetic material, DNA.

7. Transmission electron microscopy yields much more detail than light microscopy because the wavelength of electrons is much shorter than that of visible light. The shorter the wavelength, the better the resolution, i.e., the smallest distance between two objects when the objects can still be distinguished as separate objects.

8. Mayr is an idol for me due to his lifetime commitment of communicating to nonscientists about his biology profession through books like *This Is Biology: The Science of the Living World* (Cambridge, MA: Belknap Press, 1997), published when he was ninety-three years old, and *What Makes Biology Unique* (Cambridge, MA: Cambridge University Press, 2004), published when he was ninety-nine. I treasure a personal note from him encouraging my writing of *Brutes or Angels: Human Possibility in the Age of Biotechnology* (Tuscaloosa: University of Alabama Press).

9. Fossil evidence indicates that early modern humans were present in sub-Saharan Africa about 200,000 years ago. Human fossils from Morocco were recently dated at about 315,000 years old, and DNA analyses of fossils and living humans indicate that the human line diverged from Eurasian Neanderthals and Denisovans over 500,000 years ago (Stringer and Galway-Witham 2017).

10. *Race* is an eighteenth-century social construct designed to rationalize European attitudes and treatment of conquered and enslaved people and later to justify social, economic, and political inequalities between groups of persons (American Anthropological Association 1998). Modern human genomics reveal more genetic diversity within groups identified by geographical, linguistic, and cultural boundaries than between them (Rosenberg et al. 2002), suggesting that the term *race* has very limited biological significance, if any at all.

11. By *tribalism*, I mean loyalty, trust, and sacrifice toward one's group(s) such as family, club, community, sport's teams, or nation and distrust, aggression, indifference, or antipathy toward groups not one's own.

12. According to a 2013 Pew Research Center Poll, 33 percent of Americans believe that humans and other living things were created and have always existed in their present form, and 24 percent believe that humans and other living things are results of God-directed evolution (Pew Research Center 2013).

13. Two well-known and highly respected biologists, Francis Collins (2006), director of the National Institutes of Health, and evolutionary biologist Kenneth Miller (1999), have written books describing how their acceptance of evolution is compatible with their personal religious faith.

14. Ethical issues associated with proprietary health-related databases and the problem of maintaining patient privacy when submitting genetic and health history information to public databases, particularly in the context of Myriad and the *BRCA* genes, are thoroughly discussed by Conley et al. (2014).

Sources for Additional Information

Cudmore, L. L. Larison. 1977. *The Center of Life: A Natural History of the Cell.* New York: QuadrangleBooks. This book, written by an English professor enthralled with cell biology, provides an overview of how cells work.

Feinberg, Gerald. 1969. *The Prometheus Project.* Garden City, NY: Doubleday.

ScienceDaily. 2016. "Wealth of Unsuspected New Microbes Expands Tree of Life: Bacteria Make Up Nearly Two-thirds of All Biodiversity on Earth, Half of Them Uncultivable," April 11. https://www.sciencedaily.com/releases/2016/04/160411124716.htm (accessed May 18, 2016).

Skloot, Rebecca. 2010. *The Immortal Life of Henrietta Lacks.* New York: Random House.

US Supreme Court. 2013. Opinion of the Court No. 12–398. Association for MolecularPathology et al. v. Myriad Genetics, Inc. et al. http://www.supremecourt.gov/search.aspx?Search=Myriand ad+Genetics&type=Site (accessed May 31, 2016). In excellent prose, this opinion explains what a gene is and the relationship between genes, their protein products, and laboratory-created copies of genes used in research and diagnostic tests as well as the scientific, business, and legal history of the *BRCA* genes.

2

Embryos, Stem Cells, Genetic Enhancement, Genomics, and Synthetic Biology

What I cannot create I do not understand.
—Richard Feynman, 1988, Feynman's blackboard at
CalTech, Pasadena, at the time of his death

An earlier volume on modern biotechnologies, *Brutes or Angels* (Bradley 2013a), contains chapters dedicated to stem cell biology, embryo selection via preimplantation genetic diagnosis, cloning, genomics, genetic enhancement, age retardation, and synthetic biology. This chapter summarizes scientific and ethical topics discussed in the earlier volume and describes some major developments in those areas since the publication of *Brutes or Angels*. Specific questions addressed here include:

1. What is an embryonic stem cell (ESC)? Why is ESC research controversial? What other kinds of stem cells show promise for clinical applications and contributing to human health in other ways?
2. What is preimplantation genetic diagnosis (PGD)? What are its benefits and associated ethical concerns?
3. What are the different types of gene therapy and enhancement? Is there a clear difference between gene therapy and genetic enhancement? What is the current state of human gene therapy?
4. What is cloning? Have humans been cloned?
5. What is the connection between mitochondrial disease and three-parent babies? Why is this technology of grave concern to some bioethicists?
6. What was the Human Genome Project, and where is human genomics going from there? What is the Epigenome Project and its ethical implications?
7. What is age retardation and what ethical issues will arise if the technology is applied to humans?
8. What is synthetic biology? What are its major accomplishments so far and where is it headed?

Stem Cells and Associated Ethical Issues

Three categories of stem cells occur naturally in humans and other mammals: ESCs, adult stem cells, and embryonic germ cells. Of these, ESCs show the greatest potential for clinical applications. Adult stem cells are less versatile than ESCs, and embryonic germ cells are obtainable only from six- to eight-week-old embryos. In this chapter, we examine ESCs for their biology and clinical potential and for ethical issues their use raises. We also look briefly at induced pluripotent stem (iPS) cells, a class of stem cells created in the laboratory.

Stem cells get their name from the fact that they can generate specialized types of cells, just as the growing tip of a plant stem gives rise to root, branch, leaf, and flower. In fact, some stem cells can generate all 252 types of cells in the human body. Ancestral lines for the trillions of cells in the human body originate in a small group of pluripotent (having a plurality of developmental potentials) stem cells present in very young embryos and ultimately back to the fertilized egg cell itself.

In most cells, mitotic cell division produces two identical daughter cells. But when a stem cell divides, it produces two kinds of cells: (1) another cell like itself and (2) a progenitor cell committed to producing specialized cells via subsequent cell divisions (fig. 2.1). The stepwise specialization accompanying cell division that produces cells dedicated to specific tasks is called cell differentiation. Examples of differentiated cells include oxygen-carrying red blood cells, electrical signal-conducting nerve cells, contracting muscle cells, and insulin-producing pancreatic cells. Development of an entire plant or animal from a single fertilized egg, wound healing, and the daily replacement of worn-out blood and skin cells are examples of what stem cell division and subsequent cell differentiation accomplish.

To date, of the four major types of mammalian stem cells—ESCs, adult stem cells, embryonic germ cells, and iPS cells—ESCs show the greatest potential to relieve human suffering. Yet, ESC research provokes heated dispute in churches, Congress, classrooms, public media, bioethics committees, political campaigns, homes, and the streets. Alas, the passion in ESC debates is matched by public misunderstandings about what ESCs are and where they come from.

Human ESCs (hESC) originate from pluripotent cells in five- to seven-day-old embryos called *blastocysts*. At this age, the embryo is about 0.2 mm

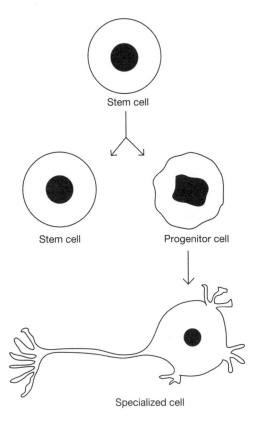

Stem cell

Stem cell   Progenitor cell

Specialized cell

Figure 2.1. Stem cell biology. A stem cell divides to produce another stem like itself and a progenitor cell that goes on to develop into a specialized cell type. Thus, stem cells regenerate themselves and also give rise to different types of cells. (Courtesy of the author)

in diameter, less than one-fifth the size of a pin head. The blastocyst is a hollow ball of cells whose interior contains a small group of undifferentiated cells, the *inner cell mass* (ICM), which has the potential to give rise to all of the cell types in a fully formed individual (fig. 2.2). The trophectoderm, a single layer of cells called trophoblasts, surrounds the ICM and gives rise to the fetal portion of the placenta if the embryo implants itself in the uterine wall. In a five-day-old blastocyst, the ICM contains about one hundred cells. All of the cells look alike; none of them are specialized as heart, muscle, liver, brain, or other cell types. Laboratory-cultured ESCs are derived from the blastocyst's ICM.

All blastocysts used for ESC research come from surplus embryos stored frozen by *in vitro* fertilization (IVF) clinics. Couples using IVF to have a child generally produce one to two dozen embryos. Physicians usually transfer only three or fewer embryos at a time to the woman's womb. Surplus embryos are early stage embryos left over after parents complete the

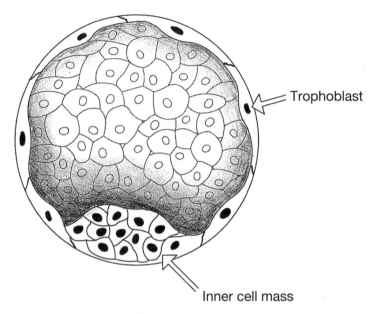

Trophoblast

Inner cell mass

Figure 2.2. Diagram of four- to five-day-old blastocyst stage embryo characteristic of humans and other mammals with its two components, the trophectoderm comprised of trophoblast cells and the inner cell mass. In blastocysts implanted in the uterine wall, the trophectoderm forms the fetal portion of the placenta, and the inner cell mass develops into the fetus. (Courtesy of the author)

IVF procedure (fig. 2.3). If the parents no longer wish to keep their embryos in frozen storage, they consent for them to be discarded, donated to infertile couples, or donated for research. At any one time over four hundred thousand surplus embryos exist in the United States alone. The majority of these are eventually discarded by clinics with parental approval.

Biologists receiving surplus embryos for hESC research first thaw and re-animate them in glass dishes. When an embryo reaches the blastocyst stage, researchers remove the ICM and disperse its pluripotent cells into a liquid nutrient solution in a laboratory culture dish. These ESCs remain undifferentiated until stimulated with hormonal and/or other chemical agents that induce them to produce specialized cell types (fig. 2.4). Many people believe that hESCs come from the wombs of pregnant women, umbilical cord blood, or aborted fetuses. Mistaken notions like these contribute to vehement opposition to hESC research. In fact, couples willingly donate surplus embryos used for the production of hESCs. These couples have completed their family or stopped trying to conceive, and rather than having them destroyed, they donate them for medical research.

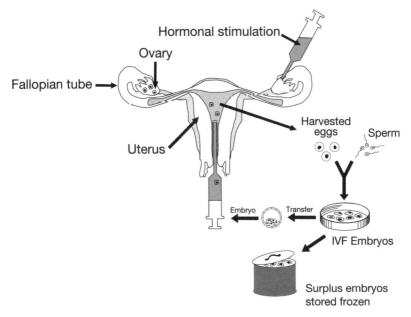

Figure 2.3. Assisted reproduction by *in vitro* fertilization. Superovulation of a woman's ovaries by hormone injection causes the release of up to two dozen eggs all at once. Collected and fertilized in the laboratory, the eggs develop into blastocyst embryos. Typically, two or three embryos are transferred to the uterus. The remaining embryos are stored frozen in liquid nitrogen for later attempts at implantation, donated for research, or eventually discarded. (Courtesy of the author)

The pluripotency of hESCs is what makes them so valuable for regenerative medicine. Eventual application of this research to repair diseased, injured, and worn-out tissues could improve, save, and lengthen the lives of millions of people. Diseases likely to respond to ESC therapy include insulin-dependent diabetes, Parkinson's disease, chronic heart disease, liver failure, kidney disease, multiple sclerosis, osteoarthritis, rheumatoid arthritis, some cancers, and even Alzheimer's disease. Human ESCs could become the preferred treatment for heart, kidney, and liver disease, replacing organ transplantation to treat these conditions. Skin and other tissues destroyed by severe burns could also be repaired using ESC therapy. Over 250,000 Americans suffer from spinal cord injuries, with about 11,000 new injuries occurring every year. Nearly 60 percent of all victims of spinal cord injuries are under thirty years old. Research during the past decade or so gives hope that ESC therapy may soon be able to return normal life to future victims of spinal cord injuries.

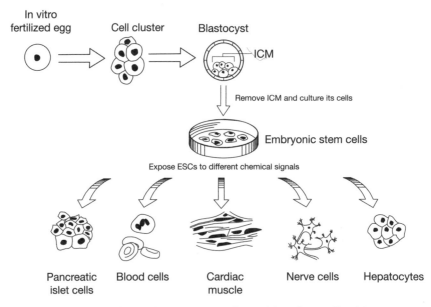

Figure 2.4. Creation of an embryonic stem cell (ESC) line from cells of the inner cell mass of a blastocyst stage embryo. Pluripotent ESCs grown in the laboratory are chemically stimulated to develop into desired specialized cell types. Current research and clinical trials aim to develop ESC therapies to repair diseased and injured tissues. (Courtesy of the author)

hESCs and the Personhood Question

Three views on personhood underpin decisions about the moral status of human embryos: the single-criterion view, the pluralistic view, and the potentiality view. The single-criterion approach identifies a single, decisive factor for granting personhood. For some people, this criterion may be conception, for others a beating heart, viability outside the womb, or another characteristic of the developing embryo or fetus. When that characteristic first becomes present, it is assumed that personhood is established. The pluralistic approach views personhood as emerging gradually during embryonic, fetal, and/or postnatal development. This view lends itself to assigning increasing levels of moral status to embryos and fetuses as they accumulate more and more of the set of traits that marks full personhood.[1] Finally, the potentiality approach assigns moral status to the embryo based on what it can become rather than what it actually is.

Single-criterion views on personhood come from both religious and secular realms. A frequent religion-inspired criterion for personhood is the presence of a soul; secular views usually focus on some biological attribute of the developing individual. Let us look at some examples of each approach.

Prior to the thirteenth century and the influence of Thomas Aquinas (1225–74), Christianity's view on the beginning of personhood is unclear. But with Aquinas, who adopted Aristotle's view on when the rational soul enters a human being, the Christian Church maintained that the human embryo becomes a person on the fortieth day after conception.[2] In addition, Christianity replaced Aristotle's rational soul with a God-given immortal soul.

No church claims to know precisely when ensoulment occurs, but Roman Catholicism chooses to give full moral status to the conceptus in order to be on the safe side. The Church now marks the beginning of personhood with conception. In the latter part of the nineteenth century, German and Belgian biologists showed that both plants and animals begin life as fertilized eggs. Responding belatedly to this biological information, the Roman Catholic Church declared in 1917 that human embryos must be treated as persons from conception onward. In August 2000, the Pontifical Academy for Life reiterated the Church's position that personhood begins at conception and condemned the use of human ESCs because human embryos are destroyed:[3] "The Church has always taught and continues to teach that the result of human procreation, from the first moment of its existence, must be guaranteed that unconditional respect which is morally due to the human being in his or her totality and unity in body and spirit: The human being is to be respected and treated as a person from the moment of conception; and therefore from that same moment his rights as a person must be recognized, among which in the first place is the inviolable right of every innocent human being to life." The Holy Synod of Bishops of the Orthodox Church in America and the Southern Baptist Convention join the Roman Catholic Church in giving the laboratory-generated conceptus personhood status and opposing ESC research. Unclear is whether these institutions consider the twenty-four-hour lag after conception (sperm fusion with the egg surface) and before fertilization (fusion of egg and sperm nuclei inside the egg) when forming positions on personhood. The distinction is important since the action of the "morning after" pill, use of which is opposed by the Catholic Church, occurs after conception but before fertilization.

Other Christian denominations take stands in the ESC debate without formal declarations on the beginning of personhood. The 2001 General Assembly of the Presbyterian Church (USA), a group of elected representatives from the denomination, endorsed the use of ESCs for research and therapy so long as they are derived from "embryos that do not have a chance of growing into personhood because the woman has decided to discontinue further [IVF] treatments and they are not available for donation to another woman for personal or medical reasons." The General Conference of the United Methodist Church adopted a similar position in 2004.

Single criteria for personhood status in human embryos also come from the secular realm. A US government panel on embryo research advised adherence to a fourteen-day rule for the latest stage at which human embryos can be used in medical research because fourteen days after conception is when some cells in the embryo are committed to develop into parts of the central nervous system, including the brain (NIH 1994). Since a pain-sensitive, rational brain is still many months away, the panel played far on the safe side with its day fourteen rule.

The viability of the fetus is a criterion for personhood with the status of federal law in the United States since the 1973 Supreme Court decision in *Roe v. Wade*. The Court's decision defines the first two trimesters as a period during which women cannot be denied an elective abortion. Six months into a pregnancy is still about the earliest that modern neonatal units can sustain fetal life outside the womb.[4]

Pluralistic views of personhood recognize an increasing progression of moral status as an embryo or fetus develops. Such views usually do not enter into debates about personhood and using human embryos for research. An example of pluralistic thinking regarding personhood is embedded in a resolution passed at the 1999 Southern Baptist Convention. The resolution declared that "protectable human life begins at fertilization." Presumably, this reflects the belief that an immortal, God-given soul is present at that time. But in the same document that declares opposition to the use of human blastocysts to obtain ESCs and describes human embryos as "the most vulnerable members of the human community," the convention reminds its denomination that abortion can be sanctioned "to save the physical life of the mother." The latter position reveals pluralistic thinking since it implies that the moral status of the mother is greater than that of the fetus. The document does not detail why the mother is given greater consideration, but one might assume that it is because she has accumulated more hallmarks of personhood, such as rationality and a life narrative, than has the fetus.

Potential personhood arguments declare that since an embryo has the potential to develop into a person, the embryo is entitled to the moral status of a fully developed person. Some bioethicists speak of "latent potential" versus "active potential," giving greater moral status to the latter. For example, an unfertilized egg and a nearby sperm cell have only latent potential to give rise to a person, compared to the active potential of a fertilized egg to do the same. The problem with distinguishing between latent and active potentials is that the line between them is necessarily arbitrary. For example, one could say that the potential of an embryo frozen at an IVF clinic is "latent," whereas that of one transferred to a uterus is "active." Or once inside the uterus, a preimplantation blastocyst is latent in its potential for personhood and an implanted blastocyst is active.

For bioethicist John Harris, assigning personhood to a cell or group of cells on the basis of its potential to develop into a person leads to absurdities. For example, all human sperm and eggs are potential progenitors of persons. Should we strive to help every egg and sperm cell realize its potential? Put in the hands of skilled cloning technicians, potential persons also reside inside the thousands of somatic (body) cells that we kill and rinse down the drain after bathing or brushing our teeth. Harris argues for conferring personhood to entities based on what they actually are, not on what they might become. Next, we consider some recent advances on the path toward the eventual clinical use of pluripotent stem cells.

Clinic-Ready hESC Repository

In 2017, the UK Stem Cell Bank created the first hESC repository designed specifically for clinical use. Glyn Stacey, who oversees the repository, explained that every hESC line offered abides by stringent ethical standards set by the European Tissue and Cell Directives "right from the donor selection, procurement of tissues, the consent process, the storage of the cells, and the transfer of the cells to the lab where they were derived" (Grens 2017). Establishment of this hESC repository is a huge boon to researchers and clinicians working toward clinical applications of hESCs. Previously, most hESC lines were derived solely for research purposes and often not with the use of clinical-grade manufacturing protocols. Getting a research cell line in shape for clinical use and then obtaining details of the cells' procurement procedures needed for clinical applications is challenging and very time consuming. With the new hESC repository, the time required to ready a cell line for clinical applications is greatly reduced. In the United States, a major funder of stem cell research is the California Institute for Regenerative Medicine (CIRM). Geoff Lomax, a program officer at CIRM, stated that the UK bank's procedures for procuring hESC lines satisfies all of CIRM's standards, so the new repository will be invaluable to clinical researchers in the United States as well as in the EU.

ESC Clinical Trials to Treat Parkinson's Disease

Plans were announced in mid-2017 to use hESCs in a clinical trial in China to treat patients with Parkinson's disease (Cyranoski 2017), and the Chinese workers' protocol passed pretrial muster in 2018. This will be the first clinical trial using ESCs derived from embryos grown from fertilized eggs. Similar trials are planned in the United States by researchers at the Scripps Research Institute in La Jolla, California, and the Memorial Sloan Kettering Cancer Center in New York City. An earlier clinical trial in Australia used cells derived by parthenogenesis from unfertilized eggs. The Chinese researchers will inject hESCs that have been treated to develop into brain cells into the region of the brain associated with Parkinson's disease in hopes that these cells will become dopamine-producing brain cells once inside the brain. In the US trials, cells already committed to developing into dopamine-producing neurons will be injected into patients' brains. So by the end of this decade, there should be concrete information about the efficacy of ESCs for treating Parkinson's disease.

Pluripotent Stem Cells without Using Embryos

In 2006, Japanese researchers stunned the stem cell research community by showing that pluripotent stem cells could be obtained without using human embryos. The new form of stem cells was derived from already specialized skin cells called fibroblasts. By giving these cells extra copies of four different genes, the workers got the cells to dedifferentiate to an embryonic state in which their developmental potential was similar to that of cells in very early embryos and also of hESCs derived from early embryos. The new class of pluripotent cells was called induced pluripotent stem cells or iPSCs (fig. 2.5). Four areas of research stand to benefit greatly from iPSCs: disease modeling, drug development, regenerative medicine, and gene regulation (Papapetrou 2016). In addition, clinical applications are in the foreseeable future for treating diseases and conditions including macular degeneration and other eye diseases, diabetes, myocardial infarction, and spinal cord injuries. Japanese researchers are poised to use iPS cells in a clinical trial for patients with Parkinson's disease (Normile 2018).

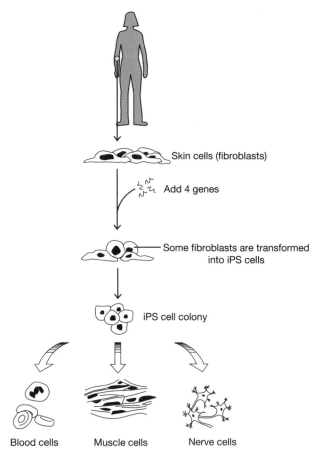

Skin cells (fibroblasts)

Add 4 genes

Some fibroblasts are transformed into iPS cells

iPS cell colony

Blood cells   Muscle cells   Nerve cells

Figure 2.5. Induced pluripotent stem cells created by adding four genes to the genome of skin cells called fibroblasts. The added genes include transcription factors that activate numerous other genes. Although only a small fraction of treated fibroblasts develop into iPS cells, the iPS cells that are produced appear to have the developmental potential of normal ESCs. (Courtesy of the author)

Minibrain Organoids from Stem Cells

Organoids are miniature organs grown in laboratory culture dishes from iPS cells or ESCs that are induced to develop into specific cell types. The cells spontaneously organize into three-dimensional cell clumps that exhibit some of the functions of mature organs. Since organoids are only a few millimeters in diameter and lack vascularization, they cannot perform all of the functions of a mature organ. Still, they are very useful for studying normal and diseased organ function and for testing the effects of pharmaceuticals and environmental toxins on specific organs. Researchers now study human organoids of thyroid, heart, liver, stomach, intestinal, pancreatic, kidney, testicular, and brain cells. As researchers coax human brain organoids or minibrains closer to mimicking brain function, some very interesting ethical questions arise.

Recently, seventeen neuroscientists, bioethicists, and law professors authored an essay identifying eight categories of ethical issues related to the future use of human minibrains. These include concerns about how to recognize whether a minibrain is conscious, human-animal blurring when human brain cells are incorporated into nonhuman animal brains, issues of informed consent and ownership when one's cells are used to create a minibrain, and the ethical handling of functional brain tissue at the end of an experiment (Farahany et al. 2018). Currently, the small size of minibrains, presumably due to lack of blood flow to the cells, and the imperfect replication of normal brain organization seem to preclude the development of consciousness within a minibrain, but this could change as the technology advances.

Preimplantation Genetic Diagnosis and Embryo Selection

What is preimplantation genetic diagnosis (PGD)? PGD is a technology that allows parents to assess the genetic health of embryos generated by IVF (fig. 2.3). When embryos are at about the eight-cell stage, clinicians remove a single cell from each embryo that a couple has produced and subject it to tests for various genetic diseases and conditions. The remaining clump of cells from each embryo continues to grow and retains the potential to give rise to a fetus if transferred to a woman's uterus (fig. 2.6). Depending on the results of DNA testing, certain embryos are selected for uterine transfer or stored frozen for future pregnancies. As for the genetically abnormal embryos, parents decide whether to discard them or donate them for research.

Ethical issues for PGD. PGD is generally used only to diagnose genetic content for couples that know they carry disease-causing mutations. Some clinics opt to test embryos for gender when a couple desires family balancing. No laws prohibit using PGD for family balancing in the United States, but clinics are not required to perform PGD for that purpose. Some persons adamantly oppose gender selection on grounds that it may lead to valu-

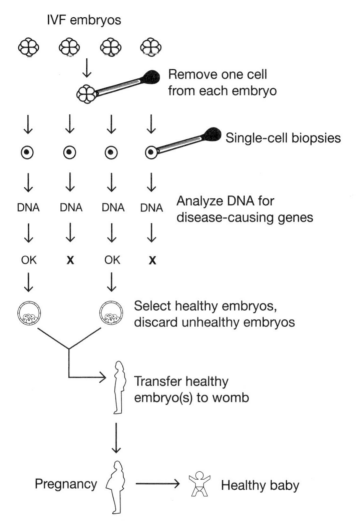

IVF embryos

Remove one cell
from each embryo

Single-cell biopsies

DNA   DNA   DNA   DNA   Analyze DNA for
                       disease-causing genes

OK   X   OK   X

Select healthy embryos,
discard unhealthy embryos

Transfer healthy
embryo(s) to womb

Pregnancy   Healthy baby

Figure 2.6. Preimplantation genetic diagnosis by single-cell biopsy. Testing
embryos created by IVF for genetic disease involves analyzing DNA extracted
from a single cell removed from each embryo at the six to ten cell stage. Embryos
develop normally after removal of one or two cells for biopsy. Disease-free embryos
transferred to the uterus and/or stored frozen for later transfer can develop into a
healthy baby. Embryos with genetic defects are discarded or donated by the parents
for research. (Courtesy of the author)

ing one gender over another and involves discarding perfectly healthy embryos of the other gender.

Creating a savior sibling is another controversial use for PGD. A savior sibling is a baby created using IVF and PGD for the purpose of providing an immunologically matched donor of stem cells for a sick, existing child. A famous savior sibling case is that of the Nash family from Denver in which parents used IVF and PGD to obtain a child whose umbilical cord blood was used as a source of hemopoietic stem cells (HSCs) to save the life of their daughter who had a rare form of leukemia, Fanconi anemia. HSCs give rise to all of the blood's red blood cells, white blood cells, and platelets. Genetically aberrant white blood cells cause Fanconi anemia, and replacement of the diseased cells with healthy cells from a sibling having histocompatibility with the sick child can cure the disease. In the case of the Nash family, PGD was used to select an embryo that was disease free and had the proper histocompatibility. Some critics of the procedure claim that using PGD to create a savior sibling is unethical because a child is being created for the purpose of saving another child and not simply because the family desires another family member. But in the few highly publicized savior sibling cases, it appears that the savior sibling is loved and cared for just as much as every other family member. In my earlier book, I discussed ethical aspects of gender selection and savior sibling creation in detail (Bradley 2013a).

Use of PGD for eugenic purposes is also a major concern for the future of the technology. Human values and lack of information about genetic determinants, not laws, are what presently prevent using PGD for selection of embryos to obtain children with certain hair or eye colors, physiques, cognitive abilities, or other non-disease-related traits. However, advances in human genomics and gene-editing technologies such as CRISPR (which is discussed in chapter 4) will eventually make eugenic uses of PGD possible. In order to use PGD to select embryos with particular genes or gene combinations, one must first have detailed knowledge about human genes, gene variants, and their functions. An impressive display of what is already known about human gene location and function is available at the Online Mendelian Inheritance in Man (OMIM) website.[5] Among the more than fifteen thousand human genes catalogued on OMIM are ones relating to dwarfism, deafness, short sleep habit, perfect pitch, epilepsy, early cataract formation, susceptibility to migraine headaches, psoriasis, schizophrenia, many cancers, age-related macular degeneration, eye and hair color, skin pigmentation, baldness, physique, and mathematical skills. Genetic testing for any one of these genes could be accomplished, but we do not yet have enough knowledge about gene function to use this information to create designer babies. For example, we have only a paucity of information about how the protein products of the approximately twenty thousand human genes interact with each other. Simply because a gene is known to be associated with perfect pitch does not mean that it does not have a dozen other unknown functions, some considered beneficial and some not. The same is true for known disease-causing genes. Single copies of genes that cause or predispose persons to sickle cell anemia, cystic fibrosis, and Alzheimer's disease may protect people from malaria, help prevent tuberculosis, or actually improve cognitive abilities, respectively, under certain environmental conditions (Velasquez-Manoff 2017). So, we are a long way from being able to safely and accurately design the genomes of future generations even if we wished to do so. But it is not too early to be thinking and discussing how our burgeoning information about the human genome ought to be used to test, genetically edit, and select human embryos.

Human Gene Therapy and Enhancement

Kinds of human genetic engineering. Human genetic engineering involves manipulating DNA in living, human cells. Potential types of human genetic engineering can be grouped into categories depending upon the kind of cell being altered and the motive for doing so. One way to classify genetic engineering is by the cell type: somatic (body) cells versus germ cells (eggs and sperm). Genetic alterations to somatic cells last only for the lifetime of the person receiving the genetically engineered cells be they liver, nerve, or white blood cells. Somatic cell alterations are not passed to a person's progeny. By contrast, germ line genetic alterations are potentially immortal since they can be passed on to future generations.

A second classification of genetic engineering is based on motive: therapeutic versus enhancing. Gene therapy aims to cure or treat a genetic disease or some other genetically based health problem. Genetic enhancement, not yet performed in humans, would aspire to augment non-health-related traits. For example, providing a person with perfect pitch, a particular hair color, increased empathy, or improved long-term memory would be an enhancement.

In a 1997 book *Ethics of Human Gene Therapy*, bioethicist LeRoy Walters and molecular biologist Julie Palmer use the following matrix to illustrate how combining the above categories yields four types of human genetic en-

Table 2.1. Types of human genetic engineering

Engineered cell type	Somatic	Germ line
Therapeutic	Type 1	Type 2
Enhancing	Type 3	Type 4

Note: There are two classification schemes for human genetic engineering: gene therapy versus genetic enhancement and somatic cell engineering versus germ line engineering. Together, the two classification methods yield four types of human genetic engineering, each with unique biological characteristics and ethical implications.

gineering: (1) somatic cell therapy, (2) germ line therapy, (3) somatic cell enhancement, and (4) germ line enhancement (table 2.1).

Somatic cell gene therapy (Type 1) is the only kind of genetic engineering currently done in humans, but that may change with the advent of mitochondrial replacement therapy, which we will observe shortly. The first clinical trial for somatic cell gene therapy in humans was in 1990, but the technology is not yet routine for treating or curing any disease or other medical condition. Still, recent advances in gene therapy technology and many ongoing clinical trials predict a promising future for Type 1 genetic engineering in humans.

Somatic cell gene therapy aims to cure genetic disease by correcting abnormal genes. Ways to do this include (1) inserting a normal gene into the genome in hopes that it will be active and compensate for the nonfunctional gene, (2) replacing the abnormal gene with a normal copy of the gene by an "add and delete" process, (3) repairing the defective gene while it is still in the genome, and (4) altering the abnormal gene by increasing or decreasing its activity. The first method is the most common approach.

Distinguishing between therapy and enhancement. The line separating genetic therapy from enhancement is not always sharp. Some writers prefer to call any genetic alteration an enhancement. For example, consider dwarfism or congenital deafness. Should these traits be considered a condition in need of therapy? Many dwarfs and deaf persons do not consider themselves handicapped. In fact, they see benefits in the qualities with which they were born and do not wish anything different for their children. Who are tall people to say that short persons should be genetically corrected? As another example of fuzziness between therapy and enhancement, consider one's immune system. If it could be strengthened beyond normal by genetic manipulations in embryos, the resulting persons may be better able to fend off viral infections and some cancers. Some argue that genetic alterations that prevent disease are therapeutic, whereas others argue that increasing any trait beyond normal is enhancement.

Many other ethical concerns about human genetic enhancement, particularly germ line enhancement, exist. These include possible resurrection of past efforts at eugenics, stigmatizing and devaluing persons with or without certain traits, unequal access to the technology would further stratify society into genetic "haves" and "have nots," germ line enhancement would violate the autonomy of unborn individuals, and the research needed to make the technology safe would require sacrificing human embryos.

Gene therapy trials. During 2013–16, nearly five hundred new somatic cell gene therapy clinical trials were approved worldwide, up from about four hundred during the previous four years. Sixty-five percent of all gene therapy clinical trials have been aimed at treating various types of cancer. Gene therapy treatments for inflammatory, ocular, neurological, cardiovascular, and infectious diseases each account for between 0.6 and 7.5 percent of all clinical trials.[6]

The kind of human gene therapy receiving the most media attention in 2016–17 was mitochondrial replacement therapy. Because it involves altering the human germ line, it is much more controversial than any of the somatic cell gene therapy trials. Mitochondrial gene therapy offers a way to remove disease-causing mitochondrial genes from a family's germ line and is discussed more fully in the next section.

Mitochondrial Gene Therapy or Three-Parent Babies

The phrase *three-parent baby*, appearing frequently in the news lately, has fueled a major bioethical controversy. What it refers to is an assisted reproduction technology used to remove genes for debilitating mitochondrial-based diseases from a family's lineage. The technology is easy to understand but very controversial. Before considering the controversy, we need to know something about mitochondria.

The mitochondrion is often called the "powerhouse" of the cell. A typical human cell has hundreds of these small organelles, and a human egg cell contains thousands of them. Mitochondria reside in the cytoplasm of the cell, that is, in the region outside the cell's membrane-bound nucleus that contains gene-laden chromosomes. In the deep evolutionary past, mitochondria were free-living bacteria. More than one billion years ago, they took up residence inside other larger cells, and those composite cells ultimately gave rise to all eukaryotic organisms, including ourselves.

The powerhouse descriptor comes from the fact that mitochondria are the cellular sites of energy transformation in a cell. In mitochondria, energy in food molecules is transformed into energy-rich molecules called adenosine triphosphate (ATP), the common energy currency for virtually all of a cell's life processes. Mitochondria are distinctive in two other ways: they possess their own small DNA molecule, and they are inherited exclusively from mothers. All of the mitochondria in your body are derived from the mitochondria in your mother's egg cell, the single cell that gave rise to you. Like the egg itself, mitochondria grow and divide, which is how the trillions of mitochondria in your body came into being. Now, what about that small amount of DNA inside every mitochondrion?

Mitochondrial DNA contains about 0.05 percent of the total DNA in a human cell and just thirty-seven genes compared to approximately twenty thousand genes in chromosomal DNA inside the nucleus. Even though mitochondria contain a relatively small number of genes compared to the rest of the cell's DNA, mutations in mitochondrial genes can cause over a dozen genetic diseases with symptoms ranging from deafness and diabetes to heart dysfunction and epilepsy (Chinnery 2014). Three-parent baby technology can ensure that babies are born with healthy mitochondria.

Suppose a couple wishes to have a baby, but the woman knows that her mother had a disease caused by a mutation in her mitochondrial DNA. This means that the woman also has the mutation even though it may not have manifested as a disease yet. By using another woman's egg containing healthy mitochondria and nuclear transfer technology during the IVF process, the woman can have a baby free of mitochondrial disease and still be the biological mother of her child.

The procedure is straightforward. A clinician carefully removes the nucleus from one of the mother's egg cells. The nucleus contains 99.95 percent of the mother's genetic information. The enucleated egg cell that contains thousands of genetically defective mitochondria is discarded. Another woman without defective mitochondria donates one of her eggs to provide healthy mitochondria. After removing the nucleus from the donor's egg, the clinician injects the mother's nucleus into it. At this point the clinician has engineered an egg cell containing normal mitochondria and the mother's nucleus with her genetic material. Next, the father's sperm fertilizes the engineered egg cell *in vitro* (fig. 2.7). Transfer of the resulting embryo to the mother's womb results in pregnancy and birth of a child containing chromosomes from its mother and father and mitochondria from the donor woman. Because the donor's mitochondria are normal, the child is spared

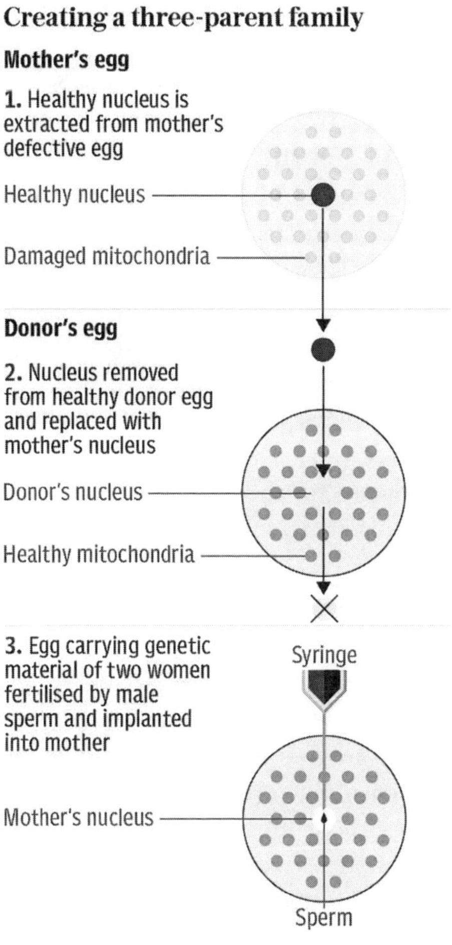

Creating a three-parent family

Mother's egg

1. Healthy nucleus is extracted from mother's defective egg

Healthy nucleus ———

Damaged mitochondria ———

Donor's egg

2. Nucleus removed from healthy donor egg and replaced with mother's nucleus

Donor's nucleus ———

Healthy mitochondria ———

3. Egg carrying genetic material of two women fertilised by male sperm and implanted into mother

Syringe

Mother's nucleus ———

Sperm

Figure 2.7. Three-parent babies. When the egg of a prospective mother has defective mitochondria containing a disease-causing mutation in mitochondrial DNA, a donor's egg can supply healthy mitochondria. The nucleus from the donor's egg, which contains all of the chromosomal DNA, is removed and discarded. Then, the nucleus from the mother's egg is removed from her defective egg and transferred to the donor's healthy egg. Finally, the donor's egg is fertilized with a sperm from the father. The resulting embryo contains chromosomal DNA from the mother and father (99.95 percent) and mitochondrial DNA from the egg donor (0.05 percent). Illustration from © Telegraph Media Group Limited 2015 and used with licensed permission. (© Telegraph Media Group Limited 2015)

the mitochondrial DNA–based disease carried by its mother. If the child is a girl, all of her eggs will have normal mitochondria, and none of her offspring will be afflicted with the disease carried by her grandmother.

In 2016, the UK's fertility regulatory office, the Human Fertilisation and Embryo Authority, sanctioned the three-parent baby method for preventing mitochondrial disease. The UK is the first nation to officially approve use of the technology. In early 2017, the authority approved use of the technique by a clinic at Newcastle University. US scientists working in Mexico, which has lax assisted reproductive laws, have reported the birth of a baby boy from the technology (Hamzelou 2016), and in 2018, a clinic in Ukraine reported having created four three-parent babies.[7] In addition to unknown safety issues for the babies, the main controversy with this technology is that it is a form of germ line DNA therapy/enhancement. Some bioethicists fear it will follow a slippery slope toward germ line enhancement procedures for nuclear DNA and eventually to eugenic practices having undesired societal consequences.[8]

Human Cloning

In 1997 came an announcement that the first cloned mammal had been born. The lamb Dolly was created by introducing the nucleus from a sheep somatic cell into an enucleated egg. This method of cloning is called somatic cell nuclear transfer (SCNT) (fig. 2.8). Since Dolly's birth, SCNT has produced many cloned animals including mice, cows, goats, pigs, a bull, an endangered Asian ox, a cat, a race mule named Idaho Gem, dogs, and monkeys.

Because humans are mammals, we share many aspects of reproductive biology with sheep, mice, and other hairy creatures. Therefore, with some refinements, the SCNT technology used to create Dolly can probably be used to clone humans. Evidence that this is so came in 2013 when researchers at the Oregon Health and Science University announced success in creating human embryos cloned from the DNA of fetal skin cells or DNA taken from the skin of an eight-month-old baby (Tachibana et al. 2013). hESCs were created from the embryo clones for use in research for disease treatment. This is an example of therapeutic cloning, which is distinguished from reproductive cloning in that the aim of the former is to obtain a blastocyst embryo from which to derive ESCs for clinical use, whereas the aim of the latter is birth of a child (fig. 2.8).

In 2014, two research groups took therapeutic cloning a step further by

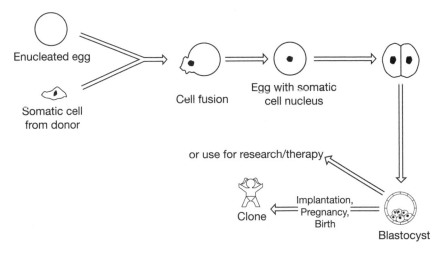

Figure 2.8. Cloning by SCNT. Fusion of a diploid somatic cell with an enucleated egg cell produces an egg that mimics a fertilized egg and develops into a complete individual. (Courtesy of the author)

deriving ESCs from blastocyst embryos obtained by cloning adult humans (Chung et al. 2014; Yamada et al. 2014). One group at the CHA University in Seoul, South Korea, cloned two men aged thirty-five and seventy-five using DNA from skin cells. Another group at the New York Stem Cell Foundation Research Institute cloned a thirty-two-year-old woman with type 1 diabetes. This second group went on to produce insulin-secreting cells from the ESCs. Derivation of ESCs from adult humans shows that therapeutic cloning is a feasible technology for obtaining immunologically compatible, pluripotent stem cells to treat injuries and disease.

For many persons, the thought of cloning humans elicits a gut-level repugnance. When pressed for specifics, objectors offer a range of reasons that they believe make human cloning immoral. Among these are that cloning is "playing God," it infringes on human dignity, it removes reproduction from the realm of human sexuality, and it violates a right to a unique human constitution. Upon analysis, none of these objections present an argument strong enough to justify banning human cloning. Currently, the strongest argument against human reproductive cloning is that the research needed to develop the technology would pose health risks to the pioneer fetuses. In addition, unrealistic and misinformed expectations of clones by parents or others could compromise a clone's freedom to develop her/his own life plan. The latter argument is not against the procedure per se; rather, it reflects the incorrect notion that a clone would be identical to the cloned individual in all regards including personality and life choices.

Genomics, Epigenomics, and Personal Lives

The Human Genome Project. The deciphered human genome is called by some the Holy Grail for medicine and cell biology. Just what is a genome, and why is it important to learn about the human genome? As described in chapter 1, the combined DNA from one copy of each type of chromosome in an organism constitutes that species' nuclear genome (fig. 1.4).[9] Determining the base (A, T, C, and G) sequence in an organism's DNA, identifying the location of genes, and analyzing the organization of the DNA in chromosomes is called genomics. Knowing the precise base sequence of an organism's DNA allows genome researchers to identify the sequences coding for proteins and to deduce from the genetic code (appendix 1) the amino acid sequence of every protein required for the development and life of that organism.

In 1988, the United States made a national commitment to determine the sequence of all 3.1 billion base pairs of the human genome. The venture was dubbed the Human Genome Project (HGP). Soon after the United States committed to an HGP, other nations were brought on board as collaborators in an international public consortium called the International Human Genome Sequencing Consortium (IHGSC) to sequence the entire human genome. The project was completed by 2003 with an error rate of only about one in every one hundred thousand bases.[10] The genomic information, licensed technologies to private companies, and research grants to the private sector that emerged from the HGP spawned a multibillion-dollar, international biotechnology industry.

Five percent of the HGP's budget approved by Congress in 1989 was devoted to mandatory studies of ethical, legal, and social issues (ELSI) associated with human genomics. Contributors to this subproject were theologians, ethicists, scientists, and philosophers in study groups, at conferences, on panels, and as authors of books and articles. Their work fulfilled the ELSI mandate and addressed a variety of societal concerns arising from advances in genomic science, including (1) use of genetic information in reproductive decisions, (2) possible stigmatization and negative psychological effects due to genetic differences, (3) designing appropriate policies on patents and commercialization of gene-based products, (4) distinguishing between gene therapy and genetic enhancement and a resurgence of eugenics, (5) fairness, privacy, and confidentiality in the handling of genetic information, (6) assessing the role genes play in human behavior and perceptions of genetic determinism, and (7) making informed decisions based on genetic test results with uncertain health implications.

Human gene patenting. A major development for human genomics since completion of the HGP in 2003 was the 2013 ruling of the US Supreme Court banning the patenting of human genes in their naturally occurring form (Liptak 2013; Marshall 2013).[11] As discussed in chapter 1, at issue was a fifteen-year monopoly by Myriad Genetics on a $3000 test for mutations in the *BRCA1* and *BRCA2* genes that put women at high risk for breast and ovarian cancer. A consortium of doctors and researchers sued Myriad over its patents on the two naturally occurring genes, arguing that Myriad's discovery of an object in nature does not justify a patent. The Supreme Court agreed and in its unanimous opinion declared that "Myriad did not create anything. To be sure, it found an important and useful gene, but separating that gene from its surrounding genetic material is not an act of invention" (US Supreme Court 2013). The decision allowed other gene-testing companies to devise tests for *BRCA* genes and lower the cost of testing, which made obtaining vital information possible where it had previously been out of reach for many women. But as discussed in chapter 1, an ethical issue remains in Myriad's exclusive use of its proprietary database developed over many years from patients undergoing its BRACAnalysis test.

The pangenome. Before 2015, most genome researchers thought that sequencing the genome from one or a few individuals would give an accurate picture of the genome for an entire species. Researchers of course recognized that a particular gene may differ in different individuals. For example, eye color gene(s) for blue eyes in one person and brown or green eyes in other person represent diversity in eye color gene(s). But few persons suspected that major portions of the genome of an individual or of a strain of bacteria could be unique to that individual or strain. Now, we know differently.

During the years following completion of the HGP, advances in DNA sequencing allowed thousands of genomes from each of many different species to be sequenced. A surprising outcome of this work is the discovery of the pangenome. The pangenome is "the entire set of genes possessed by all the members of a particular species," and the total (estimated) size of a species' pangenome is called the supragenome (Offord 2017). Pangenomics is still in its infancy, and so far, most work is with bacteria and a few crop plants.

What is unfolding though is that individuals (or strains of bacteria) possess three types of genes: core genes present throughout the species, variable genes present in some but not all individuals, and unique genes restricted to just one individual. Surprisingly, core genes account for only a small fraction of the pangenome of many species. For example, the photosynthetic marine bacterium *Prochlorococcus* has a core genome of one thousand genes but an estimated supragenome exceeding eighty thousand genes. Pangenome analysis is expected to allow production of universal vaccines against viral and bacterial pathogens and to give insights into causes and treatments for human disorders including Parkinson's disease, Alzheimer's, and autism.

The epigenome and the epigenome project. Now that the human genome is deciphered, many genome researchers are turning their attention to the human epigenome. *Epi-* means "on the outside of." Thus, the epigenome is a layer of genetic information superimposed over the genome itself; that is, over the sequence of the A, T, G and C bases in the DNA. Specifically, the epigenome consists of patterns of chemical modifications (epigenetic factors) to certain bases in the DNA or on some of the proteins associated with DNA molecules (fig. 2.9).[12]

One science journalist likens the epigenome to "chemical amendments that dangle like charms on a bracelet from the linear string of letters that spell out the genetic code" (Nemade 2005).[13] To decode the meaning of those chemical amendments, a consortium of institutes and researchers recently organized the Human Epigenome Project (HEP).[14] Why is learning about the human epigenome so important? A cell's epigenome regulates the activity of the genome. Since the genome's activity is different in different cell types, there must actually be many types of epigenomes. For example, there is an epigenome peculiar to liver cells, heart cells, nerve cells, gut cells, and to each type of cancer cell.

The stated goal of the HEP is "to identify all the chemical changes and relationships . . . that provide function to the DNA code, which will allow a fuller understanding of normal development, aging, abnormal gene control in cancer and other diseases, as well as the role of the environment in human health" (Jones and Martienssen 2005, 11241). In 2005, the editor-in-chief of the journal *Cancer Research* strongly supported establishing a HEP, writing that "therapies using the epigenome as a target have the capacity to deliver on the promise of genomic/molecular medicine" (Rauscher 2005, 11229). Soon thereafter, a HEP consortium became reality with current members including the Wellcome Trust Sanger Institute, Epigenomics AG, and the Centre National de Génotypage.

The Epigenome

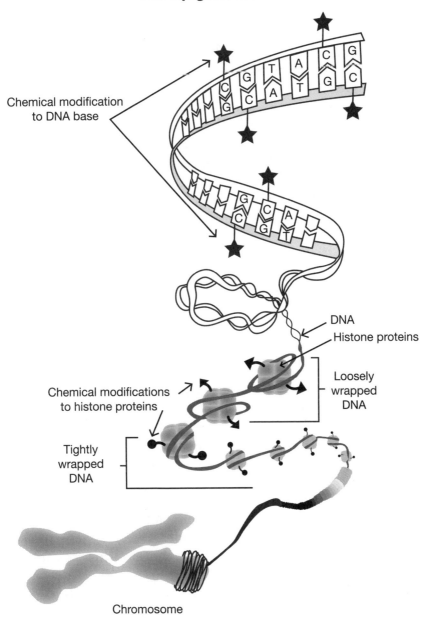

Figure 2.9: The epigenome consists of chemical modifications to some of the C bases that lie next to Gs in the same DNA strand and of chemical modifications to some of the histone proteins around which double-stranded DNA is wrapped. Some histone modifications loosen DNA's association with the histones and facilitate gene activity. Other histone modifications tighten DNA's association with the histone proteins and repress gene activity. (Courtesy of the author)

Since 2005, impressive advances relating the epigenome to cancer causes, diagnosis, and therapy presage a revolution in understanding and treating cancers (Esteller 2011). For example, by 2010 approximately three hundred epigenetically modified genes had been associated with cancer cells. Some of these genes are called tumor-suppressor genes whose activity normally prevents noncancer cells from transforming into cancer cells. An epigenetic modification called methylation can inactivate genes.[15] When a tumor-suppressor gene is methylated, cancer can ensue. Examination of the methylation pattern in a certain gene is now a powerful diagnostic tool for prostate cancer. Analyzing the epigenetic modifications on certain genes can indicate the most effective chemotherapy for certain brain tumors and breast cancers. It appears certain that learning more about the epigenome of cancer cells will significantly improve the early detection and treatment of many types of cancer.

Epigenomic changes induced by environmental factors, including a person's experiences, can influence human behavior (Berreby 2011). For example, connections exist between childhood abuse, epigenetic changes in the brain, and suicidal behavior. Behavioral epigeneticists also study connections between the epigenome and long- and short-term memory, drug addiction, trauma, childhood aggression, and depression.

Precision medicine. Personalized or precision medicine "is an emerging approach for disease prevention and treatment that takes into account people's individual variations in genes, environment, and lifestyle" (Precision Medicine Initiative 2015). Announced in President Obama's 2015 State of the Union address, the Precision Medicine Initiative (PMI) aims to tailor medicine to the individual. As it is, the efficacy of a medical treatment or drug is not the same for everyone. Some people have adverse side effects from certain drugs, which makes treatment with an alternative drug or drug combination a better choice. Personal variation in responses to and tolerance of specific pharmaceuticals is often at least partly genetically based. One hoped-for benefit of the HGP is the efficient personalization of medicine based on one's genomic constitution, and the PMI's objective is to realize that benefit. The PMI is housed in the NIH and overseen by NIH director Francis S. Collins, who also directed the HGP. The initiative contains an All of Us Research Program enlisting at least one million volunteers to share information about their medical histories, lifestyles, environments, and genetics. The All of Us website invites anybody living in the United States to participate in the program.[16] The goal of the All of Us program is to enhance health by extending precision medicine to the treatment of all diseases.

An example of precision medicine in action, separate from the US's PMI research program, is the Human Heredity and Health in Africa (H3Africa) Initiative established in 2010 by the NIH and the London-based Wellcome Trust (Nordling 2017). H3Africa promotes collaborations between scientists in Africa, the United States, and Europe to apply genomics to improve health care for native Africans and others of recent African descent. This program intends to change the present situation in which 81 percent of global genome-wide association studies, which link genetics to disease propensities and drug tolerance, are on persons with European ancestry and only 3 percent on persons with African roots. Opportunities for H3Africa projects include testing HIV patients for their tolerance to the anti-HIV drug efavirenz, testing persons for a gene variant causing a rapid conversion of the painkiller codeine into the dangerous opiate morphine, and investigating a gene variant that increases the risk of developing kidney disease but also confers resistance to the parasite that causes sleeping sickness.

Genetic basis for amyotrophic lateral sclerosis (ALS). More than six thousand people in the United States are diagnosed with ALS (Lou Gehrig's disease) each year. Recent genomic studies of families with a hereditary form of ALS revealed an abnormal number of repeats of a short base sequence (GGGGCC) within a particular protein-coding gene on chromosome 9. Normally, this sequence is repeated no more than twenty-three times, but in persons with hereditary ALS, it is repeated up to thousands of times. How this causes ALS is not clear. Some evidence suggests that the repeated base sequences result in the production of proteins with toxic effects in nerve cells that control muscle movement. This new knowledge about the genetic basis for ALS and results from ongoing studies of other mutations associated with the disease should ultimately lead to new approaches for treating ALS. Once causative genetic variants are clearly identified, a difficult decision that carriers of the variants may face is whether to use PGD to prevent the birth of babies with ALS. On the other hand, perhaps precision medicine for various types of ALS will offer treatments so effective that the ethics of using eugenic measures to prevent the disease will become a moot issue (Taylor et al. 2016).

Age Retardation

Age retardation involves slowing down the rate of undesirable changes to our minds and bodies as they move through time. It is different from life prolongation, which simply protracts life regardless of its quality. Age retardation is one approach to extending the period of vigorous, enjoyable living.

Four major lines of age retardation research are hormone therapy, caloric restriction, antioxidant treatment, and genetic. Each is based on a particular hypothesis about aging, and each shows potential for slowing the aging process in humans. Together, these approaches may eventually extend vigorous life spans well beyond one hundred years. Restricting calorie intake by 60 percent can increase the life span by 50 percent in organisms ranging from yeast and worms to dogs and monkeys. In 2002, a research project investigating antiaging effects of calorie restriction in nonobese humans began with funding from the National Institute of Aging, but it is still too early to tell the effects of the regimen on life span.[17] The genetic approach also has dramatic effects in nonhuman animals. Mutating just one gene associated with aging extends the life of roundworms by two to six times that of unaltered worms and by 50 percent in mice.[18] The potential for the new gene-editing tool CRISPR (discussed in chapter 4) to alter similar genes in humans makes public discussion of the pros and cons of age retardation in humans a pressing matter. For example, who will have access to the procedures, are there unintended consequences for lives in which all stages are extended by 50 to 100 percent, how would family and spiritual life be affected by 150- to 200-year life spans, and what about effects on world population dynamics and the work force?

Synthetic Biology

Synthetic biology is an interdisciplinary endeavor emerging from the convergence of molecular/cellular biology, genetic engineering, physics, chemistry, biochemistry, electrical engineering, nanotechnology, and information technology (fig. 2.10). Many researchers who call themselves synthetic biologists are trained in physics, electrical engineering, or computer science. Synthetic biology is a highly collaborative venture, drawing on the expertise, creativity, and goals of a cadre of scientists, engineers, and technicians. For convenience, we will use the term *synbiologists* for persons who practice synthetic biology. In a way, we are all synbiologists because the products of synthetic biology will reflect the research we support and the commodities we wish for, purchase, and allow to be created.

Some commentators claim that synthetic biology and genetic engineering are basically the same thing and that regulations already in place for genetic engineering are sufficient for synthetic biology (Parens et al. 2008). Others refer to synthetic biology as "genetic engineering on steroids" or "extreme genetic engineering" to emphasize a qualitative difference between the two

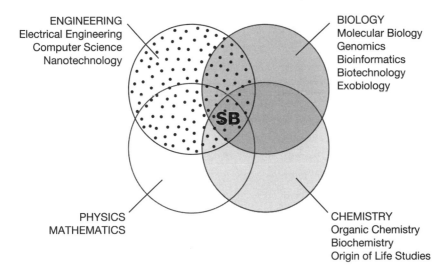

Figure 2.10. The interdisciplinary nature of synthetic biology. Synthetic biology thrives on an amalgamation of ideas and methodologies derived from the life sciences, chemistry, physics, other natural sciences, engineering, and mathematics. (Courtesy of the author)

technologies.[19] In this section we compare the two technologies, and the reader is invited to decide whether synthetic biology deserves to be in a scientific and/or ethical category by itself.

Genetic engineering and synthetic biology are similar in that both produce organisms with genomes different from the genomes made by nature. On the other hand, the two technologies differ in at least three ways: (1) the character of the life forms produced, (2) the underlying objectives of the technologies, and (3) the views of nature reflected and advanced by the technologies.

Life forms constructed by synthetic biology differ from those fashioned by genetic engineering. Genetic engineers tinker with the genetic material of organisms but do not fundamentally change the organisms themselves. Usually the character of a genetically engineered organism differs from its nonengineered counterpart by just a single trait. For example, a corn plant given bacterial genes to render it resistant to insect pests is still a corn plant and still does corny things such as tassel, produce pollen, and develop ears. By contrast, synthetic biology aims to create brand new genes, gene circuits, genomes, and organisms with traits vastly different from anything invented by nature during life's 3.5 billion years on Earth. Currently, synthetic biology focuses mainly on single-celled organisms (bacteria), but its

Top-Down Synbio

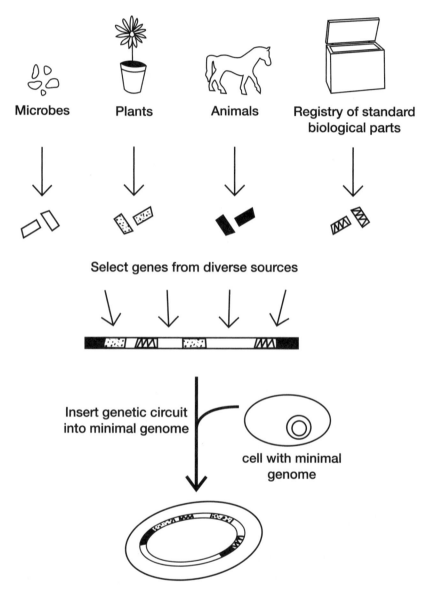

Figure. 2.11. Top-down synthetic biology uses genes or other components of existing organisms to generate radically new forms of life. For example, a "minimal genome," the bare essential genetic information needed to maintain microbial life, serves as a chassis upon which to add genetic elements from diverse sources that work together to direct the activities of a new life form. (Courtesy of the author)

vision for the future encompasses multicelled organisms. Stanford University synbiologist Drew Endy claims that "there is no technical barrier to synthesizing plants and animals, it will happen as soon as anyone pays for it" (ETC Group 2007, 7; Ball 2004). Finally, synthetic biology's integration of designer genes into specially prepared genomes is much more precise than gene transfer via genetic engineering where the number and location(s) of foreign gene(s) in the host genome are difficult to control.

There are two basic approaches to doing synthetic biology: top-down (fig. 2.11) and bottom-up (fig. 2.12). The top-down approach uses DNA segments or genes with known functions and arranges them in creative ways to design new life forms with properties useful to at least some humans. Top-down synbiologists insert extensive swaths of uniquely combined pieces of DNA into existing cells. The inserted DNA may be a mixture of genes from several other organisms, or it may be completely artificial—envisaged by humans and made by a DNA-synthesizing machine. One active area of top-down research consists of developing host cells whose own genomes are pared down to facilitate acceptance of the new DNA and acquisition of correspondingly new cellular life styles.

Bottom-up synthetic biology aims to create new life forms from scratch using readily available chemicals. Bottom-up synthetic biology is not yet a reality, but if and when it arrives, both the genetic information (DNA or some new type of genetic information-carrying molecule) and its cellular containers will be wholly human-made. Both top-down and bottom-up synthetic biology make humans the creators of brand new forms of life that are dramatically different from nature's products created via biological evolution.

Top-down synthetic biology. Assembling a large collection of "standard biological parts" (SBPs) began in 2003 as the project of a research group at MIT, and in 2012, it morphed into the independent, nonprofit iGEM (International Genetically Engineered Machine) Foundation. An SBP is a well-defined segment of DNA with a known function in living cells. Some twenty thousand SBPs now stored in iGEM Foundation super-cold freezers comprise iGEM's Registry of Standard Biological Parts available to synbiologists worldwide.[20] SBPs are categorized under headings that reflect collaborations between biologists, physicists, and engineers: for example, measurement, reporters, inverters, protein generator, protein coding, regulatory, tags, signaling, composite devices, and terminators. The idea is to mix and match the parts to construct artificial gene circuits that function inside living cells and

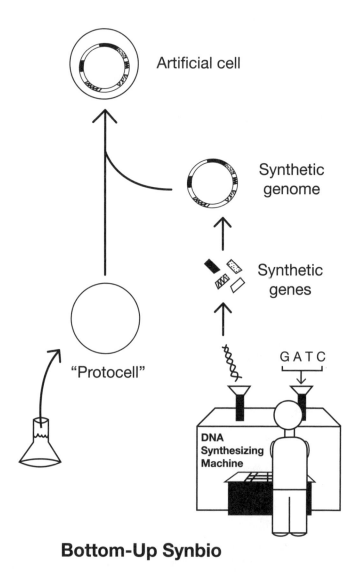

Bottom-Up Synbio

Figure 2.12. Bottom-up synthetic biology aims to build living cells from scratch. Here a human-made cellular container, the "protocell," receives human-designed genes made by a machine. The result is a hypothetical, artificial, living cell designed to perform a particular task or produce a specific product. (Courtesy of the author)

instruct them to do new things. It is like putting new software into a computer. Combinations of SBPs comprise the "software," cells themselves are the "hardware," and inserting the former into the latter commandeers cells to do the bidding of the software creators.

Likened to LEGO bricks, SBPs are dubbed BioBricks.[21] Each BioBrick is customized to join to other BioBricks in certain conformations. Some Bio-Bricks are derived from living cells, while others are of purely human design. The ultimate aim is to assemble a collection of BioBricks large and diverse enough to allow the construction of almost any imaginable genetic outcome.

Synlogic, a company using synthetic biology to create living medicines, recently reported positive results from the first clinical application of synthetic biology (Synlogic News Release 2018). Researchers at Synlogic in Cambridge, Massachusetts, tested the efficacy of genetically redesigned bacteria to relieve patients' symptoms of a liver enzyme deficiency. Patients in the clinical trial had a urea cycle disorder (UCD) in which ammonia produced in the gut during food digestion accumulates to dangerous levels in the body. High-protein foods such as meat, dairy products, and eggs contain lots of nitrogen. Normal metabolism ensures that excess nitrogen is excreted in urine as urea, but in persons with a UCD, that nitrogen ends up as toxic ammonia. To remedy the condition, the Synlogic researchers gave the common gut bacterium, E. coli, a new genetic circuit that removes ammonia from the gut as soon as it is produced and converts it into the non-toxic amino acid arginine.

Other recent products of synthetic biology still undergoing research and development include microbes designed to combat sepsis and cancer, human cells that produce insulin, bacteria with a revised genetic code that require synthetic amino acids for growth, and yeast cells that synthesize the opiate pain killer hydrocodone.[22] The latter involved giving yeast twenty-three new genes from four species of plants, rats, and a bacterium. In the case of cancer, researchers converted the normally pathogenic bacterium Salmonella into an organism that can home in on cancer cells and release toxic drugs locally to kill the cells.

Synthetic biology seems poised to transform medical care, and most people welcome this application of synthetic biology. But vigilance is in order as synbiologists become more adept at creating new life forms to do new things. Possible misuse of the technology for nefarious ends, such as biological warfare, bears monitoring. Also, methods in synthetic biology combined with the gene-editing technology CRISPR will eventually make feasible the creation of novel multicelled eukaryotes. Plants and animals with traits not invented by the nearly four-billion-year-long process of biological evolution, if released into naturally evolved ecosystems, could disrupt and compromise the integrity of interdependent webs of life on the planet.

Bottom-up synthetic biology. Two areas of research that focus on creating new life forms from scratch show tremendous progress in recent years: gene synthesis and self-replicating protocell formation. New gene-synthesizing technologies enlist microfluidic, automation, and silicon chip-based technologies that dramatically decrease the cost and time required to create "designer" genes from off-the-shelf chemicals. Researchers at industrial centers in the Bay Area, La Jolla, California, and Boston work to design genes that could create meat and milk without involving animals, renewable fuels, and fabrics stronger than steel but soft as a kitten. Accompanying risks are that gene-synthesizing technology may be used by bioterrorists to create deadly new microbes and that synthetic organisms escaped from laboratories could threaten human health or damage ecosystems.

As for protocell formation, Nobel Laureate Jack Szostak at Harvard Medical School is the undisputed leader. His research aims to understand how life could have originated on Earth via natural chemical processes by replicating feasible early Earth events in the laboratory. Szostak and his colleagues have created vesicles with membranes similar to living cells. These synthetic protocells grow and divide in an environment containing the raw materials for growth, and they can encapsulate genetic information in the form of RNA. The RNA, which represents a primitive genome, distributes itself to daughter vesicles when the protocells divide. A goal of Szostak's laboratory is to create a synthetic cellular system that undergoes Darwinian evolution, which will provide insight into universal properties of modern cells.[23]

Balancing benefits and risks of synthetic biology. In a December 2010 report on synthetic biology, President Obama's Presidential Commission for the Study of Bioethical Issues analyzed the benefits and risks of synthetic biology.[24] The commission recommended a policy of ongoing "prudent vigilance" to monitor, identify, and mitigate risks and potential harms from synthetic biology. Among the commission's eighteen specific recommendations are: establishing a coordinating body or mechanism to regularly review developments in synthetic biology, assessing the appropriateness of regulations relevant to the field, interacting with the international community, and keeping the public informed about its work. Regarding responsibility and accountability for safety and risk management, the commission emphasizes corporate responsibility and self-regulation by the research community, application of the existing *NIH Guidelines for Research Involving Recombinant DNA Molecules*, and adherence to the national Environmental Policy Act for the field release of research organisms.[25] As of mid-2018, President Trump had not established a similar commission to study and make recommendations on bioethical issues.

Other knowledgeable and responsible persons take a view different from President Obama's commission and argue that "prudent vigilance" is not an adequate response to synthetic biology. A letter undersigned by fifty-eight organizations from twenty-two countries written to Amy Gutmann, past chair of President Obama's Presidential Commission for the Study of Bioethical Issues, maintains that the precautionary principle (Raffensperger and Tickner 1999) not "prudent vigilance" should guide synthetic biology regulations.[26] One component of the precautionary principle shifts the burden of proof of the safety of an action to the proponents of the action and away from those who are concerned about safety (Kriebel et al. 2001). In my earlier volume on biotechnology and human values, I argue that synthetic biology needs an ethical framework beyond that for existing bioethical issues (Bradley 2013a, 274–81). One project for a "synbio ethics" is to assess the values we place on naturally evolved organisms compared to organisms created by humankind. Would a human-created plant or animal ever deserve to be protected on an endangered species list? So far, we have been curators of life on Earth. How should we behave as actual creators of life? One thing perhaps everybody can agree upon is that decisions about creating and using new life forms need to involve both scientists and nonscientists. As put by MIT synthetic biology researcher Kevin Esvelt (2017, 51), "if we are to engineer life, let us all decide how to do it together."

Chapter Summary

Recent advances in biotechnology raise diverse ethical and social issues including privacy, criteria for personhood, eugenics, environmental integrity, and how inviolable human nature and life itself ought to be. The cost and time required to sequence a human genome has plummeted so dramatically as to make precision medicine, treatments tailored for one's specific genetic constitution, a likely reality in the foreseeable future. But along with detailed genetic knowledge about individual patients come risks to genetic privacy and of genetic discrimination despite the 2008 Genetic Information Non-discrimination Act. Gene editing by CRISPR technology promises improvements in agriculture and energy sectors but also poses potential to alter human nature itself and to dramatically change the Earth's biosphere. How to use CRISPR technology safely and for the benefit of the planet and humankind deserves broad public discussion. The same applies to our future use of age retardation technology and synthetic biology, the latter of which is on the verge of making humans creators of life itself.

Questions for Reflection and Discussion

1. Do you believe that using three-parent mitochondrial gene therapy to prevent mitochondrion-based disease is a slippery slope leading to human germ line enhancement and "designer babies"? If not, why not? If so, do you have a problem with germ line enhancement?
2. If it became possible, would you wish your children or grandchildren to be genetically enhanced to retard the aging process so that they live productively and healthfully for several centuries? Why or why not?
3. Are you in favor of human therapeutic cloning to obtain stem cells to cure debilitating diseases such as cancers, Parkinson's disease, and sickle cell anemia? If not, why not?
4. What ethical questions do you believe arise from synthetic biology?
5. How important do you believe it is for the general public to be aware of the material in this chapter? Government officials?

Notes

1. Specific traits used to recognize full-fledged personhood vary among ethicists and religious thinkers. They may include brain activity distinctive to consciousness, living free of the mother's body, having the sense of a life narrative, and possessing the ability to reason morally.

2. Aristotle proposed three souls for humans: a nutritive soul that guides basic activities essential for life, including food intake, growth, and reproduction; a sensitive soul possessed by all animals that permits perception of the environment and movement; and a rational soul responsible for our abilities to reason and symbolize. According to Aristotle, all animals possess nutritive and sensitive souls, but only humans have a rational soul. Interestingly, Aristotle and Aquinas taught that ensoulment at forty days applies only to male em-

bryos and that female embryos require ninety days of development before the inviolable event occurs.

3. The Pontifical Academy for Life is an institution within the Roman Catholic Church founded in 1994. It promotes the Church's position that all human life is sacred and should be protected. It also acts as a think tank on bioethics and Catholic moral philosophy.

4. The onset of viability may vary a week or two, depending upon the fetus and the neonatal care given to it, but there are biological constraints on how far back technology can push viability. For example, one limiting factor is the development of the lungs to a point at which they can exchange oxygen between the atmosphere and the bloodstream. It is highly unlikely that technology to sustain fetuses much younger than six months will be developed in the foreseeable future.

5. The Online Mendelian Inheritance in Man (OMIN) site offers free public access to a compendium of information about over fifteen thousand human genes, their known functions, and their chromosomal locations. Established in 1996, the site receives new and updated entries daily. By using the "statistics" section of the site, one can easily monitor the number of genes and associated traits on the site, https://www.omim.org/about (accessed April 12, 2017).

6. *Journal of Gene Medicine*, Gene Therapy Clinical Trials Worldwide, John Wiley and Sons. http://www.abedia.com/wiley/ (accessed October 15, 2018). This site gives numbers of clinical gene therapy trials by year from 1989 to the present, continent, country, type of gene transferred, condition treated, vector used for the gene transfer, and clinical phases I–III.

7. A transcript of a National Public Radio report on the procedure used in Ukraine is at https://www.npr.org/sections/health-shots/2018/06/06/615909572/inside-the-ukrainian-clinic-making-3-parent-babies-for-women-who-are-infertile (accessed June 8, 2018).

8. The Center for Genetics and Society website (http://www.geneticsandsociety.org/index.php) is a rich source for thoughtful essays on this topic and other ethical issues associated with biotechnology. A site search for "mitochondrial replacement" yields dozens of essays.

9. Specifying the genome derived from eukaryotic chromosomal DNA as the *nuclear genome* distinguishes it from the mitochondrial genome also present in animals and plants and from the chloroplast genome present in green plants. It is called the nuclear genome because chromosomes are located in the cell's nucleus.

10. International Human Genome Sequencing Consortium 2004. This paper

reports on the high accuracy sequencing of 99 percent of the portion of the genome containing active genes, i.e., 2.85 billion bases. This classic paper also reports the surprising finding that the human genome contains only 20,000–25,000 protein-coding genes.

11. The Supreme Court ruling does still allow patenting of laboratory-altered forms of human genes. Specifically, complementary DNA (cDNA) prepared by copying the messenger RNA (mRNA) for a protein-coding gene to obtain a complementary DNA copy of the mRNA is allowed. One of cDNA's many uses is to detect and measure the level of specific gene expression in a cell or tissue.

12. The most common DNA modification is methylation of the base C when it occurs just before a G base. Generally, DNA methylation inactivates genes in which it occurs. Histone proteins function to package eukaryotic DNA within chromosomes. Histone modification by methylation, acetylation, or phosphorylation can loosen or tighten DNA packaging and thereby make genes accessible or inaccessible to being expressed. Generally, histone methylation and phosphorylation are associated with inactive DNA and acetylation with active DNA.

13. The metaphor appeared in Rashmi Nemade's article "Human 'Epigenome' Project" on December 21, 2005, in HUM-MOLGEN news, an international internet resource on HUMan MOLecular GENetics: http://hum-molgen.org /NewsGen/12-2005/000025.html, accessed October 19, 2018.

14. Organization of a Human Epigenome Project was proposed in 2005. Current members of the consortium are the Wellcome Trust Sanger Institute, Epigenomics AG, and the Centre National de Génotypage. The consortium's website is at http://www.epigenome.org.

15. DNA methylation involves replacing a hydrogen (H) atom with a methyl group (CH_3) on several C-bases that lie in a region of DNA just in front of the start point for a gene. Methylation of this region of DNA inactivates the nearby gene by preventing it from being ready to produce messenger RNA molecules needed for protein production. A clear, concise, nicely illustrated description of DNA methylation and other epigenetic modifications is at S. Kubicek, "Epigenetics: A Primer," *Scientist* 3: 3–3.

16. https://www.nih.gov/allofus-research-program/participation.

17. The project is called CALERIE (Comprehensive Assessment of Long-Term Effects of Reducing Intake of Energy). http://calerie.dcri.duke.edu/index .html (accessed September 16, 2009).

18. The roundworm species *Caenorhabditis elegans* is the worm used in genetics of aging studies. It is a favorite experimental animal for developmental

and cell biologists because it is easy to grow in the laboratory, every cell division from the fertilized egg through the adult worm is mapped in time and place, and its entire genome containing about twenty thousand genes (roughly the same number as in humans) is sequenced.

19. ETC Group, 2007. This sixty-four-page review of Synbio and its implications contains 281 endnotes and is written in nontechnical language.

20. The iGEM's Registry home page introduces the registry like this: "The iGEM Parts Registry is a growing collection of genetic parts that can be mixed and matched to build synthetic biology devices and systems. As part of the synthetic biology community's efforts to make biology easier to engineer, it provides a source of genetic parts to iGEM teams and academic labs." http://parts.igem.org/Main_Page (accessed October 20, 2018).

21. BioBricks are actually molecular entities: DNA segments like promoters and terminators, genes coding for proteins like repressors and activators of gene expression, and DNA that codes for RNA with regions that tell ribosomes (cellular protein synthesizing machines) how many proteins to make, how quickly to make them, when and where to stop protein synthesis, and how long the lifespan of a protein will be.

22. The purpose of this research was to create a microbe that cannot survive outside the laboratory, providing an intrinsic biocontainment mechanism for microbes used in synthetic biology.

23. The website for Jack Szostak's laboratory is a rich source of information about current work in origin of life and creation of life studies: http://molbio.mgh.harvard.edu/szostakweb/index.html (accessed April 13, 2017).

24. Presidential Commission for the Study of Bioethical Issues, 2010. *New Directions: The Ethics of Synthetic Biology and Emerging Technologies*, Washington, DC. Available at http://www.bioethics.gov/documents/synthetic-biology/PCSBI-Synthetic-Biology-Report-12.16.10.pdf.

25. http://oba.od.nih.gov/rdna/nih_guidelines_oba.html (accessed March 26, 2011). An excellent slide presentation explaining the history of recombinant DNA research and the current NIH guidelines is at http://oba.od.nih.gov/oba/IBC/ASGT_2007_Training/Overview%20of%20the%20NIH%20Guidelines.pdf. For the National Environmental Policy Act, visit http://www.fema.gov/plan/ehp/ehplaws/nepa.shtm#2 (accessed March 26, 2011.)

26. Letter to Presidential Commission for the Study of Bioethical Issues regarding its December 10, 2010, report, *New Directions: The Ethics of Synthetic Biology and Emerging Technologies*: http://www.etcgroup.org/sites/www.etcgroup

.org/files/publication/pdf_file/Civil%20Society%20Letter%20to%20Presidents %20Commission%20on%20Synthetic%20Biology_0.pdf (accessed October 20, 2018).

Sources for Additional Information

Akst, J. 2012. "Targeting DNA." *Scientist*, June. http://www.the-scientist.com /?articles.view/articleNo/32141/title/Targeting-DNA/ (accessed April 12, 2017).

Baker, M. 2014. "Stem Cells Made by Cloning Adult Humans." *Nature* News, April 28. http://www.nature.com/news/stem-cells-made-by-cloning-adult -humans-1.15107 (accessed April 12, 2017).

Conley, J. M., R. Cook-Deegan, and G. Lázaro-Munoz. 2014. "Myriad after *Myriad*: The Proprietary Data Dilemma." *North Carolina Journal of Law and Technology* 15: 597–637. https://www.ncbi.nlm.nih.gov/pmc/articles/PMC4275833/ (accessed April 12, 2017).

ETC Group. 2007. "Extreme Genetic Engineering: An Introduction to Synthetic Biology." http://www.etcgroup.org/content/extreme-genetic-engineering -introduction-synthetic-biology (accessed April 12, 2017).

Knoblich, J. A. 2017. "Lab-Built Brains." *Scientific American* 316 (1): 26–31.

Krieger, L. M. 2015. "Designing Life from Scratch: A Fledgling Field Is about to Take Off." *MercuryNews*, August 8. http://www.mercurynews.com/2015 /08/08/designing-life-from-scratch-a-fledgling-field-is-about-to-take-off/ (accessed April 13, 2017).

Offord, C. 2016. "The Rise of the Pangenome." *Scientist* 30 (12): 30–6.

Petrucelli, L., and A. D. Gitler. 2017. "Unlocking the Mystery of ALS." *Scientific American* 316: 46–51.

Pollack, A. 2013. "Gene Therapy with a Difference." *New York Times*, September 23. http://www.nytimes.com/2013/09/24/health/gene-therapy-with-a -difference.html (accessed April 12, 2017).

Service, R. F. 2015. "Modified Yeast Produce Opiates from Sugar." *Science* 349: 677. http://science.sciencemag.org/content/349/6249/677 (accessed April 13, 2017).

Smith-Spark, L., and M. Senthilingam. 2017. "'Three-Parent' Babies: UK Clinic Gets OK for Groundbreaking Technique." CNN, March 16. http://www.cnn .com/2017/03/16/health/three-parent-babies-embryos-approval-uk-bn/ (accessed April 12, 2017).

Stevens, H. 2016. *Biotechnology and Society: An Introduction.* Chicago: University of Chicago Press.

Szostak, J. W. 2016. "On the Origin of Life." *Medicina* 76: 199–203. http://molbio.mgh.harvard.edu/szostakweb/publications/Szostak_pdfs/Szostak_2016_MedicinaB.pdf (accessed April 13, 2017).

Tucker, J. B., and R. A. Zilinskas. 2006. "The Promise and Perils of Synthetic Biology." *New Atlantis*, Spring, 25–45. http://www.thenewatlantis.com/publications/the-promise-and-perils-of-synthetic-biology (accessed April 12, 2017).

3

Genetically Engineered Organisms

> All the food we eat, whether Brussels sprouts or pork bellies, has been modified by mankind. Genetic engineering is only one particularly powerful way to do what we have been doing for eleven thousand years.
> —Michael Specter, American science journalist

> Many of the genetically modified foods will be safe, I'm sure. Will most of them be safe? Nobody knows.
> —Jeremy Rifkin, American economic and social theorist, activist

> There doesn't seem to be any other way of creating the next green revolution without GMOs.
> —E. O. Wilson, preeminent biologist and naturalist

Genetically engineered food plants and animals seize at least as much worldwide public attention as cloning, stem cell research, and other biotechnologies. Concerns over the use of genetically engineered organisms (GEOs) for food and other purposes are global. In the United States, the primary issue has been lack of requirement to label genetically engineered (GE) products to give persons freedom of choice in the use of GE food products. This situation should be remedied by 2020 via the National Bioengineered Food Disclosure Law signed by President Obama in 2016.[1] By contrast, persons in many other countries are adamant about keeping GEOs outside their borders and GE material out of their food. Since the United States is the primary developer of GEOs, most of the international controversy over GEOs pits the desire of US companies to market GE seeds and other agricultural products abroad against fears that GE foods are unsafe. People oppose or embrace GE food products for diverse reasons. For some, the adage "you are what you eat" holds truth beyond the physical. They believe that when science and industry alter the DNA of food organisms, an ethical line has been crossed and that to consume GE food somehow defiles the genetic, biochemical, and even spiritual integrity of our own species. Others are troubled that GEOs threaten the balance within natural ecosystems. Still others worry that consuming GE foods threatens their physical health.

Proponents of GEO technology argue that it increases crop yields, decreases input costs for farmers, improves food quality, reduces worldwide malnutrition, and provides better and cheaper pharmaceuticals. Finally, there are those for whom a potato is a potato no matter how you cut it, dice it, grow it, or modify its genetic material. So long as it still looks, tastes, and feels like a potato, "no problem." This chapter focuses on the biology, promises, perils, controversies, and morality of genetically engineering plants, animals, and microbes for food and other uses. Among the questions we consider are the following:

1. What are GEOs and genetically modified organisms (GMOs)?
2. How are GEOs made?
3. What reasons are there to create GEOs?
4. What are the major controversies and misunderstandings about GEOs?
5. Are GEOs safe?
6. Is the rapid reshuffling of nature's genes ethically justifiable?
7. Is justice served by GEO technology?
8. Should people have the right to choose whether to use GE products?

Although *GMO* refers to a GE organism, the acronym is ambiguous because it stands for *genetically modified organism*. An organism can be genetically modified in many ways other than by genetic engineering. These include natural selection, hybridization, exposure to mutagens, and other means. To avoid such ambiguity, I use GEO to refer to a plant, animal, or microbe that has had its DNA altered by the process of recombinant DNA technology or other technologies such as CRISPR (discussed in chapter 4) that directly alter the base sequence of DNA. Likewise, I use GE as a descriptor for a product of genetic engineering, for example, GE plant, GE salmon, GE crop, GE food.

Genetically Engineered Organisms: The Biology

GMOs and the Origin of Agriculture

Just what is a GEO? Basically, a GEO is just one type of GMO. GMOs are living things that have been genetically altered by any one of a variety of human-directed processes. Most ways for genetically modifying plants and animals, such as selective breeding, cross-pollination, and artificially induced mutation, affect large and multiple regions of an organism's DNA in ways that are not very predictable. Many genetic experiments are normally required to achieve the desired result and sometimes the hoped-for outcome is never obtained. By contrast, a GEO is created by manipulating just one or a small number of genes in an organism. As a mechanical engineer designs a specialized part for a complicated machine, a genetic engineer typically adds, deletes, or alters just one of tens of thousands of genes in an organism so as to affect just one or very few traits. Still, there is imprecision in genetic engineering since the biotechnician often cannot control the number of copies or locations for the transferred gene in the host organism's genome. Nonetheless the predictability of outcomes for genetic engineering is far greater than for traditional methods of selective breeding and hybridization.

How long has the genetic modification of living things been going on? Nature has been at it for nearly four billion years. Instead of calling the results of nature's genetic experimentation GMOs, biologists call them "evolutionary survivors"—bacteria, plants, animals, and viruses so well adapted to their environment that they leave more offspring in successive generations than their competitors. The natural selection of genetic variants that spring up in populations by mutation, genetic recombination during sexual reproduction, or from other naturally occurring events has produced the wondrous variety of living things that populate the earth today. We humans also have been genetically modifying organisms for a long time, but only for about one quarter of 1/100,000th of the time that nature has been at it, that is, about 10,000 years.

The earliest genetic modification of crop plants by humans was probably the inadvertent domestication of wheat about 10,500 years ago. In the Fertile Crescent, between the Tigris and Euphrates Rivers in what is now Iraq, Neolithic people harvested wild emmer wheat from the hillsides. Most wild wheat grain heads contain seeds that dislodge easily and disperse widely in the wind, a trait that enhances the reproductive success of the plants in nature.

The earliest agriculturalists added a new dimension to seed dispersal. Neolithic harvesters sawed off bundled handfuls of wheat with polished, knifelike, stone sickles. When a mutant variety of wheat appeared containing seeds that clung tightly to the grain heads, proportionately more of these seeds per head ended up in the harvesters' baskets because more of the highly dispersible seeds were lost as the stalks were sawed during harvesting. Unknown to these early planters, their grain gathering method led to the domestication of wheat.

At some point, these gatherers became planters and saved back a portion of each season's harvest for sowing the next year. More and more of the mutant plants with the tight grain heads grew in the planted fields each year since the seeds from these plants stood a better chance of making it into the baskets and therefore of being planted in the fields the next year. After generations of planting seasons, virtually all of the planted wheat possessed tight, easily harvested grain heads. This is an example of the genetic modification of a species by simple selection—the preferential growing or breeding by humans of a plant or animal possessing particular traits. Similar actions led to the domestication of other staple food crops like rice and corn.

Once people noticed that the offspring of plants and animals tend to resemble the parents, simple selection became planned and purposeful. Wolves were reformed into dogs of diverse shapes, colors, and sizes in southwest Asia, China, and North America. And six thousand to ten thousand years ago, the five large animals upon which modern animal agriculture is based—sheep, goats, cows, pigs, and horses—were domesticated, by simple selection.

Over time, humans added to their repertoire of methods for genetically modifying plants and animals. These now include crossbreeding/pollination of variants that appear naturally within a species, cross-fertilizing individuals from different species (especially in plants), fusing cells from different species, growing the hybrid cells into new fertile individuals, and using mutation-causing chemicals or radiation to produce random genetic variants from which individuals are selected for reproduction. Human creation of GMOs by nongenetic engineering has continued without interruption for over ten thousand years. In the 1970s, genetic engineering entered the inventory of methods humans use to genetically modify organisms, and in 2013, CRISPR technology, which provides for easy, accurate, and inexpensive genome editing, debuted.

How Are GEOs Made?

Genetic engineering entails adding, replacing, or inactivating one or more genes in an organism. This commonly requires genetic engineers to insert foreign genes into the egg or sperm chromosomes of the organism being engineered. The process is called transgenesis. The modified individual is called a transgenic organism, and the transferred gene is called a transgene.

DNA transferred into an organism may be a gene from the same or a different species, or it may even be a segment of artificial DNA synthesized by biotechnologists. Specific procedures for creating transgenic organisms differ between plants and animals, but in all cases they involve four major steps: (1) putting foreign DNA into a form that can be transferred into a host organism, (2) getting the foreign DNA into host cells, (3) incorporating the transgene into chromosomes of host eggs, sperm, or pollen, and (4) demonstrating that the transgene is present and functioning in subsequent generations of offspring derived from the engineered host.

In order to construct a transgenic organism, biotechnicians must first find the gene(s) in nature that will confer the desired trait(s) to the host organism. Searching for and identifying appropriate genes to transfer to the host species and the transfer procedures are expensive enterprises. One or more teams of highly trained molecular biologists, laboratories equipped with highly specialized instruments, costly nonreusable supplies, and a measure of good fortune are required for success.

Once created, GEOs must be tested for human health and environmental safety. Research, development, and testing needed to gain federal approval to market a new GEO may cost a pharmaceutical or agricultural company tens of millions of dollars. But, as we will see, the industry has ways of recouping their investment, plus some.

To better appreciate the process of creating a transgenic organism, we need to take a look at some interesting behind-the-scenes details that led to Roundup Ready soybeans, a GE variety of bean that now accounts for over 90 percent of the annual soybean production in the United States. The creation of Roundup Ready soybeans exemplifies how most GE plants are created.

Making transgenic plants. Roundup is the commercial name for a herbicide (glyphosate) that kills weeds and most other plants by inhibiting an enzyme called EPSP synthase. Plants need EPSP synthase to produce three essential amino acids. Without these amino acids, the plants die. As the commercial name implies, Roundup Ready soybeans are ready to encounter Roundup without suffering ill effects. Roundup Ready beans are engineered to contain a gene for a form of EPSP synthase that is resistant to the herbicide. This was accomplished in 1994 by scientists at Monsanto, the chemical/agricultural product megacompany.

At the start of this endeavor, Monsanto searched for a plant that is naturally resistant to Roundup. Their search led them to the petunia, a favorite of spring gardeners and of plant cell biologists and geneticists who grow the plants regularly in their laboratories. The researchers discovered that a few petunia cells growing in laboratory culture dishes survive treatment with glyphosate. Next, they showed that resistance to the herbicide was due solely to the cells' ability to make a mutant form of EPSP synthase that is resistant to Roundup. The workers reasoned that if the gene for the mutant enzyme was transferred to and functioned in soybeans, the beans would become glyphosate resistant. The petunia gene was transferred. It worked in its new host, and the transgenic soybean plants were dubbed Roundup Ready. As soon as Roundup Ready bean seedlings emerge from the ground, bean fields are sprayed with Roundup to kill off weeds competing with the young bean plants for sunlight and nutrients with no harm to the bean plants. Now let us see how the petunia gene was inserted into the soybean genome. Also notice how basic scientific research sometimes leads later to unexpected, valuable applications.

Roundup Ready soybeans were created by a procedure called *Agrobacterium*-mediated gene transfer (fig. 3.1), a common method for genetically engineering plants. *Agrobacterium tumefaciens* is the scientific Latin name for a common soil bacterium that causes crown gall disease in certain plants. Crown gall disease provokes harmful tumor growth in the crown, the part of the stem just above the surface of the soil. Basic research in plant pathology in the 1970s showed that *Agrobacterium* causes crown gall disease by transmitting a small segment of DNA from itself into plant cells. The transferred DNA becomes integrated into the plant's chromosomes and directs the production of proteins that transform normal tissue into tumor tissue.

More research in the 1980s led to using the transmittable *Agrobacterium* DNA segment to carry foreign genes into plant cells. To accomplish this, scientists first made the *Agrobacterium* DNA harmless to plants by removing its tumor-causing genes without impairing its ability to invade plant cells. Then, using recombinant DNA techniques devised in the 1970s, they joined a transgene to the *Agrobacterium* DNA and successfully sent the new recombinant DNA into plant cells. To create a variety of transgenic plant, like Roundup Ready beans, which could grow on hundreds of thousands of acres year after year, one must somehow get the *Agrobacterium* DNA and its attached transgene into seeds.

Getting transgenes into seeds relies upon the remarkable ability of most plants to undergo vegetative reproduction, that is, to reproduce without sex. Tiny disks of leaf tissue formed with an ordinary paper punch are incubated in a special laboratory culture medium along with *Agrobacterium* cells carrying the transgene. The disks sit there long enough to let the transgene

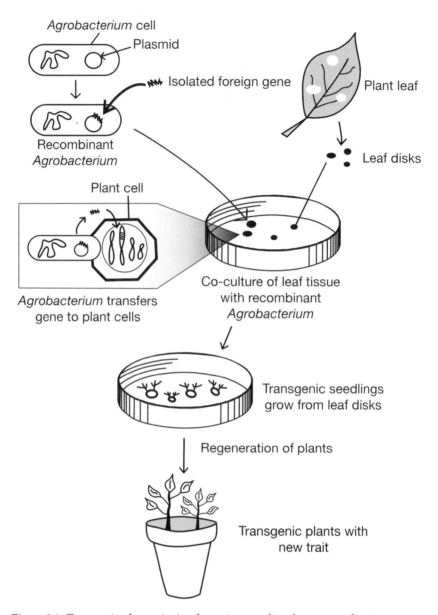

Figure 3.1. Transgenic plants via *Agrobacterium*-mediated gene transfer. *Agrobacterium* cells receive a transgene via long-established methods of recombinant DNA technology. Small tissue disks punched from a plant's leaf, when incubated with the transformed bacterial cells, incorporate the transgene into chromosomal DNA of their own cells. The transformed tissue disks then sprout roots and shoots, eventually developing into a mature transgenic plant. The seeds produced by these plants carry the transgene. (Courtesy of the author)

enter the leaf cells. Next, the transformed disks are incubated in a nutrient-containing medium along with growth-stimulating hormones that induce the disks to reproduce entirely new plants vegetatively, that is, without involving seeds. This is a remarkable phenomenon. Cells in each leaf disk produce a small mass of unspecialized stem cells, and from this clump of cells emerges new root, stem, and leaf tissue. Eventually the tiny plants mature and produce seeds containing the transgene. When these seeds are planted and themselves grow into mature plants with seeds, the transgene travels along with all of the plant's other genes into seeds and is thereby passed on to the next generation. This is how the petunia gene conferring Roundup resistance became a permanent resident in the soybean genome. Since commercialization of Roundup Ready soybeans in 1996, several other commodity plants have been rendered Roundup Ready including corn, sugarbeets, canola, cotton, and alfalfa.

Now, after nearly a half century of Roundup use and over twenty years of Roundup Ready crops, a new twist to Monsanto's Roundup story is coming to light. Although Roundup Ready GE-plants themselves pose no known health risks, close and chronic contact with Roundup may not be safe. In August, 2018, a California Superior Court jury ordered Monsanto to pay $284 million to a groundskeeper who contracted non-Hodgkin lymphoma after years of working with the herbicide. The jury ruled that handling glyphosate was at least partly responsible for the plaintiff's cancer and that Monsanto failed to warn of glyphosate health risks. Monsanto countered that over 800 studies and reviews had found Roundup to be safe and plans to appeal the jury's decision (James and Ortiz 2018).

A second example of a transgenic, commodity crop produced by *Agrobacterium*-mediated gene transfer is *Bt* corn. *Bt* stands for *Bacillus thuringiensis*, a bacterium that naturally produces proteins that are deadly to many insect pests. The pesticide proteins are called *Bt* toxins. To produce *Bt* corn, researchers transferred a bacterial gene for *Bt* toxin into corn plants. Now, all tissues in transgenic *Bt* corn plants produce *Bt* toxin, which protects corn from one of its most destructive pests, the European corn borer. So when an insect eats any part of a *Bt* corn plant, the *Bt* toxin kills the insect. This happens because ingested *Bt* toxins are partly digested into protein fragments that attack cells lining the gut. In 2016, 79 percent of the corn and 84 percent of the cotton grown in the United States were genetically engineered to resist insect pests with *Bt* toxins. *Bt* tobacco and soybeans are also grown in the United States.

Agrobacterium-mediated gene transfer is not the only way foreign genes can be introduced into a plant. Another method is called biolistics, a procedure developed in the 1980s that employs a "gene gun." Quite literally, the leaves or embryos of a normal plant are shot with tiny, gold "buckshot" coated with copies of a foreign gene. Some of the DNA-bearing projectiles explode into the tissue at close range, pierce the plant cell wall and cell membrane, surge through the cell's cytoplasm, and enter the nucleus. Within the nucleus of each cell are chromosomes, bearers of the genes required to produce an entire plant. Once inside the nucleus, copies of the foreign DNA dissociate from the surface of the tiny gold beads and incorporate themselves into the chromosomal DNA. GE plants and seeds are then obtained by growing individual transgenic cells into new plants (fig. 3.2).

Neither *Agrobacterium*-mediated gene transfer nor biolistics is very accurate. It is nearly impossible to control the numbers and locations of transgene insertions into host plant genomes. This situation is changing though with discoveries in 2009 and 2013 for accurately targeting transgenes into plant genomes (Porteus 2009; Sander and Joung 2014).[2]

Making transgenic animals. Animal cells cannot be genetically transformed by *Agrobacterium*-mediated gene transfer or by biolistics. Three other methods are commonly used to create transgenic animals. Two methods can be applied to any type of animal including birds, insects, amphibians, and mammals. The third method applies only to mammals. A fourth recently developed method called CRISPR-mediated gene editing is briefly introduced below and thoroughly discussed in chapter 4.

The first method is to inject viral DNA carrying copies of foreign gene(s) into the nucleus of a fertilized egg. In this approach, the viral DNA is first disarmed; that is, it is modified so that it cannot produce new virus particles inside a host cell. By contrast, the property of the viral DNA that allows it to integrate itself into host DNA is left intact. Once inside the nucleus of a fertilized egg (zygote), the viral DNA along with its transgene cargo integrate themselves into one or more chromosomes of the zygote nucleus. Since each cell division is preceded by chromosome replication, every cell of the individual that grows from the transgenic zygote, including eggs and sperm, contains the foreign gene(s). Transgenic animals made in this way pass their engineered genomes on to future generations when they reproduce in the normal sexual way.

The second method is called somatic cell nuclear transfer (SCNT), a method of cloning described in chapter 2 and illustrated in figure 2.8. In

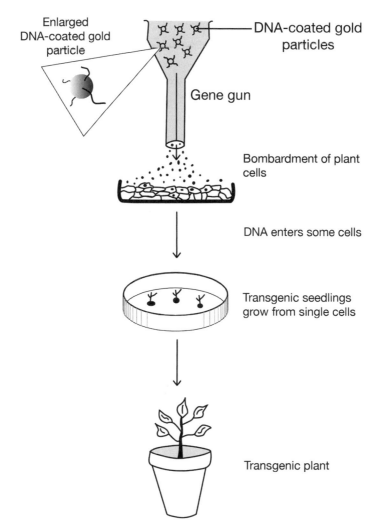

Figure 3.2. Transgenic plants via biolistics. Tiny gold particles, coated with copies of a transgene, are shot through a "gene gun" at a layer of laboratory-cultured plant cells. Some of the gene-carrying gold particles enter plant cell nuclei, where the transgenes dissociate from the gold and become incorporated into the cells' chromosomal DNA. Individual transformed cells then grow vegetatively (asexually) into mature plants with seeds containing the transgene. (Courtesy of the author)

SCNT, the DNA-containing nucleus of a body cell is transferred into an egg cell that has had its DNA-containing nucleus removed. Using SCNT to produce a transgenic animal relies on the fact that several cell types commonly grown in the laboratory will incorporate foreign genes into their chromosomes to produce transgenic cells. When the nuclei from transgenic cells are transferred into enucleated eggs, some of the eggs behave like normal fertilized eggs and develop into transgenic embryos. Transgenic embryos grow into transgenic individuals whose cells, including eggs and sperm, all carry the transgene(s). Thus, a transgenic animal is created that can pass its transgene(s) and transgenic qualities on to future generations.

The third method for creating transgenic animals is to produce chimeric embryos (fig. 3.3). This procedure makes use of the pluripotency of embryonic stem cells (ESCs) and researchers' ability to grow blastocyst stage mammalian embryos in the laboratory. ESCs incubated with copies of foreign genes incorporate the transgenes into their chromosomal DNA to produce transgenic stem cells. Transplanting these cells into laboratory-grown blastocyst stage embryos obtained by IVF produces chimeric embryos in which some cells carry the transgene(s). Chimeric embryos transferred to the wombs of surrogate mother animals develop into chimeric individuals whose organs, including ovaries and testes, contain transgenic cells. The transgenic eggs and sperm from chimeric individuals give rise to animals that are wholly transgenic.

Transgenic cows, goats, sheep, mice, pigs, and monkeys have all been produced via chimeric embryos. One practical product of this technology includes pigs with organs immunologically compatible with humans that may be used as organ transplants in the future. Another application involves transgenic goats that produce large quantities of spider silk protein in their milk. The spider silk is prized for its strength and elasticity and is used to construct material for eye suture thread, artificial tendons and ligaments, bullet proof vests, and car crash air bags.

A gene-editing technology called CRISPR/Cas9 and named "The Breakthrough of the Year" for 2015 by *Science* magazine is rapidly becoming an important means for creating GE agricultural crops and animals. Animals already obtained using CRISPR technology include dairy cows with no horns, tuberculosis-resistant cows, and pigs with human pancreases (GEN News Highlights 2017; Staropoli 2016; Wade 2017). The latter are also a form of human-animal chimera. Chapter 4 examines CRISPR technology and its many applications.

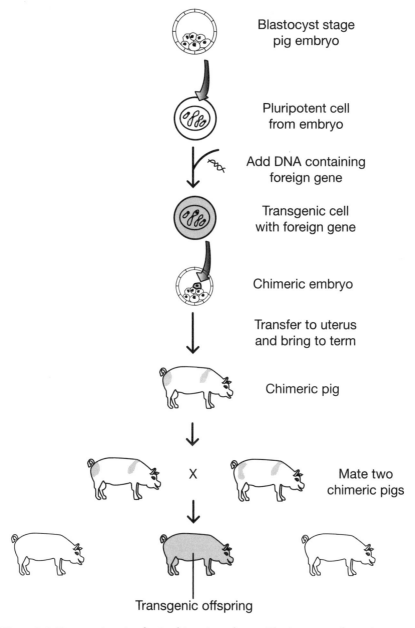

Blastocyst stage
pig embryo

Pluripotent cell
from embryo

Add DNA containing
foreign gene

Transgenic cell
with foreign gene

Chimeric embryo

Transfer to uterus
and bring to term

Chimeric pig

X   Mate two
chimeric pigs

Transgenic offspring

Figure 3.3. Transgenic animals via chimeric embryos. Pluripotent embryonic stem cells created from early embryos of an agricultural animal are genetically modified by adding one or more transgenes to their chromosomal DNA. The transformed stem cells are injected into a normal blastocyst stage embryo and eventually contribute to the tissues and organs of a mature animal. The animal is a chimera that contains some transformed cells and some normal cells. When two chimeric animals mate, some transformed sperm fertilize transformed eggs resulting in fully transgenic animals that propagate the genetic modification(s) into succeeding generations. (Courtesy of the author)

Why Create GEOs?

Aside from creating crops impervious to insect pests and weed killers, what is the point of creating transgenic organisms? The answer to this question comes from viewpoints of two types of scientific research: applied and basic.

An applied researcher asks, "What are the immediate and practical benefits of working to insert foreign DNA into the genome of an existing life form?" A basic researcher asks, "How can I manipulate the DNA of this organism to gain new insights into how living cells work?" Let us look at some examples of GEOs created for their practical benefit to humankind and some GEOs created to gain basic knowledge about life itself.

Practical benefits from GEOs. The public is interested primarily in applied research. This is because of its tangible effects on the quality, cost, and availability of food, medical treatments, and other products that directly affect our daily lives. In agriculture, the benefits of GEOs are similar to the benefits sought for thousands of years through simple selection and crossbreeding—more and better food and other agrarian products for less expense, time, and energy. The US National Academies of Science (2004) recently declared that biotechnology, including genetic engineering, has "been developed to improve the shelf life, nutritional content, flavor, color, and texture of foods, as well as their agronomic and processing characteristics."

In medicine, the motivation for creating GEOs is to improve human health and add years of high quality to our lives. For example, since the 1970s GE bacteria have been medicinal sources of human insulin and growth hormone. The microbe-produced hormones benefit diabetics and children whose growth would otherwise be stunted by low levels of their own growth hormone. Earlier these hormones had to be laboriously isolated from cattle tissue obtained at slaughter houses, and the cattle hormones did not work as well in humans as the bacteria-produced human hormones do. Other examples of GEOs created specifically to improve the quality of human life are transgenic pigs and transgenic rice.

Thousands of persons die every year due to insufficient numbers of immunologically compatible organs for transplantation. Pig physiology is similar to human physiology, and pig organs are similar in size to human organs. So why not use pigs as organ donors? A major hurdle to overcome before swine can become a routine source of organs for humans is foreign tissue rejection by the recipient's immune system. The attacking cells of the immune system recognize "foreignness" or "sameness" by proteins on the surface of other cells. If the cell surface proteins in pig organs were made compatible for humans, many lives could be saved. Transgenic pigs may be the answer to the shortage of transplantable organs.

A few genetically engineered pigs, whose cell surface proteins were replaced by human cell surface proteins, have already been produced. But creating transgenic pigs is time consuming and difficult, and the success rate is low. Creating a transgenic pig for every person who needs an organ transplant is not feasible. The best approach may be to create a small number of transgenic pigs, one for each of the major groups of human histocompatibility cell surface proteins. These pigs could then be cloned by SCNT to produce as many pigs as needed for organ donation. Transgenic pig organs are not yet available for people needing organ transplants, but they surely will be in the foreseeable future. Pig heart valves have been used to replace worn-out human heart valves for over thirty years (Pick 2008).

Turning back to plants, what about the benefits of transgenic rice? Rice is the food staple for much of the world's population, but rice lacks many nutrients and vitamins needed by growing children. Vitamin A is one of these. The World Health Organization (WHO) estimates that 250,000 to 500,000 children go blind every year from vitamin A deficiency, and over half of those children die within a year of losing their sight.[3] Our bodies cannot make vitamin A from scratch. Instead, we make it from dietary beta-carotene, also known as provitamin A for its role as an essential metabolic precursor to vitamin A. Beta-carotene is the compound that gives carrots their orange color. It is needed for healthy retinas required for good night vision.

In a humanitarian project, funded mainly by the Rockefeller Foundation and by funds from Switzerland and the European Union (EU), Ingo Potrykus at the Swiss Federal Institute of Technology in Zurich developed transgenic rice containing beta-carotene. Potrykus led a research team that in the year 2000 created GE rice containing a bacterial gene and a daffodil gene that work together to give rice the ability to produce beta-carotene. The rice is called Golden Rice because the beta-carotene gives the grain a yellow hue.

Potrykus's Golden Rice could save the lives of millions of children and improve the health of millions of others in countries where vitamin A deficiency is high and rice is a staple food. Unfortunately, fears of the biotechnological origin of Golden Rice and regulations based on the "precautionary principle" have so far prevented it being used by people who could benefit the most.[4]

It is a long and difficult journey from lab to farm field. In the case of Golden Rice, arrangements had been made with the holders of dozens of patents on the procedures and materials used to develop it to allow commercialized farming of the rice in India and the Philippines, where rice is a staple food. All appeared to be going well until time for the field tests arrived. In India, test plots of Golden Rice were burned by critics of GE food, and in the Philippines, field tests were halted by a court injunction. Of course, it is impossible to know whether the rice would now be reaching the mouths of malnourished children even if the field tests had been successful. Bureaucracy and selfishness have prevented surplus food from reaching starving people for decades. In the case of Golden Rice, we at least have an example of biotechnology producing a product that could help millions of people enjoy more healthful lives. Due mainly to lengthy regulatory processes, more than a decade will have elapsed between its development and its first release to farmers in the Philippines or India where it is most needed.[5]

On a positive note, the Rockefeller Foundation announced in late 2008 that it would provide funds to help Golden Rice through regulatory approval processes in Bangladesh, India, Indonesia, and the Philippines.[6] This resulted in some field tests of the rice in those countries, but the rice is still not planted on a large scale. Since vitamin A–deficient children have not yet eaten the rice as a staple food item, it cannot be shown that the rice will correct the deficiency or that its yield will match non-GMO varieties of rice (Everding 2016).

GEOs and basic research. Less widely known than the practical reasons for creating GEOs is the fact that scientists create them to help answer basic questions about how cells work. The central purpose of basic biological research is to discover principles that govern the processes of life. GE mice provide an example.

For three decades, basic researchers have used GE mice to learn about the function of particular genes. A method commonly used in this research is called gene knock-out. In gene knock-out, a normally functioning gene is replaced by an inactive, mutated copy of the same gene. By knocking out the activity of a specific gene and then observing the effect(s) on the appearance, physiology, or biochemistry of the mouse, researchers gather clues about the normal function(s) of the gene. The knock-out is performed at the egg stage or the blastocyst stage of development in order to obtain animals whose cells all possess the nonfunctional version of the gene.

Gene knock-ins can also be created by replacing a specific gene with an active copy of the same gene that is altered in some small and specific way. Gene knock-ins give researchers information about the function(s) of specific regions of genes.

Using knock-outs and knock-ins in similar research in plant biology and microbiology employs GEOs to discover new principles governing life's processes and to uncover new details about previously discovered principles. For example, a principle in biology since the mid-nineteenth century is that all cells arise from preexisting cells. In humans and other nonbacterial cells, this occurs by *mitosis*, a cellular process involving the duplication of chromosomes and their subsequent separation during cell division so that each new daughter cell receives a complete set of genetic instructions from the dividing mother cell. Although the overall behavior and movement of chromosomes during mitosis was described in the 1870s, cell biologists still do not fully understand how mitosis is regulated and choreographed so accurately. Every year new genes that regulate mitosis are discovered. Information about how they do so is gathered by knock-out experiments using GE yeast cells and other GE organisms.

These are some reasons for creating GEOs as communicated to us by scientists and the biotechnology industry. In reality, the motivating forces behind biotechnological research are more numerous and complex than I have described. There is not always a clear distinction between basic and applied research. For example, uncovering fundamental principles about insect reproduction or cell senescence may lead to innovations in pest control or human age retardation, respectively. Similarly, basic research on enzymes called zinc finger nucleases made possible the accurate targeting of transgenes into food plants, facilitating development of new food sources for malnourished persons around the world.[7]

Some GEOs now marketed and a look at the future. So far, marketed GEOs are mainly for medicine and plant agriculture. Earlier, I cited engineered bacteria that produce human insulin and growth hormone. The most common agricultural GEOs are soybeans, canola, corn, potatoes, cotton, and tobacco that are engineered for herbicide resistance or to produce natural insecticides. Less than a handful of GEOs have been developed specifically to offer consumers more nutritional, flavorful, or attractive food, and these were not very successful.

Crops are susceptible not only to insects but also to attack by bacteria, fungus, and viruses. Some genetic engineering is aimed at protecting crops against these plant pathogens. For example, a tomato variety was engineered to resist bacteria that cause a tomato disease called "bacterial speck." Moving a gene from a resistant tomato variety into a sensitive one is all it took to accomplish this. Other genetic engineering research led to virus-resistant squash and grapevines and also fungus-resistant squash.

With two notable exceptions, most genetic engineering of agricultural plants is directed at crop protection rather than toward increasing the nutritional value or desirability of food plants for consumers. Golden Rice is one of those exceptions, and the FlavrSavr tomato developed in the late 1980s is the other. As with Potrykus's Golden Rice, distrust of the biotech industry and its products fated the FlavrSavr tomato and foreshadowed the controversy that now engulfs GE products.

A frustrating experience for lovers of fresh garden salads is to buy a nice red tomato at the supermarket and discover at home that once again you have been fooled by the outer appearance of an unripe, tasteless, fruit masquerading as a tomato. This happens because the tomatoes are picked green (before the flesh has ripened) in order to avoid bruising during transport to market. Later, when the flesh is still hard, the fruit is treated with ethylene gas, which makes the unripe fruit turn red for presentation to consumers before it goes mushy. To solve the problem of turning a tomato red before it actually ripens, the now defunct Calgene Company of Davis, California, recognized that ripening to produce good flavor and the inevitable tissue softening are two separate processes. The two events normally occur simultaneously, but by genetic engineering, Calgene separated them in time by shutting down the activity of the gene responsible for tissue softening during the ripening process. The result was the FlavrSavr, a tomato that turned red and developed good flavor while still on the vine but without softening. Finally, attractive, tasty red tomatoes could be harvested and shipped to consumers without bruising.

So where is this FlavrSavr? Why must we still buy tomatoes that taste like hard pears? The problem is that the process used by scientists to develop FlavrSavr resulted in consumer distrust of the product's safety. GE-food opponent Jeremy Rifkin and others proclaimed that FlavrSavr biotechnicians had given the plant a gene for resistance to the antibiotic kanamycin and attached it to the gene that prevents tissue softening. The implication was that people eating FlavrSavr would develop resistance to kanamycin, making the antibiotic useless for later treatment of infection or disease. The fact is that other research showed that eating FlavrSavr tomatoes does not make people resistant to kanamycin; moreover, scientists used the kanamycin-resistance gene for a good reason—to help them identify which tomato cells they had successfully engineered during the initial stages of developing FlavrSavr. A simple screen for kanamycin resistance revealed which of the thousands of cells exposed to the foreign genes had actually been transformed and could be grown into tomato plants with fruit that would remain firm during ripening. But the public relations damage had been done. Trust in FlavrSavr never developed, consumers shunned the tomato, and marketing ceased soon after it began. Scientists now have ways of identifying genetically transformed plant cells without using genes for antibiotic resistance, but so far, no new generation of FlavrSavr has appeared.

What seems clear from the FlavrSavr episode is that GE products of the future will need to have consumer trust from the start in order to gain public approval. Perhaps if Calgene had anticipated public concern over use of the kalamycin-resistance gene and openly and effectively explained its use and the supporting information about its safety from the beginning, Flavr-Savr would have succeeded. In addition, the GEO industry will need to diversify beyond the herbicide and pest-resistant food crops that now dominate the world of marketed GEOs to provide tangible benefits for the public. Real advantages that easily can be seen, tasted, felt, or smelled resulting from GE products will probably gain more interest and acceptance from consumers than pest-resistant corn.

Some biotech companies seem to realize this and are giving attention to research and development aimed at the future marketing of GE products with improved nutrition, product quality, and benefits to the environment. Time will tell whether the industry follows through on this and whether the products will gain public acceptance. Among the many new GE products we can anticipate are:

—hypoallergenic peanuts, wheat, and soybeans,
—decaffeinated coffee that retains the flavorful oils of regular coffee,

—microbes that decompose plastic,

—vegetable oils with greater proportions of unsaturated and monosatu-
rated fatty acids, perhaps in just the right ratios to enhance cardiovascular health,

—potatoes with a higher starch content that absorb less fat during cooking,

—wheat and corn containing protein with a complete complement of essential amino acids,

—leafy green vegetables with increased iron content and other vegetables
with increased levels of micronutrients such as essential vitamins,

—leafy green vegetables engineered to lack vitamin K for persons on the
blood-thinner, Coumadin,

—cotton and corn that produce bioplastics for use in manufacturing wrinkle-
free clothing with hybrid cotton-plastic fibers and nonpetroleum based
plastic goods, respectively,

—edible (plant-based) vaccines against a range of viral and bacterial diseases,

—edible plants containing antibodies to fight disease and infection, and

—trees and other plants engineered to remove or inactivate toxic, industrial
pollutants like chlorinated organic hydrocarbons, mercury, cadmium,
and copper that have accumulated in soil and wetlands.

All of these GE-based products and many more could appear within the
next ten to forty years and maybe sooner. Today, we can imagine possible
benefits and risks of GE technology and the related CRISPR gene-editing
technology. What remains to be seen is the extent to which our imaginings
become reality.

One might think that the possibility of future GEOs like the ones listed
above would convince all but the staunchest Luddites that we should forge
ahead aggressively into a GEO future. But no; instead, there is widespread
and vehement opposition to GE technology. Why is this?

GEO Controversies and Misunderstandings

Several incidents involving GE-plant technologies fuel controversy and
criticism of GEOs and widespread public distrust of biotech companies
and their motives for developing these products. Some reveal real risks for
the public or the environment, while others reflect media and/or public
misunderstandings of the technology or worries spawned by irrationality
and ignorance. What is most lacking in the GEO debate, especially over GE
plants, is knowledge about the biotechnology itself and how it bears upon

human health and environmental quality. To help judge whether certain concerns over GE crops are legitimate or misplaced, we will look at a few of the most contentious issues.

The "Terminator," seed saving, and the "enforcers." The Technology Protection System (TPS), aptly dubbed the "Terminator" by its critics, is a three-gene module that renders plants sterile. No viable seeds are produced by plants containing the Terminator. The plan by Monsanto Corporation, which purchased the right to use the TPS from its developers at the USDA and the Mississippi-based Delta and Pine Land Company in 1998, was to incorporate the TPS into seeds for crop plants that had already been engineered for insect and/or herbicide resistance. Monsanto's interest in the TPS is that it would help them recover their financial investment in research and development for GE plants by forcing farmers to purchase new seed every year. Presently, farmers pay a technology fee to Monsanto every year they buy GE seeds. Farmers must also sign an agreement not to save seeds from their GE crops for future replanting and allow Monsanto investigators on their property to assure compliance. The TPS offered Monsanto a foolproof guarantee that farmers would purchase new seeds and pay technology fees every year, no investigators necessary.

Within a few months of the announcement by the *Wall Street Journal* that the TPS had been patented, an international outcry arose against the technology. Environmental groups considered the TPS especially scandalous because of the perceived possibility for TPS genes to escape from crop plants into populations of native plant species.

In defense of the TPS, the president of Delta and Pine Land announced that the motive for developing the technology was to protect American farmers against a market weakened by growers overseas who avoid paying annual "technology fees" by saving their seeds. But this statement, along with Monsanto's buy up that same year of Delta and Pine Land and also the nation's second-largest seed corn company, did little to quell public concern. Finally, in late 1999, Monsanto yielded and announced that for the next five years it would not seek marketing approval for any seeds containing the TPS. That period has expired, so Monsanto is now faced with a dilemma: (1) use the TPS to protect its profit margins, its investments in GE crop development, and perhaps the market for American commodity crops or (2) refrain from commercializing crops with the TPS in the interest of environmental stability and helping poor farmers in developing nations.

Engineering and gaining approval for a new GE plant is expensive. Monsanto's need to devise an effective way to recover its investment in GE products while still turning a profit is understandable. Yet, when GE plants are grown by poor farmers in developing nations where traditions of seed saving and sharing are millennia old, there is a strong argument for allowing those traditions to continue. Disrupting long-standing cultural practices can have destructive effects on the structure of societies. Also, the economic realities in some regions of the world preclude the purchase of GE seeds and the payment of seed use fees every year. As it stands, all farmers throughout the world must agree to allow inspectors from Monsanto or other sellers of GE seeds on their fields up to three years after the purchase of GE seeds. This is to allow the industry to monitor compliance with its no seed-saving policy.

That Monsanto is serious about protecting its investments is evidenced by the case of Percy Schmeiser, a farmer of rape (a European herb of the mustard family whose seeds yield canola oil) in Saskatchewan, Canada. In the fall of 1997, Monsanto's field inspectors found their patented Roundup-resistance gene among canola seeds harvested from Mr. Schmeiser's farm. Then, the real trouble began.

For generations, the Schmeiser family saved non-GE seed from its own crops for the next year's planting, and Schmeiser claimed he had done this in 1997. To Monsanto's accusatory questioning, he replied that the GE genes discovered in his harvest must have grown from seed blown into his field off passing trucks carrying other farmers' harvests or that his crop had become contaminated by pollen carrying the gene from his neighbors' fields. So in 1998, he stuck by the family's practice and sowed his fields with seed from the controversial 1997 harvest.

Monsanto promptly sued Mr. Schmeiser and a six-year patent battle ensued. The case went all the way to Canada's Supreme Court, which rendered its decision in the spring of 2004. The court found Schmeiser guilty of infringement of patent law but leveled no penalties against him. He was not even required to pay an estimated $11,000 technology fee for the use of Roundup Ready seeds in 1998.

The court's decision confused patent lawyers across Canada. On the one hand, it upheld an earlier ruling that banned patenting of "higher life forms" like whole animals or plants, while on the other hand, it permitted enforcement of patents on genes wherever the genes are used, including inside plants or seeds. The decision leaves in ambiguity the question of the legal difference between a patented gene and a seed or plant carrying a patented gene. Interestingly, the court noted that Mr. Schmeiser earned no more profit from his GE rape seed crop than from previous non-GE crops and that he did not use Roundup on his fields in 1998.

The monarch butterfly. I am charmed by the beauty and grace of the monarch butterfly and its striking green, yellow, and black striped caterpillar stage that feeds almost solely on milkweed leaves. The adult animal's lovely orange and black pattern, large size, extreme wariness, and airborne agility made me ambivalent about capturing one of the beauties as a boy working on an insect collection for a Boy Scout merit badge. Over five decades later, I still have that butterfly, pressed gently beneath glass into a background of white cotton in a small frame purchased from Woolworth's Dime Store. Now, as then, I know in my heart that this animal is a part of nature best left in its natural habitat rather than pressed under glass.

Reports that the specific destination in Mexico for the monarch's annual three-thousand-mile southward migration is being rapidly and illegally deforested only strengthens the appreciation most of us have for this creature. Monarchs undoubtedly rank with bald eagles, bluebirds, and giant sequoias as universally loved symbols of North American nature. So it is easy to understand why any suggestion that biotechnology threatens this species would rain public ire down upon the perceived perpetrator. This is exactly what happened in May 1999 when a group of researchers at Cornell University reported a 40 percent mortality rate for monarch caterpillars fed high doses of pollen from *Bt* corn in their indoor laboratory.

The report alarmed environmental activists and nonactivist citizens alike in North America and all over Europe who feared that *Bt* corn pollen settling on milkweed plants could spell the monarch's demise. In the United States, it resulted in a mandate from the EPA requiring farmers who plant *Bt* corn to include buffer zones of non-GE corn at the perimeter of their fields. Less well-known is that further experimentation outdoors concluded that *Bt* corn poses little or no danger to the monarch butterfly. Reasons for this conclusion include: (1) caterpillars in nature would virtually never encounter the doses of pollen that they were fed in the laboratory, (2) corn pollen is so heavy that it rarely falls more than a few yards from the edge of the corn field, (3) milkweed plants are rare around corn fields, (4) the butterfly's migration schedule would rarely bring it into contact with corn pollen at all, (5) milkweed leaves are so smooth that corn pollen does not cling to them very well, and (6) caterpillars fed milkweed leaves dusted with *Bt* pollen in experimental field plots showed far less mortality than in the university laboratory.

The *Bt* monarch butterfly story was a false alarm in the *particular*, but it points to an important problem in *principle* and should be taken seriously by biotech companies and their regulators. The principle is that living organisms and ecosystems are extremely complex, and prudence requires humility and caution in the face of our ignorance about how they work. Just because we can manipulate genomes does not mean that we fully understand how genomes work to produce and sustain organisms. Prudence demands that we acknowledge the limits of our knowledge. Introducing foreign genes into genomes at locations we cannot control and growing transgenic plants in enormous acreages adjacent to and within natural ecosystems carry risks. We cannot yet even list with confidence all of the components in a relatively simple natural ecosystem such as a grassland prairie or a freshwater marsh, let alone comprehensively understand the interactions between those components. Our incomplete understanding of ecosystems coupled with an exuberant confidence of biotechnicians is a prescription for unforeseen and sometimes undesirable consequences of introducing transgenic organisms into nature. For example, we do not fully understand the mechanism(s), extent, and frequency of horizontal gene transfer in the movement of genes between species in nature. Until we do, we ought to assume that the genes and gene products we introduce into our crop plants *will* make their way into genomes of other organisms, creating the potential to destabilize complex networks of interdependent organisms that are products of hundreds of millions of years of evolution.

Brazil nut allergen. Some persons are deathly allergic to shellfish, peanuts, Brazil nuts, and other foods. A case in which the Brazil nut allergen was inadvertently transferred to soybeans shows that the allergenicity of GE food is a legitimate concern. Fortunately, biotech industry's self-imposed safeguards worked in 1995. Researchers at Pioneer Hi-Bred examined the protein content of soybeans they engineered for enhanced protein content by transferring a gene from Brazil nut plants into the soybean genome. An erroneously transferred allergen gene was detected, and the project was immediately terminated. But news of the incident gave fodder to detractors of GE foods, who said that the error simply proved that GE foods are unsafe. By contrast, the biotech industry claimed that the incident proved that the industry's internal safeguards work.

Clear labeling of foods containing GE products, including the plant or animal species that contributed the transgene(s), would relieve many consumers. Until then, it is little wonder that shoppers with food allergies buy groceries with some angst even though biotech companies assure us that they take great care to screen GE food plants for allergens before their seeds ever reach farmers' fields.

StarLink corn and Bt10 corn. In another incident, a transgene protein product not approved for human consumption entered the human food supply. StarLink is a variety of insect-resistant corn developed by Aventis CropScience. The transgene product is a protein called Cry9C, one of the several insecticidal genes carried by the bacterium *B. thuringiensis*. The presence of Cry9C is approved for corn eaten by farm animals but not for corn destined for human consumption because studies show that CRy9C is not degraded under conditions mimicking the human digestive tract.

In 2000, StarLink corn seed was bought and grown by about three thousand US farmers who produced over eight hundred million bushels of the corn. Later that year after harvest time, Cry9C was detected in taco and tostada shells manufactured by a company with distribution serving seven western states. Obviously, during the corn harvest and/or bin storage of the 2000 crop, StarLink and non-StarLink corn were mixed and sold to a food manufacturer. Recall of StarLink corn-containing products cost the food industry about $1 billion.

In March 2005, the scientific journal *Nature* reported another unintended release of a GE-crop. Between 2001 and 2004, Syngenta, one of the largest agribusinesses in the world, had inadvertently produced and distributed seed for a type of GE corn not yet approved by US governmental agencies. The unapproved plant, known as *Bt*10 corn, contains a gene for the *Bt* toxin that differs slightly from the *Bt* toxin gene approved earlier for *Bt* soybeans, cotton, and corn. How the mix-up occurred has not been made public, nor has the list of countries to which *Bt*10 corn was sold.

Not surprisingly, Syngenta and critics of GE crops voice very different perspectives on this event. Syngenta spokespersons point out that the company informed its regulators, the Environmental Protection Agency (EPA), the US Department of Agriculture (USDA), and the Food and Drug Administration (FDA), as soon as the company discovered its mistake. Immediately, US governmental scientists examined *Bt*10 corn and determined that it posed no threat to human health. Syngenta claims that their handling of the situation shows that the system regulating GE crops works. By contrast,

Michael Rodemeyer, who directs a Washington, DC, think tank on food and biotechnology, contends that the error reflects an inadequate monitoring system for GE products in the food supply.

Crying "wolf" on GE potatoes. A fear of some GEO critics is that the very act of genetically engineering a food plant makes it harmful to people. Actually, the evidence is that just the opposite is true. GE technology for food crops is safer than conventional procedures for developing new plant varieties. The latter pose a greater risk of producing unintended, harmful effects on human health. Why this is so is taken up shortly after a look at one example of why this particular fear of GEOs exists.

In 1998, a debacle within the scientific community itself fueled unwarranted fears that genetic engineering per se poses human health risks. In a project aimed at creating a strain of potatoes resistant to insects, an insecticide-producing gene from a flowering plant called snowdrops was transferred into potato plants at a government-funded institution in Scotland. When the snowdrop potatoes were fed to rats, some of the rats appeared to develop organ abnormalities but not due to the natural insecticide itself.

This was an alarming finding because it suggested that the very act of genetically engineering the potatoes had made them toxic to mammals. If this were true, it would mean that the entire enterprise of genetically modifying food plants was doomed and that the whole array of products already on the market could threaten human health. The researcher rushed to the press with this explosive story. His big mistake was that he went public with his findings before his work could be properly reviewed by other researchers, the standard practice in all fields of science. When the work was reviewed by peers within his own research institution and also by Great Britain's premier scientific body, the Royal Society, the data were found not to support the researcher's premature conclusions. The incident only worsened when the work was subsequently published by a top-notch medical journal, *Lancet*, much to the later embarrassment of the journal's scientific advisory board.

The snowdrop potato fiasco was unfortunate for both sides of the GE food debate. It was unfortunate for the biotech industry because perceptions resulting from bad press are difficult to correct. It was also unfortunate for public health watchdogs concerned about the safety of GE food. Crying "wolf" on snowdrop potatoes may mean that less heed is given to future warnings. This is a bad situation because unpredictable secondary effects of genetic engineering do remain a possibility, as evidenced by an honest revelation and honorable actions in 2002 by molecular biologist Paul Lurquin (Lurquin 2002).

While performing federally funded research in the mid-1980s, Lurquin's research team identified five bacterial genes whose coordinated activity can degrade the herbicide 2,4-D. In research aimed at transferring these genes into plants to make them resistant to 2,4-D, Lurquin discovered that one breakdown product of 2,4-D in transgenic plants is a suspected cancer-causing agent in humans. Moreover, the likely carcinogen was bound tightly to plant tissues, making it highly likely to accumulate to dangerous levels in any plant tissues containing the five bacterial genes and also exposed to 2,4-D. Recognizing the hazard in continuing to pursue this strategy for creating herbicide resistance, Lurquin discontinued the research project.

GE salmon almost makes it to the dinner table. As a final example of a GEO controversy, we consider the case of GE salmon. Here is an example of how politics and business interests can thwart the acceptance and use of a perfectly good and safe source of protein derived from GE technology. In November 2015, the FDA approved the sale of AquAdvantage salmon to US consumers. That decision was foiled a month later by a rider attached to a federal spending bill and signed into law by President Obama requiring the FDA to develop a plan to label the salmon as GE (Dennis 2015). Interestingly, the rider's author, Sen. Lisa Murkowski (R-Alaska) and Sen. Maria Cantwell (D-Washington), an outspoken supporter of the rider, were both from states with large commercial salmon industries. Alaska's 2015 commercial salmon harvest was worth $414 million (ADF&G 2015). The rider mandate was designed to be difficult for the FDA to fulfill. The FDA requires additional labeling of foods when the products are materially different from their conventional counterparts. In the case of AquAdvantage salmon, the relevant material difference would be its nutritional profile. But the FDA had already determined that the nutritional profile of the GE salmon is no different from its non-GE, farm-raised or wild-caught counterparts.

Development and FDA approval of AquAdvantage salmon took twenty years. AquAdvantage salmon have two genes from two other species of fish, Chinook salmon and ocean pout.[8] The transferred genes give the AquAdvantage salmon a much greater growth rate than conventional non-GE salmon. In just eighteen months from fertilized eggs, AquAdvantage salmon are ready to harvest, compared to three years for conventional salmon. AquaBounty, the company that developed AquAdvantage salmon, promoted its GE fish as a means to reestablish a domestic US salmon industry being overwhelmed by farm-raised salmon from Chile and Norway. Critics, including Murkowski, referred to the GE salmon as a "Frankenfish," implying that it was freakish and dangerous to humans. This unfortunate rhetoric is not based on scientific evidence. In fact, the contrary is true. The two transferred fish genes are comprised of DNA made of the same chemical stuff as the DNA of any other organism on Earth. And humans digest DNA every day from a plethora of plant, animal, and microbial genomes. That digestion process acts the same way and releases the same nutrients regardless of the source of DNA. Another health concern raised was that the fish may contain an allergen. If true, the allergen would also occur in other fin fish to which some persons are allergic. Nothing about the addition of the two genes from other fish creates a new allergen in AquAdvantage salmon.

Environmental concerns were also raised—that if AquAdvantage salmon escaped into the ocean, it could mate with native salmon species and disrupt finely balanced ecosystems with rapidly growing fish. AquaBounty foresaw this risk and preemptively did two things to prevent GE salmon from mating with native fish. First, the developers made their GE fish triploid; that is, the fish have three sets of chromosomes instead of the normal two sets. Triploid fish are sterile. Then, to counter the small chance that a fraction of a percent of the GE fish might not be triploid, they raised the fish in landlocked ponds in a mountainous region of Panama. Moreover, AquaBounty specified that all future farmers of their fish would be required to raise them in landlocked ponds.

AquAdvantage salmon was the first GE animal approved for US consumers by the FDA. But politics, business interests, and misinformation about GE products may prevent it from ever reaching dinner tables.

Why Europeans are so opposed to GE foods. The enmity of Western Europe to GE crops and food products is an interesting phenomenon that reflects Europeans' views of the corporate United States and also their images of themselves. A diversity of causes for Europe's hostility toward GEOs cited by several different authors include:

1) a gut-level association of the term "genetic engineering" with "eugenics," conjuring up memories and images of the Holocaust,
2) ambiguous feelings about US influence on European culture,

3) European distrust of their own elected governments, and
4) European pride in their own agrarian and culinary traditions and lifestyles.

One can understand why many modern-day Germans and other Europeans resist any signs that their governments are attempting to force "genetically improved" organisms upon them, especially covertly. Although it is not rational to connect the horrific extermination of six million Jews, Gypsies, and homosexuals with the genetic engineering of corn, rationality seldom wins the day when strong emotions run deep within large populations of human beings. In fact, a lesson that former US secretary of defense Robert S. McNamara implores us to learn about dealings between nations in the film *The Fog of War* is that "rationality cannot save us."

Europeans have a love-hate relationship with the United States. In countries like France and England, there remains widespread and overwhelming gratitude for American sacrifices on their soil in defense of individual liberty and national freedom during World War II. On the other hand, there is widespread European resentment toward what they perceive as a nation that has become an arrogant, uncultured economic and military bully across an ocean no longer wide enough to provide protection against cultural invasion. While American pop music, blue jeans, wireless communication, and information technology are enthusiastically embraced, especially by Europeans born after World War II, Europeans do not relish being told what to do in international affairs, nor do they wish to become "McDonaldized."

Architectural permanence and traditions of behavior such as the evening promenade of Italian families along the streets of their villages and the old sections of their cities are highly valued by Europeans of all ages and professions. "New" is not always "better" for Europeans. The tendency is to view GE crops and food products as a means for wealthy corporations from a "throw-away" culture to profit at the expense of people who value frugality and preservation. It is easy to understand why ships bearing GE products to European shores are not welcome. In 1999, one such boat was the target of a "reverse Tea Party," where protesters dumped American GE corn and soybeans into an English harbor.

Actions of elected officials in European nations do not always reflect the majority opinion of the public. Europeans viewed pressure by the World Trade Organization upon European governments in the 1990s to allow marketing of GE products as an example of the United States controlling a world organization for its own economic benefit. Not trusting that elected officials could or would resist the pressure to import GE products, and not believing corporate declarations about the products' safety, the well-organized and politically effective International Greenpeace organization turned its attention from saving whales to opposing GE foods and crops in Europe and Asia. The end result was a temporary ban in 2002 by the EU on the import of GE foods from the United States. The basis for the ban was potential hazards of GE food to human health.

One prominent member of English royalty sided with Greenpeace. In May 2000, Charles, the Prince of Wales and an organic gardening enthusiast, voiced his concerns about introducing GE plants into the environment: "Nature has come to be regarded as a system that can be engineered for our own convenience . . . and in which anything that happens can be 'fixed' by technology and human ingenuity. . . . We need, therefore, to rediscover a reverence for the natural world . . . to become more aware . . . of the 'relationship . . . between God, Man and Creation'" (Charles, Prince of Wales, 2000).

Writing about GEOs, plant biotechnologist Paul Lurquin (2002) emphasizes the culinary pride and traditions in Europe as important reasons for Europe's hostility toward GE foods. My experiences in Italy and Switzerland confirm Lurquin's views. In Rome, Florence, Pisa, Siena, Venice, and Zurich, I observed people from all walks of life shopping daily for fruits and vegetables in city center farmers' markets. Europeans value relaxed mealtimes more than we do in the United States. In Italy, businesses close every day from one o'clock to three thirty. I do not know what the majority of Italian business persons do during their daily afternoon break, but I do know that many families reunite during midday for a freshly prepared meal and a glass or two of local wine. And in the evening, remaining around the dinner table for two hours or more is not uncommon. "Faster" is not "better" for Europeans when it comes to meals. Sauces, grilled vegetables, salads, and fruit desserts prepared from fresh farm products purchased daily at the market are what generations of Europeans have known as the best foods on Earth. How could transferring a foreign gene into beans, corn, or tomatoes possibly improve what is already the best soup, polenta, and pasta primavera in the world?

Lurquin also cites a 1998 survey showing that European public knowledge about basic biology, genetics, and the nature of biotechnology is superior to that of the US public. Could it be that Europeans sense things about

GE food that we do not or that their value placed on tradition is too under-rated by some of us? Just what is the safety status of GE foods?

Are GEOs Safe?

This question has two aspects: (1) safety of GEOs for human consumption and (2) hazards of releasing GEOs into natural ecosystems.

GEO safety and human consumption. Most opinions on this issue are polarized and controversial. Here, I present information discovered, gathered, and written by scientists. The upshot is that GEO food products are safer and have gone through more regulatory hoops than food derived from plants and animals genetically modified by traditional, nongenetic engineering methods. This is not to say that GEOs never pose a risk to human health. Some of those risks and how they have been countered are described. Also cited are two well-known studies purporting to demonstrate dangers of GEO-derived food and a critique of the validity of these claims. We begin with the conclusions of a study performed by the US National Academies.

The National Research Council and the US Institute of Medicine of the National Academies published an extensive report on the safety of GE foods (National Academies 2004). A portion of the report describes how genetically modifying food organisms by any means can produce unpredictable and unintended effects potentially harmful to human health. However, the report is careful to point out that genetic engineering is just one way to genetically modify food organisms and that other traditional means of modification produce a greater number and more varieties of unpredictable effects than genetic engineering does. The authors of the report make three noteworthy points:

1) Living organisms consist of complex webs of biochemical pathways whose many interconnections are still largely unknown to us. What we do know is that altering just one component in a single pathway can produce unpredictable, rippling effects throughout an organism's metabolism. Some such effects may include production of substances that are harmful to humans or the inhibition of production of substances that are beneficial or essential to human life.

2) All modes of genetic modification, including simple selection, breeding, cross-hybridization, induced mutagenesis, and targeted genetic engineering are capable of altering the metabolism of food organisms in ways that could be deleterious to human health.

3) The chances of inducing alterations in food GMOs that are harmful to humans are greatest with the traditional methods of genetic modification (e.g., crossbreeding and induced mutagenesis) and least with targeted genetic engineering. This is because genetic changes induced by the traditional methods of gene modification are largely uncontrollable.

Living cells and organisms have efficient mechanisms to correct chemical imbalances and to compensate for other perturbations in their internal and external environments. Maintaining a stable internal milieu is called homeostasis, and it calls into play multitudinous, interdependent chemical processes in the cells of every organ system. For example, temperature changes can induce corresponding changes in blood flow to dissipate or conserve heat. Shifts in the electrolyte environment within tissues due to dehydration or dietary salt intake translate into alteration of blood volume and kidney function. Exposure to toxic chemicals induces the synthesis or activation of enzymes in the liver to degrade them.

Still, extreme external environmental conditions can sometimes cause so much havoc inside cells that homeostasis breaks down. This is why most cells die when exposed to temperatures or acidities outside the range they normally experience. Cells and organisms may also die from traumas for which they have no correcting mechanism or protection as evidenced by the disastrous effects of chemicals like chlorine, cyanide, or certain nerve toxins on humans and other living things. What this has to do with human health is that a transgene in a food organism could activate processes to maintain homeostasis that also produce substances harmful to humans.

Consider herbicide resistance. Some plants' resistance is due to a transgene that produces an enzyme that chemically changes the herbicide into a chemical that is nontoxic to the plant. Crop resistance to the herbicides Liberty, 2,4-D, Bromoxynill, and Dalapon is conferred by bacterial transgenes in just this way. There is no problem with this unless the inactivated form of the herbicide happens to be toxic to humans. Being innocuous to plants, such toxic derivatives could accumulate in food plants and pose a health hazard to persons consuming them. We saw an example of this earlier with the carcinogenic derivative of the herbicide 2,4-D found in plants being engineered for 2,4-D resistance by Dr. Lurquin and his coworkers.

A related phenomenon could arise from lack of control over the positions at which foreign genes insert themselves into host genomes. It is possible that the location or simply the activity of a foreign gene could subvert the functioning of a normal metabolic pathway in the host and transform it into a toxin-producing pathway.

Molecular biologists have learned much about the genomics of plants and animals, including humans, but they still have only a fledgling knowledge about the proteomics of these organisms.[9] It may take the rest of the century to learn the significance of the potentially hundreds of millions of life-sustaining interactions that occur between the tens of thousands of proteins encoded in the genomes of humans and the plants and animals that we consume. Even that knowledge alone may not be sufficient to predict the occurrence of unusual metabolic products in genetically altered food organisms and their effects on human health.

As it is, GE organisms are tested extensively for toxic substances before they are ever grown for food. The question is whether biotechnicians can accurately screen for all possible toxins, given the complexity of the system they are trying to protect—the human organism. The degree of this uncertainty is a source of controversy.

Human ESCs and iPS cells are very valuable for testing GEOs, drugs, and other substances for harmful effects on human health. Because virtually all 252 types of cells in a human being are derivable from ESCs and iPS cells, the effects of chemicals in genetically altered organisms on every tissue in the body can be tested in a laboratory culture dish. This minimizes the need to use human-derived tissue and nonhuman primate subjects to assess the safety of GE products.

Despite present difficulties in assuring absolute safety of all GE food organisms for all persons, the likelihood of genetic engineering producing a toxic substance in food is much less than it is for conventional selection and breeding practices. Examples of this little-known and underappreciated fact are detailed in the aforementioned government publication on the safety of genetically modified foods (National Academies 2004).

Crossbreeding practices and chemical or radiation-induced mutagenesis procedures randomly mix or mutate thousands of genes not just one or two targeted genes as in genetic engineering. There are some common food plants that provide examples of unpredicted, unintended, hazardous genetic modifications not caused by genetic engineering. Consider tomatoes and potatoes. Both have been selected and bred for increased yield and insect resistance. Both acquire the latter by elevating their production of chemicals that are toxic to insects. Unfortunately, some of these chemical defenses are toxic to some humans. Similarly, celery naturally produces elevated amounts of a chemical called psoralen when it is bruised. Psoralen is toxic to insects and therefore protects the plant from further damage by pests. Some celery cultivars are bred to contain high levels of psoralen for the advantage it gives growers in getting pristine-appearing bunches of celery to market. But psoralen adversely affects human health by sensitizing the skin to long-wave ultraviolet radiation that can cause severe dermatitis. Celery pickers chronically exposed to bruised celery stalks and working in direct sunlight are especially susceptible. Produce packers and grocery workers who handle celery on a regular basis also report higher than normal incidences of dermatitis.

Finally, consider two highly publicized studies claiming to show that particular types of GE corn (Séralini et al. 2012) and GE potatoes (Ewen and Pusztai 1999) pose serious health risks. Objective critiques of these claims by biotechnologist and science communicator Alison Van Eenennaam and others reveal them both to be unsubstantiated (Van Eenennaam 2013). Séralini's study performed on two-year-old laboratory rats claimed that feeding Roundup Ready corn and a formulation of Roundup herbicide over a two-year period caused organ damage and tumors. This study was poorly designed, is contradicted by many other peer-reviewed long-term studies, and has been thoroughly discredited by regulatory agencies worldwide.[10] The 2012 article was retracted in 2014, yet GEO critics continue to cite the article as evidence of a health risk from Roundup Ready corn.[11] The other study by Ewen and Pusztai claimed that rats fed GE potatoes that expressed a particular insecticidal lectin suffered harm to their gastrointestinal tract and immune system. The latter was claimed to be unrelated to the lectin but simply due to some effect of the genetic-engineering process itself. A careful examination of the data by the UK's Royal Society found no evidence that the lectin or the genetic-engineering process had adverse effects on the rats' immune systems. Regarding the reported harm to the gastrointestinal tract, the Royal Society's report noted the lack of control animals fed a standard rodent diet and also the fact that the GE potatoes were not intended for human consumption and would never have obtained regulatory approval anyway because of the widely known and documented toxic and allergenic features of lectins. Despite the many problems with the study, many GEO critics continue to cite it as evidence for danger of the genetic-engineering process.

In summary, although GE plants deserve to be thoroughly tested for secondary metabolites harmful to humans or to the environment, it is wrong to believe that GE food is more dangerous than non-GE food developed by conventional means of genetic modification. Just the opposite is true. Ironically, new varieties produced by conventional methods such as hybridization require no evaluation by federal regulatory agencies of their safety for human consumption or the environment before commercial release. By contrast, GEOs undergo intense scrutiny by the FDA, EPA, or USDA, and the company responsible for their development before their release.

GEO safety and the natural environment. Commonly voiced concerns about the impact of GEOs on the environment include: (1) harming of non-target insect species by insect-resistant plants such as *Bt* corn, (2) development of *Bt*-resistant insect pests, and (3) transfer of genes from GE crops into native plant species.

As we saw, reports that *Bt* corn pollen harms monarch butterfly larvae were based on incomplete studies and proved to be a false alarm. This communication debacle is not evidence that GE plants producing *Bt* toxin could not harm other desirable insects. Rather, it is a warning to be vigilant and foresighted in the development and use of GE plants.

As years of use of *Bt* crops increase, so does the likelihood that *Bt*-resistant strains of insect pests will emerge—the virtually certain outcome of an ancient evolutionary arms race perennially played out between plant and herbivore, predator and prey, host and pathogen. Should we worry about this? Farmers have been encouraged to plant strips of non-GE plants in their fields of GE crops in order to keep a healthy number of *Bt*-sensitive insects in the breeding population. Some commentators suggest that biotech companies may not really care whether pests become resistant to *Bt* since there are other natural insecticide genes to be found and transferred into commodity plants when the *Bt* era is over.

Horizontal gene transfer between species is very different from the vertical transfer of genes that occurs between generations within a species or closely related species that can hybridize. I have genes from my parents, grandparents, great-grandparents, and on back. I have acquired these genes from my ancestors by vertical gene transfer from generation to generation. Horizontal gene transfer refers to the direct nonsexual movement of genes from one species into another. How commonly this occurs is not known, but it does happen. For example, viruses carried by sap-feeding insects from one plant to another are thought to be vectors for horizontal gene transfer in the plant world.

Ethicist Gregory Pence argues that horizontal gene transfer is not a legitimate concern because genes do not move between plant species unless they are closely enough related to hybridize and produce fertile offspring (Pence 2002). This argument fails to recognize the established role of viruses and insects in moving genes between species. It is reasonable to be concerned about horizontal gene transfer. The worry most often expressed is that genes for herbicide resistance, such as those in Roundup Ready crops, may move from crop plants to weeds, creating super weeds resistant to conventional herbicides. Something akin to this has happened in Alberta, Canada, where Roundup Ready rapeseed (canola) is a commodity crop. There, volunteer rapeseed plants themselves have become pesky, herbicide-resistant weeds in wheat fields.[12]

Although herbicide resistance by native plants could become a headache for farmers, it would not be nearly as disruptive to the natural environment as would their acquisition of genes for certain other traits. Enhanced rates of photosynthesis, insecticidal properties, drought tolerance, and earlier or later flowering are all traits of great interest to plant genetic engineers. If genes for these traits entered natural ecosystems, disruption of the species composition and balance of ecosystems could be devastating. Whether transgenes are subject to the same mechanisms that apparently limit the promiscuous transfer of naturally occurring genes for traits like these between plant species is unknown; therefore, the movement of transgenes within natural ecosystems remains a legitimate concern.

GEOs: Human Values

Three main ethical issues come with creating and using genetically modified plants and animals: (1) the unprecedented, rapid mixing of genes between species that acquired their genomes over eons by natural selection, (2) justice and the benefits derived from GEO technology, especially in the context of the developing world, and 3) the right of individuals to make choices in using GEOs.

Rapid, Human-Directed Gene Mixing

This issue manifests itself in at least two different forms: first, in the admonition by some that we should not "play God" with nature, and second, in a land ethic that values the stability of natural ecosystems. Let us consider some views on these issues, beginning with the charge that biotechnicians are wrongfully playing God when they create transgenic organisms.

GEOs and playing God. Implicit in the opinion that biotechnologists are "playing God" is that traditional breeders of plants and animals are not playing God. This suggests that it is not the human manipulation of an organism's genetic material per se that is problematic but rather something about the process of genetic engineering that distinguishes it from the conventional methods of breeding and developing new varieties of plants and animals. What might this be?

The violation of the sanctity of species by transferring genes between them then seems to be the contentious point for those who believe it is immoral for science to create transgenic organisms. This brings up the questions: What is a species? What is the basis for species sanctity? Consider first what qualifies as a species.

Volumes about the concept of species are written by systematists (biologists who classify organisms), comparative morphologists, evolutionary biologists, and others, and there is plenty of disagreement among them. Philosopher of biology David L. Hull (1997) observes that species are whatever one chooses to make them.

An introductory biology course is likely to define a species as a group of organisms within which individuals can breed to produce fertile offspring. The problem with this definition is that sometimes individuals classified as belonging to the same species, but which inhabit distant regions of the species' range, cannot successfully mate. Conversely, individuals, especially plants, classified as belonging to different species can often crossbreed to produce fertile hybrids. Domestic wheat owes its origin to several such hybridization events between wild grasses.

Other definitions of species are tailored to specific uses of the species concept. Two primary uses biologists make of the notion of "species" are as fundamental units of evolution and for classification purposes (Hull 1997). Evolutionary biologists are interested in how the genetic compositions of populations change over time and how subpopulations become reproductively isolated from larger groups. Systematists, on the other hand, are mainly interested in using morphological traits to distinguish between groups of organisms. For example, the length of antennae or the number of bristles on the underside of an abdominal segment may be critical for assigning an insect to a particular species. Although overlap exists between these two ways of thinking about species, there are also areas of disagreement. For example, a systematist may view two organisms as belonging to different species or subspecies based on morphological differences; whereas, an evolutionary biologist may place them together within the same species based on interbreeding compatibility.

For this discussion, I propose we consider a species to be a group of organisms that (1) has members with the potential to interbreed and produce offspring belonging to that same interbreeding group, (2) originated from a different, preexisting group of interbreeding organisms, (3) may give rise to one or more new groups of interbreeding organisms in the future, and (4) is relatively stable over time periods measured by historical, human events. Is there anything about a species described in this way that deserves the sanctity of remaining "uncontaminated" with even one or a few genes from a different species?

My description of species is premised on the common ancestry of all living things, a reality recognized by biologists worldwide for more than 150 years. However, another reality is that about half of all Americans reject the validity of science's observations showing that all of life is related by common descent. Instead, they believe that the array of living things we see around us today and observe in the fossil record is the product of a single, supernatural, creation event. These persons are likely to view the ethics of GEOs with transgenes differently from those who accept the biological principle of evolution by common descent. My objective here is not to enter the evolution versus creationism (intelligent design) debate. Rather, I wish to point out that how one views the origin of species may influence her view about the morality of creating GEOs.

Scripture from the Judeo-Christian tradition states that "God created great whales, and every living creature that moveth, which the waters brought forth abundantly, after their kind, and every winged fowl after his kind. And God blessed them, saying, Be fruitful, and multiply, and fill the waters in the seas, and let fowl multiply in the earth" (Genesis 1: 21–22). A literal reading of this passage encourages the notion that each and every species of plant, animal, and microbe was created by God in its present form or in the form it had when it became extinct. Furthermore, it fosters the view that the assemblage of living creatures we see around us now has always existed and that the kinds of organisms in this assemblage have not changed over time. Does this interpretation of scripture make moving a gene from one species to another species disrespectful of God? Does it lead inexorably to the view of Prince Charles of Wales, who wrote in the *Daily Telegraph* that scientists are usurping "realms that belong to God and to God alone" (Bara and Zidenga 2003)?

Do these questions have absolute answers? One may believe that scientists "play God" when they transfer a gene from a petunia plant into a bean plant or from humans into bacteria. Whether this is out of place for humans to do from the perspective a scriptural literalist is not for me to say. A literal reading of Christian scripture also supports the belief that humankind was created rational so that it could exert "dominion" over Earth's other creatures. One could interpret "dominion" as including power over organisms' genomes. However, the notion of mankind exerting divinely ordained dominion over the biosphere is out of vogue now even among many believers because of the harm caused to the Earth's environment by that attitude. In the last chapter of this book, I discuss the urgent need for better and more expansive public education in the realms of evolution and ecology. Perhaps a more widespread acceptance of transgenic organisms will be a by-product of broader and deeper biological literacy.

Transgenic Organisms, Ecosystem Stability, and a Land Ethic

The second ethical issue associated with rapid, human-directed gene mixing between organisms concerns the potential effects of foreign gene products on nontarget organisms and ultimately on the stability of natural ecosystems. Will the planned or unplanned entry of transgenic organisms into natural ecosystems upset the incumbent natural balance? Limited understanding of how ecosystems actually operate handicaps our decisions when attempting to place value on natural ecosystems and also determines how one views this issue. For those who value the preservation of biodiversity and natural habitats, the possibility that transgenic organisms may ingress into the natural environment is alarming. Recall the ecological mayhem that could be caused by horizontal movement of transgenes into native species. In addition, there is a legitimate concern that a GEO itself might take up residence within established ecosystems. Like an invasive species of plant or animal, a GEO might displace existing species or destabilize the system in unpredictable ways. We take care not to transfer weeds from one freshwater lake to another in the Upper Midwest in order to prevent spreading purple loosestrife, an exotic plant now choking many Wisconsin wetlands. It may be prudent to take similar care with GEOs. Most domesticated, agricultural cultivars could never survive in native habitats, even with an added transgene or two, but creating new plant varieties with transgenes for enhanced photosynthesis, drought or salt resistance, flood tolerance, or other traits that give them a competitive advantage over native species could be disruptive to native assemblages of plants and animals.

For people who see little or no value in protecting and preserving native ecosystems, the release of GEOs into native habitats, which could possibly upset the balance between native species, is not of much concern. A similar position might be reached by persons who take literally the scriptural admonition for human dominion over other living things. What both fail to appreciate is the tightly woven nature of the web of life. Humankind depends upon the stability of the entire system for its very existence. A simple example illustrates the point. The oxygen we breathe comes from green plants and oceanic plankton. If we kill the plankton with pollution and climate change or remake the planet's surface into parking lots and plowed fields, we, along with all other oxygen-breathing creatures, will die.

The morality of introducing GEOs into the environment can also be considered from a secular point of view that asks, "What is the proper relationship between humankind and nature?" Whether biological reality carries with it imperatives or prohibitions on how humans ought to treat the natural world is a much-debated question in philosophy. The roots of the question extend back to the seventeenth century when David Hume questioned the legitimacy of using objective observations about how the world is to derive principles of moral behavior. For example, consider the fact that the stability of natural ecosystems depends on the well-being of species that comprise them. Does this observation carry a moral obligation for species preservation? We will not settle this debate here. Rather, consider the thoughts of Aldo Leopold, a twentieth-century Thoreau-like figure, on what a healthy relationship between humans and nature looks like.

Leopold, Wisconsin conservationist and author of *A Sand Count Almanac*, pointed out in the 1940s that we have ethical systems that set rules for interactions between individuals and between nations but no ethics to describe our proper relationship with the land. "Obligations have no meaning without conscience, and the problem we face is the extension of the social conscience from people to land," wrote Leopold in his essay "The Land Ethic" (Leopold [1949] 1970, 209). Here, Leopold proposed an ethical system for human treatment of the land. Recall that the upshot of Leopold's land ethic is to judge the morality of our actions by their effects on the biotic community: "*A thing is right when it tends to preserve the integrity, sta-*

bility, and beauty of the biotic community. It is wrong when it tends otherwise" (224–25, emphasis added). What is meant by the "integrity" and "stability" of living communities? Ecologists have talked to us for decades about this.

The integrity of a community refers to the level of intactness of its interconnected parts. Just as the internal physiology of a plant or animal is designed to maintain homeostasis, so too are the parts of an ecosystem. A few disruptions can be tolerated, but wholesale disorganization of the interconnections that have evolved over hundreds of millions of years will result in ecosystem collapse. As for stability, it is a biological axiom that the more diverse an ecosystem is, the more stable it is. This makes sense because an ecosystem with a large number of diverse interconnected plants, animals, and microbes will have a more extensive framework of interdependencies. It will be better able to tolerate a discontinuity introduced here or there due to the large number of compensatory relationships within the system. If a rabbit had only foxes for predators, the removal of foxes would result in the place being overrun by rabbits, which would consume more than their share of leafy greens, and the system would collapse. But if hawks, eagles, and snakes also eat rabbits, the effect of fox removal is dampened. As more components in an ecosystem are damaged, less compensatory dampening occurs when the next component is damaged.

As a nation and as a world community we are either dreadfully ignorant about what constitutes the integrity and stability of biotic communities, or we are terribly shortsighted. Wetlands are drained for parking lots and subdivisions, diverse deciduous forests are decimated and replaced with pine monocultures. Rivers are diverted and dammed to irrigate desert ecosystems, flooding others with expansive reservoirs. And we continue to expel greenhouse gases into the atmosphere in the face of overwhelming evidence for their disruptive effects on climate and biotic communities worldwide. Ignorance and obliviousness also reign at the local level. In the small college town where I live, the first things to go when developers transformed a forest into a city park were the decades old oaks, sycamores, and hickory trees—all uprooted, cut into pieces, and burned. Then, the site was scoured with bulldozers to provide a "clean" landscape for the planting of ornamental trees and bushes. True, there are points of enlightenment scattered over the land—communities that dedicate themselves to preserving beauty in its natural forms—but these are disappointingly sparse.

What are we to do? Leopold wrote that "conservation is a state of harmony between men [humans] and land" (Leopold [1949] 1970, 207). We too are part of the biotic community. We deserve a place in it. Some say that since we are part of nature, then all that we do is, by definition, natural, and we should therefore not worry about what we do. This is a shortsighted view of our relationship with the other components of the biotic community. Because humankind is rational and can ponder the consequences of its actions, it has an obligation to preserve the integrity and stability of the biological community upon which it depends. We can acknowledge that both the rational and the self-serving components of our behavior arose by natural selection and that neither one is intrinsically bad. Our challenge as Earth's rational creature is to integrate knowledge and rationality with our self-serving nature and realize that to be truly self-serving we must be farsighted, not shortsighted. Our sight must focus beyond the actions that might increase profit margins or GNP for the current year or the next decade.

What about the third criterion for Leopold's ethical behavior toward the land—beauty? Is beauty not subjective, purely in the eye of the beholder? Some feel that strategically planted clumps of exotic azalea bushes are more beautiful in a city park than tall, expansive, naturally spaced oaks and sycamores. And as Leopold points out, the stark "beauty" of the American Southwest is the result of a collapsed ecosystem brought on by the overgrazing of livestock in a biotic community too fragile to absorb such treatment and retain its equilibrium. The "beauty" in Leopold's land ethic criteria is a beauty informed by knowledge about integrity and stability. Leopold challenges us to learn to recognize the appearance of healthy, stable biotic communities and then feel their beauty when we look upon them and walk lightly through them. Appreciating this kind of beauty requires education. Appreciation comes as an intellectual process as well as an emotional response.

We all must eat, of course, and there are lots of us. Taking Leopold's land ethic seriously does not require that we turn every corn field in Illinois back into deciduous forests or natural prairie land. Rather, we ought to make farsighted decisions about our use of the Earth's remaining natural ecosystems, our lifestyles, and the export of our lifestyles.

What does Leopold's land ethic mean for GEOs? There is no simple answer. We have seen how some GEOs might actually contribute to the integrity and stability of the land. For example, pollutant-eating bacteria could restore natural habitats. No-till farming, made possible by herbicide-resistant crops, lessens soil compaction and its destructive effects on soil ecosystems. On the other hand, there are the potential deleterious effects of widespread use of insecticide-producing crops and the unknown consequences of horizontal gene transfer between GEOs and indigenous species. Education of our children and grandchildren to see beauty in the integrity and stability of nature, together with a farsighted approach to our use of GEOs, are both critical for humankind's survival and for that of the planet's biosphere.

Once again, from the pen of Leopold, "examine each question in terms of what is ethically and esthetically right, as well as what is economically expedient. . . . Abraham knew exactly what the land was for: it was to drip milk and honey into Abraham's mouth. At the present moment, the assurance with which we regard this assumption is inverse to the degree of our education" (Leopold [1949] 1970, 204–5).

Justice and Benefits from GEO Technology

As with all new technologies, distributive justice is a difficult but important issue for GE technology. Just who benefits the most from GEOs? Is the technology helping persons in greatest need of its benefits? In searching for answers to these questions, we may discover whether we are on track toward a just distribution of benefits from GE technology. Take GE commodity crops as an example.

Farmers, seed companies, and GE crops. GE crops are mixed blessings for farmers and the environment. The rationale for creating Roundup Ready corn and beans was to make the plants resistant to herbicides so that weeds can be killed by spraying the fields soon after crop plants emerge from the ground. Greater yields were expected by reducing weed competition for nutrients, but most herbicides applied to fields are expensive chemical fertilizers. By contrast, non-herbicide-resistant crops must be left to fend for themselves after emergence, or weeds must be controlled by tilling, a time-consuming method with the undesired side effect of soil compaction.

Before planting, farmers must estimate whether the cost of seeds and technology fees for herbicide-resistant crops offsets increased yields and profit. Higher yields result in lower market prices. If everybody strives for higher yields, prices are forced down even more. The vicious cycle encourages farmers and seed companies to do whatever they can to get even larger yields in order to make a profit. Additionally, farmers must do a lot of guessing and gambling. Will rain at the wrong time require replanting; will it be a bad weed year, a dry year, or a year of insect pests? Finally, there is the question of whether the market for GE crops will be as good as the market for nontransgenic crops. For example, Japanese soybean buyers pay premium prices for non-GE beans to produce tofu.

Purchasing and growing transgenic crop seeds can therefore be an iffy venture for farmers. It is much less so for GE seed companies, especially when they control the amounts and types (transgenic or not) of seeds, establish prices, and manufacture the herbicide to which their GE plants are resistant. Recently, I asked a farmer in Ohio who he believes benefits the most from *Bt* corn and Roundup Ready beans. His emphatic, one-word answer was, "Monsanto." Later the same day, I visited the manager of the local grain elevator responsible for keeping stored GE soybeans separate from non-GE beans. I asked him the same question. He responded immediately with the same one-word answer.

What about the land? Are there environmental benefits to growing GE crops? I have mentioned that growing Roundup Ready crops allows farmers to control weeds without having to till. No-till farming is good for the soil because it lessens compaction by heavy machinery. The destructiveness of soil compaction cannot be overestimated and reducing or even reversing it is an essential component of a sustainable agriculture. A Midwestern farmer recently explained to me that due to generations of farming with heavy machinery, crops in most fields grow in just the top two inches of soil. Below that the soil is brick hard, devoid of oxygen, lacking in microorganisms that release nutrients, and impermeable to water. An innovative deep-tilling procedure using eight knifelike blades that penetrate two feet into the soil and are pulled through the field by a 450 HP tractor shows promise for revitalizing the compacted layers of soil if practiced regularly for several years. Beneficial soil bacteria are usually injected into the soil while deep tilling. Tractors do contribute to compaction, but by guiding vehicles and machinery in the same tracks, soil damage is minimized.

Biotech companies like to point out that their pest-resistant seeds have environmental benefits since they remove the need for the use of pesticides. What they talk about less is that their herbicide-resistant seeds have just the opposite effect, lessening the need for prudence in the application of herbicides and fostering continuation of the deleterious effects of toxic chemicals entering the groundwater and freshwater ecosystems. Increased herbicide exposure to farmers, farm workers, and consumers is also a downside to increased use of herbicide-resistant crops. Granted, herbicides Roundup and Liberty are deemed safe for humans because their actions are on plant metabolic processes that do not occur in humans, but this is not necessarily true for every herbicide. Also, unknown, long-term, harmful effects of human exposure to herbicides cannot be ruled out.

Another liability that comes with growing GE crops is what farmers call yield-drag. This is decreased productivity in fields where Round-Up Ready crops have been grown for several seasons, compared to fields where they either have not been grown or have been grown less frequently. Although Monsanto denies that yield-drag exists, most farmers agree that it is a real phenomenon. The cause is unknown. One possibility reportedly being tested is that glyphosate, the active ingredient in Round-Up, when taken up by Round-Up-resistant crops, induces the plants to expel toxins into the soil, adversely affecting beneficial soil bacteria.

Although the genetic engineering of commodity crop plants has so far mainly benefitted GE seed companies, this situation may be on the cusp of change. Innovative GE products and plants modified with the gene-editing tool CRISPR promise direct benefits for the consumer, although as we see in chapter 4, the EU may not be too keen on CRISPR crops. Also encouraging are efforts of public sector research to provide benefits from GE technology to developing countries.

GE technology and the developing world. The relationship between GE technology and the developing world raises two important questions. First, does GE technology have the potential to relieve malnutrition and starvation in poor countries? If the answer to this question is "yes," the second question is will GE technology actually be used to benefit poor and malnourished persons. Examining effects of an earlier overhaul of the germplasm (hereditary material) of certain third-world commodity crops may help answer these questions. The Green Revolution of the 1960s and 1970s was a humanitarian-motivated project that introduced new varieties of wheat and rice and new farming practices into third-world countries in Asia and Latin America. The result was dramatically increased grain yields. To what degree and in what ways the revolution actually helped poor and hungry persons in these countries is controversial, as are the effects the revolution had on the environment.

Norman Borlaug, farmer's son, scientist, world humanitarian, and Nobel Laureate, is the father of the Green Revolution. The revolution had its roots in Borlaug's recognition in the 1940s that the productivity of wheat farmers in Mexico lagged far behind that of US farmers, primarily because of an airborne disease called wheat rust. Borlaug was confident that through intensive plant breeding and selection, rust-resistant varieties could be developed. He was right.

The outcome of thousands of crosses between different wheat plants grown at different geographical sites with varying climates and soil conditions was an array of semidwarf varieties adapted to a broad spectrum of environmental conditions and with greatly increased yields. Growing on shorter, sturdier stalks, the grain heads of dwarf varieties remained upright during wind and rain. Harvesting the new grain is easier and more complete than for the old varieties, whose bent stalks reduce the proportion of harvestable grain.

Where the new varieties were used, grain production increased dramatically. India, for example, which was not producing enough grain to feed its own citizens before 1970, began exporting grain in the mid-1980s. Both India and Pakistan increased their annual production of wheat by more than five times in about twenty-five years after beginning to use Borlaug's semidwarf varieties. Bioethicist Gregory Pence credits Borlaug's work on wheat and the work he inspired on rice for saving the lives of one billion people over a period of two decades (Pence 2002).

There is another side to the Green Revolution. Since the new varieties came with chemical fertilizer and irrigation requirements, many of the small, subsistence farmers could not afford to continue farming. They were forced off their land by wealthier, conglomerate farmers and moved into cities, where they lead lives of poverty worse than before the Green Revolution. In many areas, the gap between rich and poor widened, and the hunger problem worsened. Unemployment increased in many regions because farmers of large acreages replaced manual labor with mechanical technologies in their fields. Rural women, who had previously enjoyed respect as tenders of the crops, their family's main source of income, lost esteem within the family and society when their labor became unneeded on industrialized farms.

The land also suffered. The combined application of chemical fertilizer and irrigation increased the salt content of the soil as the water evaporated. Natural ecosystems were harmed when water from rivers and wetlands was diverted into irrigation pipes and also when plots of land containing a diversity of fauna and flora were turned over to crop monocultures.

Nearly all writers on the Green Revolution note that it bypassed Africa. One reason may be that food needs in famine-plagued Asia during pre-AIDS and pregenocide times in Africa were viewed as more urgent. Another is that indigenous crops and growing conditions in Asia and Latin America are well suited for the new varieties of wheat and rice, whereas conditions in Africa are not. Of course, neither of these reasons explains the current level of neglect by developed nations of the unspeakable human misery now rampant on the African continent.

Some criticisms of the Green Revolution have been challenged by counter-claims that (1) movement from manual to mechanical labor allows children to move from fields into schools, (2) more acres of uncultivated land would have had to be converted to farmland to feed citizens in Asian countries under traditional agricultural practices than the acreage converted under the new agriculture, and (3) farmers and their families have more to eat and can build better houses as a result of increased profit from their farms.

Whether the estimated one billion lives saved by increased yields from the Green Revolution's dwarf varieties of wheat and rice is accurate can probably never be known, but few deny that world hunger was greatly reduced as a direct result of Norman Borlaug's vision and steadfast work, for which he was awarded a Nobel Peace Prize in 1970.

What can we learn from the Green Revolution that is relevant to the prospect for GE crops easing world hunger in the coming decades? First, we know that farm land in developing countries can produce more food when new crop varieties designed for increased yield and durability are grown. Second, we learned that just increasing crop yields does not guarantee that poverty will be wiped out or even that it will be relieved. People need food to live, but they also need good jobs to emerge from poverty. Third, the Green Revolution taught that introduction of new agricultural technologies into a developing nation needs to be preceded by a holistic understanding of the traditions, values, and structures of the society. How new technologies will positively and negatively impact society must be anticipated and planned for before they are introduced. Ideally, people that will be most affected by the new technologies should have the most influential voices in whether and how they will be implemented.

So what is the prospect that GEOs can really help citizens in the developing world? It is good so long as elected officials and private corporations of the developed world take seriously the fact that the needs of agriculture in the world's poor countries are very different from the needs of agriculture in their own countries. People in North America and the EU spend 10 to 15 percent of their income on food compared to the 50 80 percent spent by poor people in developing countries. New crops engineered for increased yields do not need to be high priorities for nations like the United States, where raw food material is in surplus. Instead, the United States needs crops that leave shallower footprints on the natural environment and promote sustained soil productivity. Subsistence farmers in developing countries, on the other hand, will benefit most from increased yields and improved nutrition from staple crops like rice and sweet potatoes.

The West devotes itself mainly to producing high-yield GEOs grown in great abundance by large-scale farmers. This strategy provides biotech companies with huge profits. So-called "orphan crops," grown by small farmers in developing countries for their own family's use and to sell at the local market, get little attention by corporate giants in the GEO industry. The expense of research and development to improve yields and the nutritive value of orphan crops cannot be recovered from farmers with small acreages and unable to pay annual technology fees. Among such orphan crops are cassava, beans (nonsoy), sweet potatoes, barley, lentils, millet, sorghum, some fruits, and certain multipurpose trees. Fortunately, some of these are now receiving attention by researchers in the public sector, including public universities and international research institutions.

Right of Choice in the Use of GEOs

The importance of respecting the autonomy of people in countries that might benefit from GE technology from the West was expressed by P. Chengal Reddy, president of the Federation of Farmers' Associations in Andhra Pradesh, India. In a personal communication to the authors of the book *Seeds of Contention* (Pinstrup-Anderson and Schiøler 2001), Reddy declares concern about the influence environmental activist groups may have on the introduction of GE crops into his country and suggests leaving the choice of selecting modern agricultural technologies up to Indian farmers. Reddy's advice applies equally well to biotech companies seeking new markets for their engineered seeds.

Respect for individual choice is also an issue when it comes to labeling GE foods. Unlike European countries, the US government did not require labeling of GMO-containing food products for years. Finally, in July 2016, the US Congress passed legislation that requires most GMO-containing foods to be labeled as such. The legislation gives food companies three options for labeling: a text label, a symbol, or an electronic code readable by smartphone that indicates whether the food contains GMO ingredients (Associated Press 2016). As of late 2018, rules for GE food labeling were not finalized. The USDA suggests using *bioengineered food* or *bioengineered food ingredient* in labels and aims to require labeling by early 2020. Some writers argue against labeling of GE foods because doing so may send the mistaken message that there are dangers to consuming them (Pence 2002). Indeed, there is overwhelming evidence for the safety of GE foods. However, the argument that people deserve to have the knowledge necessary for making a choice about purchasing GE foods eventually won the day. The means for segregating GE and non-GE crops are already in place since exporting crops to countries where GE products are banned or where labeling of GE products is required has demanded it.

Chapter Summary

Genetic engineering is one of many ways that humans modify the genetic constitutions of plants, animals, and bacteria. Basic biological research in the 1970s made possible our present capacity to produce GEOs that have applications within basic research, medicine, agriculture, environmental reclamation, and production of synthetic products. So far, the most widely used GEOs are crops engineered for protection against pests and weeds. The benefits of these accrue primarily to large corporations that invest in their development. It is possible for unpredicted and unintended alterations in the metabolism of GEOs to have harmful effects on human health or the environment, but the chances for this happening are less than for equally harmful consequences occurring from the genetic modification of organisms using traditional approaches such as crossbreeding, hybridization, and induced mutagenesis. The potential for GEOs to benefit the developing world will probably be realized through ongoing efforts at national and international public research institutions. Major ethical issues raised by GE technology include the creation and introduction of genetically new organisms into the environment, the distribution of benefits from the technology, and the choices people have in using GEOs.

Questions for Reflection and Discussion

1. Do you believe currently marketed GE foods are safe for human consumption? Why or why not?
2. Do you believe any currently grown GEOs pose environmental risks? If so, are the risks outweighed by benefits?

3. Can you imagine future GEOs (other than biological pathogens) that would be unethical to release into the environment? Would any of these be unethical to create/use even under contained conditions?

4. In the context of Aldo Leopold's "land ethic," how ought the "beauty" of an ecosystem be conceived?

5. How can GEO technology best be used to improve life for the world's most impoverished persons? Who should pay for the research and development required to develop noncommodity, GE food plants for use in agriculturally poor countries?

Notes

1. The National Bioengineered Food Disclosure Law required the US Department of Agriculture (USDA) to set regulations by July 29, 2018, for implementing the law. This did not happen. The new target date for setting regulations was December 1, 2018, and January 1, 2020, is the target date for compliance by food manufacturers. The law allows food manufacturers three different options for disclosing the GE ingredients in their products. Food derived from animals fed GE crops is exempt from labeling. Once the USDA announces the regulations, manufacturers will probably have one to three years to comply (Jaffe 2017).

2. Porteus's "News and View" article is a synopsis of two full-length research papers in the same issue of *Nature* reporting the use of specially designed enzymes, zinc-finger nucleases (ZFNs), to induce double-stranded breaks in target DNA at the desired spot for insertion of a transgene into plants. Three methods are available to insert transgenes into the host DNA at the site of the DNA break: ZFNs with desired specificity can be assembled by the researcher, purchased, or obtained from a biochemical corporation like Sangamo or Dow in a research collaboration agreement in which corporate scientists make the ZFNs. The two research articles are: V. K. Shukla, Y. Doyon, J. C. Miller, et al., "Precise Genome Modification in the Crop Species *Zea mays* Using Zinc-Finger Nucleases," *Nature* 459 (2009): 437–41; and J. A. Townsend, "High-Frequency Modification of Plant Genes Using Engineered Zinc-Finger Nucleases, *Nature* 459 (2009): 442–45. In 2013, the CRISPR/Cas9 technology was successfully adapted to eukaryotic cells, making the creation of GE plants and animals much easier, more accurate, and less expensive (chapter 4, this volume).

3. The official Golden Rice website (http://www.goldenrice.org/) reports these statistics in an article on its Home Page: "Quantum Leap: Golden Rice Accumulates Provitamin A (ß-carotene) in the Grain" (accessed May 8, 2018).

4. The official Golden Rice website (http://www.goldenrice.org/) contains a "Current Affairs" page with reports on the status of using GE crops to relieve hunger in Africa and Asia. Here, viewed on June 5, 2009, is a discussion by Temba Nolutshungu (director of the Free Market Foundation, South Africa) of the precautionary principle and how its application in the context of crops like Golden Rice may be costing thousands of lives per day. The precautionary principle requires that action be taken to avoid a risk even when there is no evidence for the existence, magnitude, or potential impact of that risk. The principle demands that new technologies be assumed harmful until they are proven safe to a high degree of certainty. In Current Affairs comments on World Development Report 2008, viewed November 17, 2008, Prof. Ingo Potrykus wrote: "It is probably safe to say that GMO regulation, in the context of *Golden Rice*, is responsible for the loss of 7 x 40,000 lives in India and, of course, of many more in the other countries."

5. In a February 2008 personal communication, Dr. Jorge E. Mayer, Golden Rice Project Manager, predicted that three more years would elapse before the first variety of rice containing the beta-carotene-producing trait could be released to farmers, due in part to the lengthy regulatory process that includes required allergenicity and toxicity testing.

6. "Rockefeller Foundation to Help Golden Rice Obtain Regulatory Approval," on Home Page at: http://www.goldenrice.org/ (accessed May 8, 2018).

7. Please see note 1 of this chapter.

8. The ocean pout is native to the northwest Atlantic and is noted for its ability to live in near freezing waters due to an antifreeze protein. A segment of DNA called a promoter keeps the gene for the antifreeze protein turned on nearly continuously. To design the AquAdvantage salmon, the DNA for this promoter from ocean pout was transferred to the Atlantic salmon genome and placed next to another transferred gene for growth hormone from Chinook salmon. This results in the AquAdvantage salmon having a nearly continuous supply of growth hormone, which gives it a higher growth rate.

9. Proteomics is the study of how the protein products of genes interact with each other in living cells.

10. Problems with the study include the fact that the rats used in the research normally have a high rate of tumor development, especially by the time they are two years old, that the study used too few control animals, and that it employed inappropriate histological and statistical analyses.

11. Please see https://www.ncbi.nlm.nih.gov/pubmed/24490213?report= abstract (accessed May 3, 2018).

12. The emergence of multiple herbicide resistant volunteer rape seed plants (*Brassica napus*) is likely due to sequential crossings between different varieties of cultivated rapeseed, each GEO engineered for resistance to a different herbicide, grown near to each other during the same growing season (Hall et al. 2000).

Sources for Additional Information

Bronowski, Jacob. 1973. "The Harvest of the Seasons." In *The Ascent of Man*, 59–90. Boston: Little, Brown.

Diamond, Jared. 1997. *Guns, Germs, and Steel: The Fates of Human Societies.* New York: W. W. Norton.

Doyle, J. 1990. "Genetic Engineering Harms Agriculture." In *Genetic Engineering Opposing Viewpoints*, edited by W. Dudley, ch. 3, sec. 1. San Diego: Greenhaven Press.

Ferber, Dan. 2004. "Microbes Made to Order." *Science* 303: 158–61.

Industrial Biotechnology Association. 1990. "Genetic Engineering Benefits Agriculture." In *Genetic Engineering Opposing Viewpoints*, edited by W. Dudley, ch. 3, sec. 2. San Diego: Greenhaven Press.

Kondro, Wayne. 2004. "Monsanto Wins Split Decision in Patent Fight over GM Crop." *Science* 304: 1229.

Lambrecht, Bill. 2001. *Dinner at the New Gene Café.* New York: Thomas Dunne Books.

Leopold, Aldo. (1949) 1970. "The Land Ethic." In *A Sand County Almanac.* New York: Oxford University Press.

Macilwain, C. 2005. "US Launches Probe into Sales of Unapproved Transgenic Corn." *Nature* 434: 423.

Mendelsohn, M., J. Kough, Z. Vaituzis, and K. Matthews. 2003. "Are *Bt* Crops Safe?" *Nature Biotechnology* 21: 1003–9.

4
CRISPR and Life's Future

The idea that you would affect evolution is a very profound thing.
— Jennifer Doudna, CRISPR gene-editing scientist

Is it more ethical to edit embryos or to screen a lot of embryos and throw them away? I don't know the answer.
— Jennifer Doudna

Since the beginnings of agriculture some ten thousand years ago, humans have altered natural genomes either inadvertently or purposefully by selectively growing plants and animals with traits beneficial to humans. In the 1860s, Gregor Mendel discovered simple laws of inheritance with his cross-pollination experiments using different varieties of peas. Nobody payed much attention to Mendel's work until the early 1900s, when it was rediscovered and its wide applicability became appreciated. Commercialization of new agricultural plant varieties produced by cross-pollination began near the end of the nineteenth century. One result was a quickened pace of planned hybridizations of commodity plants and animals by agriculturalists. Over time scientists learned tricks for creating new varieties, particularly plants. One involves fusing cells from different but related species and then growing the hybrid cells into new fertile individuals that can parent new varieties. Another uses mutation-inducing radiation or chemicals to create random genetic variants from which ones with the desired traits are propagated.

Genetic engineering or recombinant DNA technology debuted in the 1970s.[1] This technology uses inactivated viruses or tiny synthetic vesicles as vectors to deliver new genes into bacterial, plant, and animal genomes. Products of recombinant DNA technology include bacteria that produce human insulin and growth hormone, crop plants like Roundup Ready soybeans resistant to herbicide, and goats that secrete spider web proteins into their milk. Genetic engineering also offers gene therapies for human diseases including certain immune system deficiencies, some leukemias and other cancers, and liver disease.

One serious drawback of recombinant DNA technology is the difficulty in precisely targeting genetic changes to specific spots in the genome. Off-targeting can inactivate essential genes in the host cells, killing them or causing them to become cancerous. Better targeting of genetic alterations arrived in the early 2000s with gene-editing methods based on proteins called zinc fingers and TALEN. These methods require sophisticated knowledge and instrumentation to redesign certain proteins so that they recognize and bind to particular regions of DNA (Gaj et al. 2013). In 2013, researchers developed an even more accurate method for gene editing. Commonly called CRISPR/Cas9 or just CRISPR, the method uses RNA rather than protein to identify the site(s) in an organism's DNA that will be altered. CRISPR is a revolutionary biotechnology for rapidly, inexpensively, accurately, and easily altering the genome of virtually any organism, including *Homo sapiens*. The acronym will be explained shortly, and we will also see that the significance of CRISPR and related technologies for the future of all life on Earth cannot be overstated. The following questions guide our exploration of CRISPR technology:

1. What is CRISPR/Cas9, and how and when was it discovered?
2. What is gene editing, and how is CRISPR/Cas9 used to edit genomes?
3. What kinds of CRISPR-mediated gene editing are in the present and future?
4. What is a gene drive?
5. How do metaphors influence public knowledge about biotechnology and scientists' research and development in biotechnology?
6. Should we genetically edit the human germ line?
7. Should we genetically alter entire populations of wild plants and animals?
8. Should gene-editing technology be regulated? If so, when and how?

CRISPR: The Biology

CRISPR in Nature

CRISPR is the acronym for a naturally occurring DNA sequence in bacteria. Ruud Jansen of Utrecht University in the Netherlands created the acronym in 2002. It stands for "clustered regularly interspaced short palindromic repeats." That's a mouthful even for a cell biologist. CRISPR's function in nature is to help bacteria defend themselves against bacteria-killing viruses (for more detail, please see appendix 1).

Once the workings of bacteria's ingenious CRISPR defense system became understood in 2007, Jennifer Doudna, a biochemist at the University of California, Berkeley, had a eureka moment. In an interview, she reported thinking, "Oh my gosh, this could be a tool" (Zimmer, 2015, 5). She was right. Shortly after her startling realization, Doudna and her colleague Emmanuelle Charpentier and others adapted the bacterial CRISPR defense system into the most powerful gene manipulation tool ever devised. With CRISPR technology, humans can now quickly, precisely, and inexpensively alter virtually any gene in any organism on Earth (Jinek et al. 2012; Cong et al. 2013).[2]

CRISPR as a Gene-Editing Tool

In December 2015, the American Association for the Advancement of Science named development of CRISPR as a gene-editing tool the Breakthrough of the Year, which the association considers the year's most significant research finding for science and society (Travis 2015).[3] When a cell's DNA is edited by CRISPR, all of the cellular progeny of that cell carry the edited DNA. Therefore, when researchers edit genes in a fertilized egg or early embryo, all or most of the cells in the resulting individual, including its eggs or sperm, carry the genetic modification.

Eggs and sperm are called germ cells, and editing genes in eggs, sperm, fertilized eggs, or very early embryos results in germ line modifications, which are passed on to future generations. In the second part of this chapter on human values, we consider the controversial prospect of human germ line modification. Similarly, if the DNA of a stem cell is altered by CRISPR-mediated technology, progeny cells of the altered stem cell carry the genetic change. This fact has profound significance for medical applications of CRISPR technology to treat organ-specific diseases like leukemia and diabetes. For a not-too-technical look at how CRISPR technology actually works to edit genes, please see appendix 1.

Within just four years after its first use in a laboratory setting, CRISPR gene editing had been applied to bacteria, yeast, fish, tobacco, rice, wheat, sorghum, a popular research plant called thale cress, mice, rats, rabbits, frogs, fruit flies, silkworm caterpillars, roundworms, human cancer cells, human pluripotent stem cells, pigs, monkeys, and human embryos (Sander and Joung 2014; Specter 2016; Shen 2014). Researchers at the University of Chicago have even used CRISPR technology to show that genes regulating digit formation during embryonic development in mammals are the same genes regulating fish fin bone formation (Nakamura et al. 2016), giving insight into the evolution of the human hand. A later section of this chapter details the CRISPR experiments on human embryos. Although using CRISPR to create gene-edited babies is not documented in a scientific or medical journal as of December 2018, birth of twin girls from IVF embryos gene-edited in China to resist HIV infection was reported to news media in November 2018.[4] If true, this move is unethical due to lack of evidence that the procedure is safe for humans and insufficient public discussion about the pros and cons of human germ line engineering.

No CRISPR-edited animals have yet been released into the environment, but this may soon change. CRISPR-edited mosquitoes may be used to control transmission of malaria and the Zika virus. CRISPR-edited agricultural plants will undoubtedly become important for farmers in the near future. CRISPR has great potential for preventing, curing, and treating genetic-based diseases, but ethical controversy surrounds its possible use in editing the human germ line and altering entire populations of wild animals. We will examine these debates later in the chapter.

Added to the bioethical disputes over CRISPR are intense legal battles over whose work and patents ought to be honored for licensing CRISPR applications in medicine and other areas. The battlegrounds are the United States and Europe, and the combatants are the University of California (UC) system and the Broad Institute at the Massachusetts Institute of Technology. At stake are potentially billions of dollars in future commercial applications of CRISPR and a Nobel Prize.

In 2012, coworkers Jennifer Doudna (UC Berkeley) and Emmanuelle Charpentier, then at Umeå University in Sweden, were first to report using CRISPR technology to edit isolated DNA in a test tube. In early 2013, Feng Zhang's group at the Broad Institute was first to report applying CRISPR technology to edit DNA in living mammalian cells, and just a few weeks later Doudna's group published similar work. In 2014, the Broad Institute received the first patent for use of CRISPR technology on mammalian cells. The UC system contested the Broad Institute's patents, arguing that applying the CRISPR technology developed by Doudna and Charpentier in mammalian cells was an obvious next step in the fundamental work pioneered at UC Berkeley. But the Patent Trial and Appeal Board (PTAB) ruled in favor of the Broad Institute in February 2017. Then in April 2018, a lawyer for the UC system contended that a legal error had resulted in the PTAB's decision and asked a federal appeals court to reverse the decision or send the case back to the PTAB for reconsideration. In September 2018, the US Court of Appeals denied the UC system's appeal.

Meanwhile, another CRISPR patent struggle plays out in Europe. The dispute there is between the Broad Institute and several opponents including the UC system and a company called CRISPR Therapeutics, cofounded by Emmanuelle Charpentier, who has moved from Sweden to the Max Planck Institute for Infection Biology in Berlin. In early 2018, the European Patent Office (EPO) ruled against the Broad Institute and revoked an earlier CRISPR patent it had awarded to Broad. Several other Broad Institute patents face similar opposition. In Europe, the disputes hinge on the validity of Broad's claims of priority based on patents filed in the United States that included a patent author not listed on the more recent patents filed in Europe. As of mid-2018, a Broad Institute appeal of the early 2018 EPO ruling was pending. The outcomes of the CRISPR patent wars in the United States and Europe will have great and far-reaching financial ramifications for the players due mainly to the prodigious potential of CRISPR gene editing to fight disease and otherwise benefit human well-being.

Potential Medical Benefits from CRISPR Technology

At this writing in late 2018 the first clinical trials using CRISPR for gene therapy in human patients are poised to begin in Europe and the United States. Earlier promising results from experiments on mice with human disease conditions and on human stem cells set the stage for clinical trials. For example, researchers at the University of Pennsylvania have used CRISPR to correct a gene mutation in mice that causes a metabolic liver disease in humans (Yang et al. 2016). The disease results from a deficiency in an enzyme called ornithine transcarbamylase (OTC), the same condition researchers at that institution tragically failed to cure in eighteen-year-old Jesse Gelsinger in a 1999 gene therapy trial gone wrong (Sibbald 2001). In the absence of normal OTC, ammonia builds up in the blood from the metabolism of dietary protein and, if left untreated, causes brain damage and death. One of every forty thousand human births has OTC deficiency. The researchers injected newborn mice with virus particles carrying genes for the CRISPR elements necessary to find the OTC mutation in liver cells, cut out the mutated region, and replaced it with a normal segment of DNA in order to fix the OTC gene. The therapy reversed the disease-causing OTC mutation in about 10 percent of the newborn mice treated, encouraging results for such a preliminary therapeutic gene-editing experiment.

Other researchers lowered cholesterol levels in mice using CRISPR technology to inactivate a gene in humans that is associated with high levels of

LDL cholesterol and increased risk of cardiovascular disease (GEN News Highlights 2015). In another CRISPR experiment, researchers repaired a gene defect that causes an eye disease, retinitis pigmentosa, which leads to blindness (Bassuk et al. 2016).[5] Gene-editing technology also promises new cancer treatments. A research team in Boston used the CRISPR method on thirty-three types of cancer cells growing in the laboratory (Aguirre et al. 2016). They programmed CRISPR to break the DNA in these cells at regions containing multiple copies of certain genes, and they found that the more CRISPR-mediated DNA breaks that were made, the more poorly the cancer cells proliferated. The results suggest a way to use CRISPR to attack a wide variety of cancers in clinical settings.

Miniorgan or organoid technology is another exciting area of CRISPR research with revolutionary medical applications (Lewis 2016). Miniorgans are laboratory-grown cellular aggregates that mimic the structure and function of fully formed organs such as the intestine, brain, kidney, liver, and heart. Organoids are usually derived from induced pluripotent stem (iPS) cells obtained by reprogramming skin cells.[6] Recently, researchers in the Netherlands successfully created intestinal organoids directly from adult stem cells of the gut, saving the time and expense of having to go through iPS cells in order to obtain miniorgans (Lewis 2016).

Organoids make possible biomedical research on human tissue without needing to perform clinical trials on patients themselves. They also often allow bypassing expensive and ethically controversial drug and toxicity testing in nonhuman primates. For example, researchers now apply CRISPR technology to the cells of miniorgans to help discover the genetic causes of organ-specific diseases and even to correct the underlying genetic defects that cause the diseases. The Netherlands group, cited above, collected intestinal stem cells from patients with cystic fibrosis, used CRISPR methodology to correct the genetic defect in these cells that causes the disease, and then grew miniorgans from the genetically edited cells. Cells in the resulting miniorgans displayed behavior characteristic of normal intestinal cells and lacked behavior associated with cystic fibrosis. Eventually, cells collected from a patient with a diseased kidney, heart, liver, brain, or other organ may be corrected genetically by CRISPR and grown into healthy, transplantable organs in the laboratory. Miniorgans in conjunction with CRISPR techniques are also used to study cancer. By introducing cancer-causing genes or gene defects into stem cells and growing the cells into cancerous organoids, researchers can study the biology of specific cancers and test new therapies without enlisting human patients in risky trials.

So far, most CRISPR-related biomedical work is limited to "proof-of-principle" experiments designed to show the feasibility of pursuing further research aimed at eventual treatments for persons suffering from genetically based diseases or conditions. A major hurdle to overcome before CRISPR can realize its disease-curing potential is the problem of accurately and efficiently delivering CRISPR's components, a guide RNA and the Cas9 protein, to the cells in need of therapy (Cross 2017). A method called *ex vivo* gene editing, in which cells taken from a patient are edited via CRISPR in a laboratory dish and then reinjected into the patient, shows some promise. A research group at the University of Pennsylvania received NIH approval in mid-2016 for an *ex vivo* CRISPR clinical trial on cancer patients. During the trial, four genes in patients' immune systems' T cells will be edited to enhance their ability to detect and kill cancer cells (Reardon 2016).

Important to note is that except for two human embryo experiments detailed in the next section and the undocumented claim of gene-edited babies born in China described above, CRISPR editing of human genes has been restricted to somatic (body) cells and not extended to germ cells (eggs or sperm). The ethical significance of this is that somatic cell edits disappear from the human population upon the death of the individual whose body cells were edited; whereas, edited genes in eggs, sperm, or early embryos move into future generations and may remain in the species indefinitely. Now, we will take a look at recent CRISPR work in human embryos done by Chinese researchers and what is about to happen elsewhere now that CRISPR editing of human embryos is approved in the UK and the United States for particular research projects.

CRISPR Experiments in Humans: Germ Line Editing and Imminent Clinical Trials

In 2015 and 2016, two separate research groups in China attempted CRISPR-mediated gene editing in human embryos. They both got mixed results. In an attempt to blunt ethical concerns about research that could lead to "designer babies," both groups used 3PN embryos, which cannot implant in the womb or develop to term because they have three copies of every chromosome instead of the normal two copies.[7] Severe developmental problems make 3PN embryos nonviable past the early, preimplantation stage of development. Recall that three copies of just one of our twenty-three types of chromosomes, a trisomy, causes conditions like Down, Edwards, and Patau syndromes.

In April 2015, researchers reported having injected eighty-five 3PN human embryos with CRISPR components they hoped would correct a mutation responsible for the blood disorder, beta thalassemia (Liang et al. 2015).[8] Their objective was to correct every cell of the embryo without causing any other changes to the embryo's DNA. The experiment failed at this objective. First of all, only four of the eighty-five treated embryos contained the desired change in the desired gene, but the genetic change did not happen in all cells of any of those four embryos. More disturbing was that CRISPR-mediated genetic change occurred in only twenty-eight of the eighty-five embryos. All twenty-eight of these embryos, which included the four containing desired genetic changes, suffered off-target alterations to their DNA. The researchers concluded that their CRISPR procedure needs further work to improve its fidelity and specificity before clinical applications are feasible.

A year later, a second group of Chinese researchers used CRISPR gene editing in an attempt to make human embryos resistant to HIV infection (Kang et al. 2016). Their results were similar to those of the first group with many of the treated embryos suffering off-target editing to their DNA. In discussing their results, the research group strongly discourages genome editing of the human germ line before global communities have thoroughly discussed its safety and ethical implications. In the second section of this chapter on human values, we see how research and ethics communities have responded to the prospect of germ line gene editing in humans.

As for actions of regulatory bodies in the UK and the United States, CRISPR research on human embryos seems poised to move forward. In early 2016, the UK regulatory body responsible for overseeing human reproduction and embryo experimentation, the Human Fertilisation and Embryology Authority (HFEA), approved CRISPR editing of DNA in human embryos for specific research applications (Callaway 2016). Researchers at the Francis Crick Institute in London will use CRISPR technology to disable a few specific genes during the first seven days of *in vitro* human embryo development in order to assess the genes' functions during normal development. Their findings should help clinicians improve IVF success rates, aid in developing treatments for infertility, and give insight about factors needed for successful implantation after normal internal fertilization. All embryos used in the research will be surplus IVF embryos used with the parents' informed consent. The work will assess the risks and feasibility of using CRISPR technology on human embryos in clinical settings. Information gained by this research could be obtained without using CRISPR technology, but the researchers argue convincingly that with CRISPR far fewer embryos are needed for the work than would be required using an alternative technology. The decision by the HFEA is the first instance of a national regulatory board giving the go-ahead for CRISPR research on human embryos. The approved project also shows that CRISPR use for basic human biological research can be clearly separated from its clinical use. The HFEA currently prohibits using genetically altered human embryos for reproduction in the UK.

In the United States, the first proposed clinical trial using CRISPR-edited cells passed its initial regulatory hurdle in June 2016 (Kaiser 2016). A University of Pennsylvania–led team won approval from the Recombinant DNA Advisory Committee (RAC) at the US NIH to disable two genes in the immune system's T cells, so named because they mature in the thymus gland. By inactivating the two genes and inserting a new gene using the CRISPR gene-editing system, the researchers hope to greatly enhance the T cells' ability find and destroy the cancer cells. T cells will be collected from eighteen patients with myeloma, sarcoma, or melanoma and who no longer respond to conventional cancer treatments. The cells' DNA will be edited in the laboratory and the redesigned cells then will be reintroduced into the patients. The FDA and ethics boards at the researchers' institutions still must vet safety and ethical aspects of the project before the trial can begin. Another human trial using CRISPR is also imminent. In August 2018, Vertex Pharmaceuticals in Boston, Massachusetts, became the first US company to sponsor a CRISPR-based clinical trial. Like the earlier Chinese work, the trial targets the blood disorder, β-thalassemia, which results in a deficiency of hemoglobin in adults. Researchers will use CRISPR technology to allow cells to make fetal hemoglobin that is normally produced only in infancy. The work is in collaboration with the Swiss company CRISPR Therapeutics and is being performed in Germany.

Gene Drive: Gene Editing in Entire Populations and Species

In addition to editing genes of individual organisms, CRISPR can also be employed to spread a human designed genetic alteration rapidly and efficiently through an entire population. The technology is called *gene drive*.

The gene drive concept has been around for most of the current century. But the biotechnology needed to design a gene drive that actually works well did not exist until after 2012 with the advent of CRISPR as a gene-editing tool.

The first CRISPR-mediated gene drive was dubbed a mutagenic chain reaction by its developers (Gantz and Bier 2015). Like a nuclear chain reaction in a chunk of enriched uranium where more and more atoms split in exponential fashion, a mutagenic chain reaction spreads a mutation explosively outward into a population through normal acts of reproduction originating with just one or a few individuals. The shorter the generation time (the period between fertilized egg and reproductively mature adult) for an animal and the more individuals in the initial nucleus of mutagenic gene drive carriers, the more quickly the gene drive can spread a mutation.

Animals with CRISPR-mediated gene drives have not yet been released into wild populations in the environment, but the technology is poised to do so with flies and mosquitoes. In principle, a gene drive could be applied to any sexually reproducing organism. How does the spread of a genetic alteration carried by a gene drive differ from the spread of a naturally occurring mutation through a population?

Naturally occurring genetic variation is the fodder upon which natural selection acts. Individuals with mutations and other genetic changes appear regularly in natural populations. Genetic variation and natural selection have driven the origin and evolution of species for hundreds of millions of years. A naturally occurring genetic variation is normally passed on to only 50 percent of the progeny of the individual(s) carrying the variant. In order for the genetic variant to spread through a population, it must give its carriers a strong reproductive advantage; that is, it must increase the fitness of the individual(s). By contrast, a synthetic gene drive like the CRISPR-mediated gene drive ensures that virtually all offspring of carriers receive the genetic variant associated with the gene drive. In this way, a gene drive can propagate new genetic material rapidly through a population even if the genetic variant does not increase the fitness of its carriers. Moreover, the genetic variant associated with a gene drive is planned and designed by humans, not simply the result of nature's randomly occurring DNA variation. In the second part of this chapter, we consider ethical and environmental implications of CRISPR gene drives. A closer look at how CRISPR-mediated gene drives work is offered in appendix 1.

To summarize, CRISPR gene drives enable researchers to engineer the hereditary material of entire populations, even entire species, of plants or animals. Potential uses of gene drives include control of disease-carrying animals such as mosquitoes and ticks, control of invasive species of plants and animals in certain geographic areas, and reversal of pesticide and herbicide resistance in weeds and insect pests. Whether we humans ought to actually manipulate entire ecosystems by selectively removing or altering particular species is an important ethical and environmental question considered in the second part of this chapter.

CRISPR Applied to Nonhuman Organisms

Since the CRISPR bacterial defense system was redesigned as a gene-editing tool in 2012, nearly one thousand scientific articles have been published referencing the use of CRISPR to manipulate genes in plants, animals, and microbes. We will look at a few of these as examples of the imagination researchers bring to CRISPR technology use.

In late 2015 and early 2016, scientists used CRISPR to create malaria-resistant mosquitoes and male mosquitoes that transmit sterility factors to females. Malaria is a parasitic disease caused by single-celled organisms, protozoans, in the genus *Plasmodium*. A species named *Plasmodium falciparum* causes the most virulent and dangerous form of the disease, which is transmitted by bites from female mosquitoes in the genus *Anopheles*. The parasite initially infects liver cells. It propagates and is released into the bloodstream, where it infects and destroys red blood cells. Mosquitoes become infected by feeding on infected humans. These infected insects then transmit the parasite to other humans. A person dies of malaria every minute, the most vulnerable groups being children and pregnant women in sub-Saharan Africa. Although insecticide use has reduced the global incidence of malaria, mosquitoes evolve and become resistant to insecticides, and the complex life cycle of the parasite thwarts development of an antimalaria vaccine. This was the situation in 2015 when researchers at the Universities of California in San Diego and Irvine succeeded in an *Anopheles* genome-engineering project that could eradicate malaria forever.

The California researchers engineered into the mosquito genome a CRISPR-mediated gene drive containing genes that render the mosquito resistant to infection by the malaria-causing parasite (Gantz et al. 2015). If released into wild populations, the parasite-resistant trait would spread rapidly though the populations, perhaps establishing itself in the species as a whole and eradicating malaria in humans. Another research group also tested the CRISPR gene drive approach to controlling malaria-bearing mosquitoes by inactivating three genes required for female mosquito fertility and showing that the gene drive works well in caged populations (Hammond et al. 2016). As of mid-2018, no gene drive–bearing mosquitoes have been introduced into wild populations. However, the devastating worldwide effects of malaria, the spreading threat of the Zika virus to unborn babies, and the annual toll of Lyme disease raise strong arguments for using robust and creative measures to control the vectors that transmit these and similar scourges. Ethical and regulatory considerations about the intentional or unintentional release of gene drives into wild populations are considered in the human values section of this chapter.

Vectors of disease are not the only targets of CRISPR gene editing. Researchers employ CRISPR to create hens that lay nonallergenic eggs, disease-resistant and more nutritious food crops and animals, novel pets, pigs with organs suitable for transplantation into humans, and much more (Ledford 2015; Reardon 2016). A chicken egg white contains four proteins. One small portion of one of these proteins is allergenic for persons allergic to poultry. Australian molecular biologist Timothy Doran aims to edit the gene for the allergenic egg white protein so that it no longer ails people with an egg allergy but still performs its normal function in chicken egg development.

Other agricultural uses of CRISPR gene editing include creating disease-resistant goats, wheat, and rice, cattle without horns so that they do not injure each other during transport, and vitamin-enriched oranges. Animals with CRISPR-edited genomes are now sold as pets, including a petite pig that grows no larger than a dachshund. A South Korean biotech firm plans to improve the abilities of guide dogs and other work dogs using CRISPR. In 2016, Chinese researchers used CRISPR to produce autism spectrum disorder in monkeys in order to provide behavioral and neuroscientists with a nonhuman model for studying the disorder. Other scientists are intent on applying CRISPR technology to the de-extinction of wooly mammoths, passenger pigeons, and other long extinct animals.

CRISPR: Human Values

The reality of genome-editing methods and the prospect of their widespread application in humans and nonhuman organisms raise many ethical issues and arouse controversy among stakeholders in the methodology, which include scientists, bioethicists, environmentalists, biotechnology entrepreneurs, clerics, policy makers, and the general citizenry. Moral questions raised by CRISPR-related technologies range from the degree of genetic control parents ought to have over unborn children to the wisdom of creating new ecosystems populated with organisms designed and created by humans. We begin by recognizing the power that metaphors for genetic phenomena and technologies have upon our thinking and deciding about the genetic future of humans and other life forms on Earth.

Genetic Metaphors Influence Research Objectives and Public Knowledge

The rapid ascendancy of CRISPR as a gene-editing tool provides an opportunity to observe how quickly metaphors describing a new technology appear and how they can influence the direction of scientific research and mold public perceptions and expectations (O'Keefe et al. 2015). Scientists often use metaphors to describe new discoveries and technologies, both when conversing with each other and when talking to nonscientists. Journalists quickly adopt the metaphors, and their use of them invariably shapes public opinion, which in turn can influence the directions science takes.

Examples of powerful biological metaphors are those associated with the human genome, especially during the period when proponents of the Human Genome Project (HGP) sought funding for the project (Dreger 2000; Bradley 2013a, 111–12). Genome scientists described the human genome as a *genetic frontier* in need of *mapping* to expose *bad genes* that cause disease and suffering. These metaphors appealed to a caricaturized American pride in exploring frontiers and rooting out evil. To oppose funding for a project described in this way was unpatriotic. The desired result was obtained. The US Congress approved $3 billion of public money for the HGP.

Metaphors attached to CRISPR are already shaping our understanding of and expectations for the technology. Combined with earlier metaphors for DNA (*blueprint for life*, *life's code*) and the human genome, images of CRISPR action are very potent. Combining prevalent metaphors for CRISPR's action on DNA such as *edit*, *guide*, *target*, and *chop* with genomic metaphors, an understanding of CRISPR emerges reading something like this: *Guided to specific target sites within life's blueprint, CRISPR technology edits genes by chopping out bad coding and replacing it with normal, healthy coding.* Both misconception and truth are carried in this statement.

Calling DNA the *blueprint for life* suggests the dangerous falsehood of genetic determinism. Certainly, we and other organisms are restricted in how we can live by our genetic makeup, but the expression of our genetic potential is greatly influenced by environmental factors as well (Bradley 2013a,

125–27). The image of CRISPR being *guided* to specific *target* sites suggests either an unrealistic accuracy or a drone-like attack that may be very accurate some of the time but tragically off-target at others. Indeed, CRISPR has often produced unintended, *collateral damage* to untargeted genes as in the case of the two experiments by Chinese scientists on human embryos described earlier. Collateral damage to genome function may also occur even when CRISPR targeting is absolutely precise simply because geneticists do not yet understand all of the functions of specific genes and their products. Chopping up or modifying a disease-causing gene could adversely affect undiscovered functions of the gene.

Recent advances in CRISPR technology do minimize off-target modifications of DNA as suggested by the title of an article in a professional newsletter reporting on this advance: "CRISPR Moves from Butchery to Surgery" (Ferreira 2016). Time and experience will tell how error free the new CRISPR surgery is. Finally, the *gene/genome-editing* descriptor for CRISPR technology implies that our genes and genomes need editing like the rough draft of a poem or essay. True, some genes do contain disease-causing mutations, and editing these genes to relieve suffering seems viscerally appropriate. But CRISPR has many potential uses other than curing or preventing disease including the addition of human-designed genes to give organisms entirely new traits, returning long-extinct animals to the planet's biosphere, adding, removing, or altering entire ecosystems, and perhaps altering human nature itself. The prevailing and nearly exclusive use of the *editing* metaphor for CRISPR action de-emphasizes, even camouflages, the uses for CRISPR beyond eliminating genetic disease. Many of these other uses for CRISPR, such as human germ line genetic enhancement or production of human-animal chimeras, have ethical implications worthy of broad discussion.[9]

Should Germ Line Gene Editing Be Done in Humans?

The term *embryo editing* now appears frequently in the context of using CRISPR to edit genes in human embryos (Cyranoski and Reardon 2015; Kaiser 2017). Associating the editing metaphor with human embryos raises concern and controversy beyond that raised when it appears with just the term *DNA*. There are at least two reasons for this. First, editing DNA in early embryos, especially at the single-celled zygote stage of development, produces genetic alterations passed on to future generations. Germ line editing concerns discussed earlier including eugenics and violation of autonomy apply. Second, embryo editing terminology suggests that the human embryo is in need of improving or correcting, conjuring the tragic turns that eugenics took during the past century.

The prospect of altering the human germ line using CRISPR and the force of the editing metaphor are experienced differently by different persons. Even researchers in this area of biotechnology have different views about how CRISPR technology ought to be developed and used. Just weeks before the first report of human embryo editing in 2015, two groups of scientists and other professionals published statements on the ethics of doing CRISPR research on human cells and human germ line gene editing. They did not agree on the most prudent path forward.

One group (Lanphier et al. 2015) advocated for a voluntary moratorium on genetically modifying the human germ line, excluding the mitochondrial gene therapy described in chapter 2. Its reasons were both practical and ethical. The group raises concern about public confusion between somatic cell and germ line gene editing. The former promises cures or treatments for several cancers, HIV/AIDS, hemophilia, and sickle cell anemia, while the latter would introduce genetic changes into unborn generations. If research on human germ line gene editing proceeds, these authors fear a backlash against all forms of human gene editing by a public that does not distinguish between somatic and germ line gene therapy.

The Lanphier group's deeper concerns about germ line gene editing arise from safety and ethical considerations. Genome researchers still know too little about gene interactions to alter or inactivate genes in future generations with confidence that only the desired results will ensue. Even correcting a known disease-causing component of a gene in a fertilized egg may have unintended, negative consequences for the individual in later life or for her or his descendants. The group points out that means other than germ line gene editing are already available to insure having a healthy baby.

One such means is preimplantation genetic diagnosis (PGD) followed by embryo selection. Here single-cell biopsies from early embryos created by IVF determine which embryos are genetically healthy for transfer to the uterus. This is a viable option if only one parent carries a genetic disorder or if both parents carry a disorder-causing gene on just one of a pair of their chromosomes. It is not an option when both parents carry a double dose of a genetic disorder since all embryos would then receive the disorder-causing gene.

The group's final concern is that germ line gene-editing research and eventual clinical applications will slide inexorably into "designer baby" production in which genome-editing methods are used for nontherapeutic purposes to enhance or alter nonmedical traits in human embryos. Making irreversible, life-altering edits to the genomes of the nonconsenting unborn, and further stratifying society by enhancing cognitive or other traits to give advantages to the relatively wealthy, and the emergence of a new eugenics of nondisease traits are the major ethical issues linked to human germ line gene editing. Counterarguments to these and other concerns are described by myself (Bradley 2013a, 169–76) and by a second group of scientists and biomedical experts weighing in on the prospect of human germ line genome editing (Baltimore et al. 2015).

This second group met in Napa, California, in January 2015 to discuss the scientific, medical, ethical, legal, and social implications of germ line genomic engineering of both human and nonhuman life forms. Included in this discussion were Nobel laureates David Baltimore and Paul Berg, along with the developer of the CRISPR gene-editing technology, Jennifer Doudna, and fifteen other scientists and medical professionals.

In March 2015, shortly before researchers in China published the first report of experimental human germ line gene editing, the Napa group published its recommendations (Baltimore et al. 2015). It begins by noting that genome-engineering technology "provides methods to reshape the biosphere for the benefit of the environment and human society," but that "with such enormous opportunities come unknown risks to human health and well-being." Rather than specifically advocating a moratorium on germ line gene-editing research for clinical application, the Napa group discourages such work in countries with "lax jurisdictions where it might be permitted" and notes that in those countries that already have a highly developed bioscience capacity, human germ line gene modification is prohibited or tightly regulated. The group made specific recommendations focusing on the urgent need for international dialogue among all stakeholders in genome-engineering technology.

Along with the scientists, persons with disabilities have voices that need to be heard in the CRISPR dialogue. Proponents of human germ line gene editing often cite the suffering that could be avoided by editing out genetic "defects" that cause disease and "abnormalities" in early embryos. Correcting the genetic causes of diseases such as Tay Sachs and Huntington's is one thing, but identifying abnormalities for removal from the germ line is quite another. For example, many persons with conditions such as spina bifida, deafness, or dwarfism embrace who they are, enjoy life, contribute to culture and society, and may be offended that others consider their condition in need of prevention or correction.

Arguing that CRISPR germ line gene editing need not be used to prevent the birth of babies with abnormalities when preimplantation genetic diagnosis is available suffers from the same prejudice. Emily Beitiks (2016), associate director of the Paul K. Longmore Institute on Disability at San Francisco State University, writes: "We don't need CRISPR to 'solve' the disability = tragedy equation." She continues, "Social changes to the built environment and cultural changes to discriminatory attitudes are a safer bet with more widely shared impacts."

Thoughtful persons aware of the rapidly expanding use of CRISPR gene-editing technology recognize the urgent need for an inclusive dialogue about the directions CRISPR research and use ought to take. But ignorance about CRISPR and its revolutionary potential for redesigning life on Earth and how we view life is still widespread. Therefore, raising public consciousness and knowledge about CRISPR is an immediate imperative that must be followed by the inclusive dialogue worldwide. Scientists, the general public, policy makers, ethicists, persons with business interests in the technology, law professionals, clergy, and others must come together in global forums. Here, they can learn from each other, develop informed opinions, and recommend policies to guide humanity's decisions and actions in this age of genome engineering. We cannot realistically expect unanimous decisions, but we ought to work toward a goal of global consensus for our path forward. The importance of this goal is the subject of the last chapter in this book.

An important early step toward international dialogue on germ line gene editing occurred in late December 2015, when an international group of scientists met in Washington, DC, to discuss the ethics of using CRISPR on humans, particularly on early human embryos for germ line gene editing. Participating in the three-day conference were representatives from the US National Academies of Science, Engineering, and Medicine (NASEM), the UK's Royal Society, and the Chinese Academy of Science. All conferees agreed that for safety reasons, now is too soon to undertake clinical germ line gene editing in humans.

Major safety issues for germ line genome editing are risk of untargeted insertions or deletions in genetic material (collateral damage) and an incomplete picture of how gene products interact with each other. The former could inactivate essential genes, causing cancer or disrupting essential cellular functions. Because of our incomplete knowledge about how proteins interact with each other, even very precise inactivation of a disease-causing gene could have unintended consequences. The danger of interfering with the action of any gene is at least twofold. First, even a deleterious gene product may interact with other proteins to produce effects beneficial or vital to cellular health (Velasquez-Manoff 2017). Second, individual genes often code for multiple, functionally diverse proteins.[10]

Identifying all of the possible protein products of human genes and how they interact with each other to support normal cell function is an ultimate goal of the Human Proteome Project.[11] So far, that goal is far from accomplished. Despite these risks and unknowns, the 2015 summit of scientists recognized CRISPR's potential to end suffering from numerous heritable diseases that plague many families around the world. The group steered clear of recommending a prohibition or moratorium on CRISPR research on human DNA and recommended revisiting the pros and cons of clinical germ line editing on a regular basis (Reardon 2015a). In the meantime, debate over whether human germ line modification for any reason should be a goal of humankind continues.

Some advocates for social justice and health equity are adamantly opposed to any germ line gene editing. Their concern is that purposeful therapeutic modification of the human germ line to prevent genetic disease could easily slide into genetic enhancement of nondisease traits in the germ line. Marcy Darnovsky, director of the Center for Genetics and Society in Berkeley, California, fears that a slippery slope from therapy to enhancement could end in "full-out germline manipulation, putting a high-tech eugenic social dynamic into play" (Darnovsky 2013).

Similarly, bioethics essayist, Pete Shanks, warns that applying gene-editing technology to human reproduction could lead to a world of genetic haves and have-nots (Shanks 2017). Others point out that germ line genetic enhancement would violate the autonomy of future generations through their ancestors' decisions and expectations.

Potential traits targeted by eugenic uses of CRISPR include cognition, stature, physique, and ability in art, science, athletics, or other disciplines. Traits like these are multigenic, involving complex interactions between many genes and their protein products. Such interactions are not yet understood, but they are objects of study for the Human Proteome Project. Although multigenic traits will not be targets of genetic enhancement in the near future for technical reasons, it is wise to look ahead and anticipate ethical challenges from future technologies. Multiple societal problems could arise from human germ line enhancement including exacerbation of already egregious stratification of economic, educational, and professional opportunities, violation of future generations' autonomy by their ancestors' decisions and expectations, and new, dangerous eugenics efforts.

Opponents of germ line editing point out that ways other than genome editing exist to avoid having a child with a debilitating genetic disease. One of these is PGD, which was discussed earlier, but PGD has ethical issues of its own. A major one is how to view the moral status of the embryo and whether it is ethical to create embryos that one knows will be discarded in order to obtain a healthy embryo. Also, PGD would not be an option for parents who both carry two copies of a mutant gene for a disease like cystic fibrosis. Adoption of a healthy baby is another option, but this is countered with the argument that we already sanction assisted reproduction technologies like IVF, intracytoplasmic sperm injection, gamete intrafallopian transfer, along with egg and sperm donation to help parents have a child genetically related to one or both of them. Furthermore, existing assisted reproduction technologies, coupled with PGD, could also be used eugenically. So why should therapeutic germ line editing be treated differently from any of these and not considered just another form of assisted reproduction to help parents exercise their right to reproduce?

Do adult humans have such a strong right to reproduce as to justify germ line gene editing? Bioethicist and law professor John Robertson argues that procreative liberty is a right protected by the US Constitution and that technology-assisted reproduction is justified to protect that right (Robertson 1994). Others maintain that there is no inviolable right to reproduce and that we must refrain from going down the path of human germ line engineering because it manipulates the genetic essence of unborn persons without their consent.

British bioethicist John Harris counters this argument by pointing out that germ line engineering does not violate anybody's right of consent because the unborn do not exist to either give or withhold consent (Harris 2015). He reminds us that parents and others make decisions for future persons regularly without their consent. Prenatal nutrition, exposure of pregnant women to chemical and emotional stressors, and the components of children's growth and development environment offered by some parents that include music, sports, friends, attention, and quality of education all influence the future life of a child and are administered exclusively or largely without consent.

Molecular biologist and author Gregory Stock foresees the inevitability of human germ line genetic manipulations and suggests that we concentrate on making wise decisions as we proceed ahead with it rather than miring ourselves in arguments over whether to do it (Stock 2002). Stock's belief in the inevitability of human germ line intervention may sound to some like a resignation to technological determinism, meaning that technology possesses an unstoppable inertia of its own, over which humans have little or no control.

Sheila Jasanoff, author on science and technology at Harvard Kennedy School, cites HAL, the treacherous computer aboard the spacecraft in Arthur C. Clarke's 1968 novel, *2001: A Space Odyssey*, as an example of a fictional outcome of technological determinism. Jasanoff gives no real-life example of technological determinism because she believes that none exists. She argues that human values can and do direct our use of technology, rendering the concept of technological determinism fallacious. She writes, "Human values enter into the design of technology. And, further downstream, human values continue to shape the ways in which technologies are put to use, and sometimes repudiated" (Jasanoff 2016, 19).

Stock, in my view, does not believe in the false idea of technological determinism. Just because application of a technology, such as human germ line manipulation, is inevitable does not mean that the ways it will be used are beyond human control. Instead, the uses to which genome editing and other biotechnologies are put will reflect human values, just as do the uses to which atomic fission and computers have been and are being put. We return to the subject of technology, choices, and human values in chapter 9.

According to Stock, the important question is not how to prevent human germ line manipulation but rather how to design our genetic future. Initial steps will probably be aimed at relieving suffering by preventing or eradicating genetic-based diseases. Opposition to that will be difficult to mount. Ultimately though, we must confront decisions about altering human nature itself.

Volumes are written about human nature from religious, humanistic, philosophical, and biological perspectives. Being a biologist, I like to approach the question of human nature by asking what quintessential traits distinguish *Homo sapiens*. Several traits come to mind including moral awareness and assessment, forethought in the context of past and present, moral decision-making, imaginative creativity, collective hope, curiosity about existential origins and fate, reason, compassion, and empathy.[12] This is not to say that other animals do not share some of these traits to some degree. In fact, Charles Darwin noted that "the difference in mind between man and the higher animals, great as it is, is certainly one of degree and not of kind" (Darwin 1871).

The question we need to contemplate now is this: When it becomes possible to modify, enhance, or otherwise change human nature, should we do it? And if we do, in what directions, when, at what rate, and how expansively should we proceed? Each of us can contribute informed opinions on these questions. In the final chapter of this book, we examine the need and prospect for worldwide discussion of biotechnology's perils and promises for humankind's future.

Jeantine Lunshof, a philosopher, bioethicist, and registered nurse in a medical center genetics department in the Netherlands, prioritizes her concerns about CRISPR technology in a rather surprising way considering her work and background in human well-being (Lunshof 2014, 127). She writes: "The ethical issues raised by human germline engineering are not new. They deserve consideration, but outcry over designer babies and precision gene therapy should not blind us to a much more pressing problem: the increasing use of CRISPR to edit the genomes of wild animal populations." Here, Lunshof refers to proposed releases of organisms with engineered gene drives into the wild. Because the impact that the rapid spread of human-designed traits through wild populations or the demise of entire populations would have on natural ecosystems is unknown, we should proceed with caution.

Should Gene Drives Be Released into the Environment?

CRISPR-mediated gene drives are already a reality as evidenced by the work with fruit flies and mosquitoes. A gene drive can be designed for virtually any sexually reproducing plant or animal. Should this technology be deployed in an open environment? If so, for what purposes, and how should it be regulated? Answering these questions requires anticipating uses and risks for gene drives.

Proposed uses for CRISPR-mediated gene drives are to control or eradicate vectors for human diseases and to eliminate or suppress invasive species in areas where they wreak havoc with native species and ecosystems.

Two gene drives are already developed for controlling *Anopheles*, the mosquito vector for malaria. Other candidates for CRISPR-mediated gene drive control are *Aedes*, the mosquito vector for dengue fever, yellow fever, the chikungunya and Zika viruses, and the deer tick vector for Lyme disease. Especially troublesome invasive species in North America include the Asian carp, Africanized honey bees known as "killer bees," imported red fire ants, kudzu, privet, zebra mussels, Burmese pythons, snakehead fish, emerald ash borer, and purple loosestrife plants. What could be wrong with using modern biotechnology to wipe out entire populations of these organisms?

Simply put, the release of genetically altered animals with gene drives into natural populations comes with unknown risks. One is that the gene drive could spread to geographical regions where the unwanted species is a native component of stable ecosystems. Since components of healthy ecosystems form complex webs of interdependency, it is nearly impossible to predict the consequences of suddenly removing or greatly reducing populations of entire species from an ecosystem. The same is true for ecosystems where invasive species have been present for many generations and may be stably integrated into the system's web of interactions. Another risk is horizontal transfer of the genetic components of the gene drive to nontarget species. Horizontal or lateral gene transfer refers to the movement of genetic material between individuals of the same or different species as opposed to vertical gene transfer, which is the movement of genes from parents to offspring. Horizontal gene transfer is common among bacteria, but gene transfer between distantly related organisms has occurred during the evolution of multicelled plants and animals. Unintended transfer of a lethal gene drive to nontarget species could be devastating for an entire ecosystem.

The lack of comprehensive knowledge about ecosystem structure and function currently makes introduction of gene drives into populations very risky. In the end, decisions about using gene drives in the wild may distill to choices between unpredictable negative consequences to the environment and hoped-for health or economic benefits from the use of gene drives. For example, consider the emerald ash borer, an insect that costs the US ash lumber industry more than $25 billion annually. Development and use of a CRISPR-mediated gene drive to eradicate the emerald ash borer would be an economic boon for the lumber industry, but what would be the long-term environmental effects of eliminating this species? How, when, and why did the emerald ash borer become a species in the first place? What role does the practice of monoculture farming have in the establishment of insect, viral, and fungal pest species? Where does the responsibility lie in balancing the immediate, tangible economic benefits of land use against long-range benefits of maintaining the ecological stability and integrity of the land?

Wisconsin conservationist and author, Aldo Leopold, asked questions like these in the 1940s and suggested that viewing land simply as a commodity to be exploited for corporate profit or human pleasure is the "key log" that must be removed before an effective "land ethic" can be developed (Leopold [1949] 1970). Now, seven decades later, a logjam of ignorance about evolution and ecology as well as the shortsighted, profit-centered attitude of industries still thwarts development of a land ethic that recognizes the interdependency of all of nature's components.

As noted by Charles P. Snow in his 1952 essay, *The Two Cultures*, scientists are optimists. They persistently try to imagine solutions to problems and then work to make solutions reality rather than lamenting that problems exist. In this spirit, some CRISPR researchers are developing so-called reverse gene drives that could be released into the wild to undo what a previous gene drive has done. Others believe that an immunization gene drive that vaccinates certain populations against other gene drives could help solve the problem of horizontal transfer of gene drives.

Whether more technology is the best solution to dealing with risks brought on by technology, either in the general sense or in the specific context of CRISPR gene drives, is a question that spawns diverse perspectives and is therefore worthy of continued debate. Gene editing and gene drives have great power to change who we are as biological creatures, how we live our lives, and how we understand nature and our place in it. This reality commands broad, inclusive dialogue and gives urgency to obtaining tangible outcomes from the debate. Consensual decisions can steady us as we proceed into a future of life consciously redesigning itself.

Regulating Genetic Biotechnologies

Wise and informed oversight of CRISPR gene editing and the development of gene drives and their potential use in natural environments is needed, but so far, at least in the United States, it is lacking. One reason for lack of formal oversight of CRISPR is the astoundingly rapid rise and expansion of the technology. CRISPR's potential applications in biomedicine, basic research, agriculture, and environmental remediation and manipulation can only be described as explosive. While scientists and ethicists grapple with this phenomenon, many citizens, legislators, and other stakeholders remain unaware that the technology even exists.

In 2014, Massachusetts Institute of Technology political scientist Kenneth Oye, Harvard University synthetic biologist George Church, and others called attention to the urgent need for regulating gene drive technology and the inadequacy of regulatory mechanisms in the United States and globally to deal with CRISPR-mediated gene drives. They concluded what others have: that informed public discussion is required for deciding whether and how to use gene drives (Oye et al. 2014). To facilitate such discussion, they recommend that researchers who are developing technologies that could affect the global commons publish information about the technologies and their potential applications well in advance of actually constructing, testing, and releasing them into the world.

In 2015 and 2016, an opportunity for updating the biotechnology regulatory apparatus in the United States occurred. In a mid-2015 memorandum, President Obama requested that the three decades old Coordinated Framework for the Regulation of Biotechnology (CFRB) be reviewed, presumably with an eye toward updating to make it relevant to new genetic technologies including CRISPR and gene drives. Three federal agencies comprise the CFRB: the Food and Drug Administration (FDA), Environmental Protection Agency (EPA), and Department of Agriculture (USDA). All three agencies have authorities to regulate the development and introduction of products resulting from biotechnology. Every new biotechnology product is assigned to one of the agencies depending upon the nature and purpose of the product. For example, under the Plant Protection Act, the USDA must approve crops genetically modified to resist insect predation; whereas, human food products containing biochemical products derived from genetic engineering are considered to possess food additives, and the FDA must declare them safe unless the additive is already generally recognized as safe.

Responsibility for the CFRB review fell to the Office of Science and Technology Policy (OSTP). Established in 1976, the OSTP advises the executive branch of government on the effects of science and technology on national and international affairs. The OSTP works with government agencies, the private sector, state and local governments, other nations, and the science and higher-education communities to develop and fulfill sound science and technology policies (US White House n.d.).

To perform the CFRB review, the OSTP established an Interagency Working Group representing the FDA, EPA, USDA, and the White House. President Obama's memorandum tasked the interagency reviewers with three objectives: (1) update the CFRB by clarifying current roles and responsibilities, (2) develop a long-term strategy to ensure that the federal biotechnology regulatory system is equipped to efficiently assess the risks, if any, of the future products of biotechnology, and (3) commission an expert analysis of the future landscape of biotechnology products (Obama White House Archives n.d.). The group released a draft of its recommendations in September 2016 in which it addressed the first two objectives (US White House 2016). For the third objective, the group commissioned NASEM for its assessment of biotechnology products likely to appear in the future. NASEM (2016) responded with a book-length report titled *Gene Drives on the Horizon* in which it assesses human and ecological benefits and risks of gene drives, gives specific recommendations for gene drive researchers, and offers detailed suggestions for engaging communities, stakeholders, and publics in conversation about human values as they relate to gene drives. The NASEM recommendations are not yet codified into policy, and sadly, the OSTP has languished under the Trump administration.

Jennifer Kuzma, codirector of the Genetic Engineering and Society Center at North Carolina State University, is critical of the process and apparent outcomes of the CFRB review (Kuzma 2016). She sees the review as a missed opportunity to provide clear regulatory pathways for the development and application of currently expanding technologies such as CRISPR-mediated genome editing and gene drives. Instead of venturing into the unfamiliar areas of these developing technologies to clarify which agency's authority would apply to anticipated products or whether revised or new authorities need to be granted, the review team remained in familiar territory. The case study examples given to illustrate clarifications of roles and responsibilities of regulating agencies (objective 1 above) all fell unambiguously into the current authorities of these agencies. Thus, the review provided no direction for how the CFRB ought to approach the regulation of gene drive–engineered mosquitoes for release into wild populations to control insect borne diseases or the release of other GE animals.

Kuzma also faulted the review process for offering very limited opportunities for input from a wide range of stakeholders in how to regulate new biotechnologies. She writes that "the OSTP process ignored the public part of the politics stream, constraining policy agenda setting to an inner circle of policy elites through the closed-door interagency process" (Kuzma 2016,

1212). Finally, the review failed to consider how US regulatory policy might be coordinated with existing international agreements for the safe transference of living, genetically modified organisms across international borders. Currently the United States is not a signee on either of two key international agreements, the Cartagena Protocol on Biosafety or the United Nations Convention on Biological Diversity, perhaps because their provisions might adversely affect certain US trade and business interests. Kuzma (2016, 1213) concludes her illuminating article by observing that "opportunities to effect meaningful and impactful change for safe, responsible, legitimate, and appropriate use of GEOs were missed at this key juncture in the biotech revolution, although it is poised to change nearly every sector and even our conceptions of nature."

Meanwhile, the Court of Justice of the European Union in Luxembourg created a major hurdle for commercialization of CRISPR gene-edited crops in Europe (Callaway 2018). On July 2, 2018, the court ruled that any organism created by precise gene-editing technologies like CRISPR are subject to the same stringent regulations governing traditional GE organisms. These 2001 regulations have severely limited growing and selling GE crops in Europe.

Chapter Summary

In nature, the bacterial CRISPR/Cas system forms a defense mechanism against viruses that infect bacteria. In 2012, researchers developed the CRISPR/Cas system of bacteria into a powerful gene-editing tool applicable to virtually any organism on earth. *Cas* genes code for Cas nucleases, proteins that can cut DNA molecules. One of these, Cas9, from a *Streptococcus* species, combined with laboratory-designed guide RNA molecules, form the most widely used CRISPR-based gene-editing tool. Guide RNAs, designed to base-pair with specific spots in the genome to be altered, direct Cas9 to those spots. There Cas9 cuts the DNA. Depending on how researchers tailor the system, a cell's own DNA repair mechanisms act on the cut sites to inactivate, activate, temporarily silence, correct, delete, or add genes at those spots in the cell's genome. Researchers can genetically alter the germ line (eggs and sperm) of an organism by applying CRISPR technology to fertilized eggs. Moreover, a technology called CRISPR-mediated gene drive can potentially propagate a human-designed genetic alteration rapidly through an entire population of sexually reproducing organisms.

Researchers have engineered CRISPR-mediated gene drives into some organisms in the laboratory, but as of 2018, none of these have been released into wild populations. CRISPR technology promises many benefits for human and environmental health and well-being, but these potential benefits come with risks. A major controversy is whether CRISPR technology ought to be used to edit the human germ line, which would pass the genetic alteration on to future generations. Another concern is whether benefits to releasing gene drive–engineered organisms into the environment to control disease vectors and invasive species outweigh risks to ecosystems. Regulation of the use of CRISPR and gene drive technologies lags far behind their many feasible applications and continued rapid development.

Questions for Reflection and Discussion

1. Should public money fund research to develop CRISPR for safe, clinical, germ line gene editing to cure genetic diseases in human embryos before they are transferred to the womb?
2. Do you believe the term *gene editing* to describe the action of CRISPR on DNA accurately represents the technology to the nonscientist public? How about the term *embryo editing* to describe CRISPR's potential use for germ line DNA modification? Do you foresee dangers or misconceptions developing due to use of the *editing* metaphor? If so, what different descriptors do you suggest using for CRISPR?
3. Do you believe that doing CRISPR research on human embryos will lead inevitably and eventually to genetic enhancement of the human germ line? If so, does this bother you? Why or why not? If you do not think that human germline enhancement is inevitable, what do you think will prevent it?
4. Should plants or animals engineered to have CRISPR-mediated gene drives be released into wild populations under any circumstances? If so, what are some circumstances under which it would be all right? If not, why not?
5. Do you believe a time will come when a majority of Earth's plant and animal species have had their genomes edited by humans? If so, what does that world look like to you? If not, what will prevent it?
6. Imagine a time when, for several generations, the vast majority of living human beings have had their germ lines edited. What does that world look like to you? What types of gene editing will be most prevalent? Who

is most likely to be making the gene-editing decisions? Will persons with and without edited genomes be noticeably different? If so, in what ways?

7. What is your personal position on human germ line genome editing?

8. Presuming that CRISPR-mediated human germ line engineering can eventually be done safely, causing no health problems for the engineered individuals, what do you consider to be the greatest ethical concern: using CRISPR tools for human germ line genetic enhancement or releasing CRISPR-mediated gene drives into wild populations of plants and animals? Why?

9. How much detail about the way CRISPR and gene drives work do you believe conscientious, lay citizens need to understand in order to contribute informed decisions on the future use of these technologies? What levels of understanding of the technologies are appropriate for junior and senior high school science teachers? Students? Elected policy makers?

Notes

1. Recombinant DNA technology creates DNA molecules with portions from two or more sources, often different species. For example, Roundup Ready soybeans contain some genes from petunias that render the beans resistant to the herbicide glyphosate.

2. A patent war over whose work and patent ought to be honored for licensing CRISPR/Cas9 applications in medicine and other areas rages between Doudna's institution, the University of California at Berkeley, and Zhang's base, the Broad Institute at the Massachusetts Institute of Technology. Doudna's 2012 work showed that CRISPR technology can edit isolated DNA, and Zhang's work was the first to apply CRISPR gene editing to eukaryotic DNA in living, laboratory cultured cells. The European Patent Office also challenged Berkeley's patent claims because one of Doudna's collaborators at another institution was not named on one version of the patent application. In February 2017, a three-judge panel of the Patent Trial and Appeal Board ruled in favor of the Broad Institute for their patents claiming invention of the CRISPR gene-editing technology as applied to eukaryotes. An appeal of this decision from Berkeley was denied in 2018.

3. Henceforth, I use the acronym *CRISPR* to stand for the CRISPR/Cas9 system unless otherwise indicated. Cas9 designates the particular protein currently most widely used in CRISPR-mediated gene editing work. It is a double-

stranded DNA nuclease, an enzyme that makes a cut through both strands of double-stranded DNA at one site.

4. Researcher He Jiankui at Southern University of Science and Technology of China in Shenzhen and Michael Deem, a bioengineer at Rice University in Houston, collaborated on the project and reported the results to the Associated Press. They claimed to have used CRISPR gene editing in early IVF embryos to disable a gene named CCR5, whose activity allows HIV to enter cells. The work was reported as an AP exclusive by Marilynn Marchione on November 26, 2018, "First Gene-Edited Babies Claimed in China," Associated Press (https://www.apnews.com/4997bb7aa36c45449b488e19ac83e86d [accessed November 30, 2018]).

5. The investigators collected living skin cells from a patient with the disease, converted these cells into stem cells, and then genetically repaired the defective gene in the stem cells via CRISPR gene editing as the cells were growing in the laboratory. Genomes in 13 percent of the stem cells in the experiment were successfully corrected.

6. iPS cells are cells taken from differentiated tissue, often skin, to which additional copies of three to four genes for transcription factors have been added while cultured in the laboratory. The transcription factors activate other genes normally active during early embryonic development and thereby induce pluripotency, the ability of cells to reproduce and give rise to all of the body's specialized types of cells. Normally, only cells in very early embryos and embryonic stem cells (ESCs) derived from early embryos are pluripotent. Research using iPS cells avoids the ethical controversy surrounding creation and use of ESCs. Although iPS cells and ESCs behave very similarly in laboratory experiments, they are subtly different. Whether one will prove to be better in regenerative medicine applications remains to be seen. Currently there are strong arguments for continuing research with ESCs. These include the broad, undisputed developmental potential of ESCs, the great value of ESC research for understanding and solving problems in early human development, and the fact that certain applications of iPS cells raise ethical issues of their own in the area of reproductive rights.

7. 3PN stands for three pronuclei. Each pronucleus has one set of chromosomes, so a 3PN embryo has three chromosome sets. This happens if the final cell division in egg formation does not occur, leaving the egg with two sets of chromosomes instead of just one. Then upon fertilization, the egg gains a third set of chromosomes from the sperm. 3PN embryos also form when two sperm

fertilize a single, normal egg cell. Many problems in cell division and gene expression result from having three complete sets of chromosomes, making the 3PN embryos nonviable.

8. The injected embryos were zygotes, the earliest stage of embryonic development. The zygote is a single, fertilized egg cell containing chromosomes from both parents. In this case, each 3PN zygote had one set of female (egg) chromosomes and two sets of male (sperm) chromosomes due to two sperm cells having fertilized a single egg cell.

9. A chimera is an organism containing cells/tissues/organs from two or more different species. For example, pigs with a human pancreas are being developed for treatment of diabetes.

10. This happens through a process called alternative splicing in which the primary RNA transcript of a gene becomes cut and spliced in alternative ways to give rise to several different messenger RNA molecules, each of which codes for a different protein. Until the discovery of alternative splicing in the1970s, the "one gene-one protein" paradigm was widely accepted.

11. The Human Proteome Project is an international, ten-year project began in 2012. The project's objective is to decipher the function of all 20,300 polypeptides (proteins or their component subunits) coded for by the human genome and to apply this information to improved disease diagnosis and therapy.

12. There is evidence that rats can feel individual hope, but no evidence yet that nonhuman animals can participate and feel collective hope.

Sources for Additional Information

Achenbach, J. 2015. "Scientists Debate the Ethics of an Unnerving Gene Editing Technique. *Washington Post*, December 1. https://www.washingtonpost.com/news/speaking-of-science/wp/2015/12/01/historic-summit-on-gene editing-and-designer-babies-convenes-in-washington/ (accessed June 23, 2016).

Cohen, J. 2018. "Federal Appeals Court Hears CRISPR Patent Dispute." *Science*, April 30. http://www.sciencemag.org/news/2018/04/federal-appeals-court-hears-crispr-patent-dispute (accessed May 7, 2018).

Doudna, J. 2015. "My Whirlwind Year with CRISPR." *Nature* 528: 469–71. http://www.nature.com/news/genome-editing-revolution-my-whirlwind-year-with-crispr-1.19063 (accessed July 21, 2016). This is an engaging, inspiring, and personal testimony of social conscience from a world-class scientist and developer of the CRISPR gene-editing technology.

Esvelt, K. M., A. L. Smidler, F. Catteruccia, and G. Church. 2014. "Emerging Technology: Concerning RNA-Guided Gene Drives for the Alteration of Wild Populations." eLife 2014;3:e03401. https://elifesciences.org/content/3 /e03401 (accessed July 26, 2016).

Greely, H. T. 2016. *The End of Sex and the Future of Human Reproduction.* Cambridge, MA: Harvard University Press. This book provides a rigorous treatment of human-assisted reproduction technologies and their ethical, legal, and social implications.

Highett, K. 2018. "Breakthrough CRISPR Gene Editing Trial Set to Begin This Year." *Newsweek*, April 16. http://www.newsweek.com/crispr-therapeutics -crispr-cas9-gene-editing-beta-thalassaemia-887051 (accessed June 19, 2018).

Knoblich, J. A. 2017. "Lab-Built Brains." *Scientific American* 316(1): 26–31.

Knoepfler, P. 2016. *GMO Sapiens: The Life-Changing Science of Designer Babies.* Hackensack, NJ: World Scientific Publishing. This is an informative and thought-provoking read for general audiences not trained in science.

Ledford, H. 2016. "Titanic Clash over CRISPR Patents Turns Ugly." *Nature* 537: 460–61. http://www.nature.com/news/titanic-clash-over-crispr-patents-turns -ugly-1.20631 (accessed October 13, 2016).

Pennisi, E. 2013. "The CRISPR Craze." *Science* 341: 833–36. http://science .sciencemag.org/content/341/6148/833 (pdf available here; accessed July 21, 2016).

———. 2017. "Mini-livers Reveal Fine Details of Organ Development." *Science* 356: 1109.

Reardon, S. 2016. "The CRISPR Zoo." *Nature* 531: 160–63. http://www.nature .com/news/welcome-to-the-crispr-zoo-1.19537 (accessed July 22, 2016). This is an excellent, accessible, informative article about the many and diverse potential uses for CRISPR-mediated gene editing and some of the ethical issues they raise.

Sevick, K. 2018. "Broad Institute Takes a Hit in European CRISPR Patent Struggle. *Science*, January 18. http://www.sciencemag.org/news/2018/01/broad -institute-takes-hit-european-crispr-patent-struggle (accessed May 7, 2018).

Specter, M. 2016. "DNA Revolution." *National Geographic*, August, 30–59. Published online as "How the DNA Revolution Is Changing Us." https://www .nationalgeographic.com/magazine/2016/08/dna-crispr-gene-editing-science -ethics/?user.testname=photogallery:c (accessed October 22, 2018).

———. 2017. "Rewriting the Code of Life." *New Yorker*, January 2, 92.

Walsh, F. 2015. "The Promise of Gene Editing." BBC News, December 1. http:// www.bbc.com/news/health-34972920 (accessed June 30, 2016).

Zimmer, C. 2015. "Breakthrough DNA Editor Born of Bacteria." *Quanta Magazine*, February 6. https://www.quantamagazine.org/20150206-crispr-dna -editor-bacteria/ (accessed July 21, 2016).

———. 2016. "From Fins into Hands. Scientists Discover a Deep Evolutionary Link." *New York Times*, August 18. http://www.nytimes.com/2016/08/18 /science/from-fins-into-hands-scientists-discover-a-deep-evolutionary-link .html?_r=0 (accessed October 3, 2016). This is an enjoyable account of researchers' use of CRISPR technology to demonstrate a link between fish fin bone formation and the evolutionary development of mammalian digits, including the human hand. Also, see A. Saxena and K. Cooper, "Fin to Limb within Our Grasp," *Nature* 537: 176–77, for a nontechnical description of this work.

5
Nanotechnology, Life, and Nanoethics

By 2100, our destiny is to become like the gods we once worshipped and feared. But our tools will not be magic wands and potions but the science of computers, nanotechnology, artificial intelligence, biotechnology, and most of all, the quantum theory.
—Michio Kaku, American theoretical physicist, futurist

A religious college in Cairo is considering issues of nanotechnology: If replicators are used to prepare a copy of a strip of bacon, right down to the molecular level, but without it ever being part of a pig, how is it to be treated?
—Charles Stross, British science fiction writer

Imagine cancer-curing medicine traveling through the bloodstream inside unimaginably small capsules or hollow, straw-like tubes and homing in on cancer cells like a bear to honey to destroy malignant tumors without nasty side effects. Envision laboratory-grown replacement organs for humans, artificially intelligent robots that reason as well as or better than humans, or deadly soot-sized particles targeted against ethnic groups or individuals by terrorist groups or governments. These and other wondrous and chilling imaginings may become realities in our lifetimes. Some will be with us in just a few years. All are products of nanotechnology and its applications in biology. The term *nanotechnology* refers to the scientific endeavor of understanding and manipulating extremely small bits of matter—atoms and small groups of atoms—in prescribed ways. Nanotechnology's present and future products, its likely impact on our lives, and the ethical issues it raises are subjects of this chapter. Our excursion into the nanoscale world emphasizes nanobiology and includes discussion of the questions:

(1) How is biology related to nanotechnology?
(2) What natural nanoscale machines do living cells have?
(3) What are the medical applications of nanotechnology?
(4) What are some nonmedical applications of nanotechnology?
(5) What dangers lurk in our nanotech future?
(6) What does responsible development of nanotechnology require?
(7) Does nanotechnology raise special ethical challenges?

Nanotechnology and Biology: The Science

The prefix *nano* is derived from *nanos*, the Greek word for dwarf. In science and engineering, *nano* means "one billionth." So 1 nanometer (nm) equals one-billionth of a meter. Since a meter is about 39 inches, 1 nm is incredibly small. The width of a human hair is about 80,000 nm. The diameter of the HIV particle is 100 nm, and the width of the famous gene-carrying DNA double helix is about 2 nm. Imagine if a meter were expanded to stretch from San Francisco to New York City. On that scale, one nanometer would be only about as large as the width of a kernel of unpopped popcorn. Scientists define a nanoscale object as anything with dimensions between about 0.1 and 100 nm.

What Are Nanotechnology and Nanobiology?

Humans have manipulated nanoscale objects for centuries. Over seven hundred years ago medieval artisans in Europe created multicolored stained-glass windows for cathedrals by mixing molten gold and glass. By varying the proportions of each element in the mixture, the artisans obtained different colors of glass. Unknown to them, it was the nanosized gold particles that gave color to the glass. Since the eighteenth century, chemists, and before them the alchemists, combined elements that unbeknownst to them formed nanoscale compounds. Today, we would call these artists and scientists early nanotechnologists; however, *nanotechnology* and *nanoscience* are late twentieth-century terms. They describe the interdisciplinary endeavor to understand the properties of matter at the nanoscale and to use this understanding to fashion new nanoscale objects for use in widely ranging areas from medicine and electronics to cosmetics, fabrics, and space travel. Nanobiology is a subdiscipline of nanotechnology. It is devoted to understanding how a cell's own nanoscale structures work and also to creating nanoscale objects that perform specific tasks inside living things.

Nanoscientists aim to discover how and why matter behaves the way it does at the nanoscale. Already, they have learned some surprising things. For example, some substances that do not conduct electricity at larger than nanoscale sizes become excellent conductors at the nanoscale; other materials emit light of various colors depending on their nanoscale size. And some material that is harmless in the size of dust particles or larger becomes toxic at the nanoscale such as carbon-containing soot particles emitted from diesel engines. Nanoscientists believe that nanoscale particles gain most of their unusual properties due to their extremely high surface to volume ratio. In relatively large objects, the properties of the bulk material swamp out or mask special reactions occurring on the surface of the object, but as particles become smaller and smaller, events at the surface become more and more important in determining the overall properties of the particle. The increase in surface area relative to the volume of a spherical object as it becomes smaller is dramatically illustrated by comparing the surface area (A) to volume (V) ratio (A/V) for the planet Earth, a baseball, and a nanoparticle 20 nm in diameter:[1]

Earth: A/V = 0.00000000000000001 (one-hundredth of one-quadrillionth)
Baseball: A/V = 0.1 (one-tenth)
Nanoparticle: A/V = 100,000,000 (one hundred million)

As A/V increases, an object's surface interactions with the environment increasingly influence the object's qualities. This applies to objects of any shape not just to spheres.

Nanotechnology uses principles discovered by nanoscience to create nanoscale objects with special properties for human uses. Nanotechnologists received early inspiration from a talk by California Institute of Technology physicist and Nobel Laureate Richard Feynman (1959). In his lecture titled "There's Plenty of Room at the Bottom" and delivered at the annual meeting of the American Physical Society, Feynman challenged scientists and engineers to begin building things from the bottom-up, atom by atom, molecule by molecule. Building things at the atomic scale, he declared, can make possible the encoding of "all the information that man has carefully accumulated in all the books in the world . . . in a cube of material one two-hundredth of an inch wide—which is the barest piece of dust that can be made out by the human eye. So there is *plenty* of room at the bottom!"

Bottom-up engineering contrasts with top-down fabrication, the miniaturizing of large (macro) scale objects. For example, switches in electrical circuits were modified top-down to produce transistors and ultimately computer chips. The bottom-up method of nanotechnology represents a completely new approach to building things and uses properties of matter that are not present above the nanoscale.

Nanotechnology is now the most interdisciplinary field in applied science. Chemists, chemical engineers, biostructural engineers, molecular biologists, pharmaceutical scientists, mechanical engineers, textile engineers, electrical engineers, biomedical scientists, aerospace engineers, computer scientists, information technologists, and many other professionals can all claim to be nanotechnologists so long as they are working with nanoscale objects.

Nanoscale objects are invisible to the naked eye and also when the eye is aided by traditional light microscopes like those used in biology class. To see and/or manipulate nanoscale objects, specialized instruments including electron microscopes and scanning probe microscopes are required. Electron microscopes have been used by engineers and biologists since the 1960s to examine very thin fibers, crystalline structures, and the nanoscale parts of cells. The more recent invention called a scanning probe microscope, in use only since the 1980s, employs extremely fine-tipped probes to examine surfaces of objects down to the scale of individual atoms. One type of scanning probe microscope is the scanning tunneling microscope (STM). An STM gathers information about surfaces by measuring minute electrical currents passing between the tip of the probe and the surface being examined.

By contrast, another type of scanning probe microscope, the atomic force microscope (AFM), actually touches the surfaces of objects and "feels" their contours, providing the researcher with an image that includes individual atoms. Both STMs and AFMs are used to move individual atoms or molecules deliberately around on a surface to create preplanned patterns.[2] This type of building, atom by atom, is expensive, painstakingly slow, and only in two dimensions. Development of instruments that rapidly and economically fabricate three-dimensional nanoscale objects will be a major nanotechnology breakthrough.

Nanobiology includes two broad areas of endeavor: (1) understanding the workings of naturally occurring nanoscale structures made by living cells and (2) creating nanoscale structures to study cells, fight disease, or otherwise repair, enhance, or alter living things. Nanobiology directed at curing or diagnosing disease is *nanomedicine*. Let us now explore the exciting and diverse realm of nanobiology.

Nanoscale Objects and Biology

In this section, we consider both human-made and cell-made nanoscale objects. Before humans began creating nanoscale particles of matter for uses in living organisms, cells had been doing it for a very long time. We begin with examples of nanomachines that evolved by natural selection in living cells during the 3.8 billion years that life has graced the surface of the Earth.

One of nature's nanoscale machines. Examples of nanomachines created by nature are the protein called *kinesin*, a molecular motor that moves things around inside cells, and the *ribosome* (appendix 1), the cell's protein-synthesizing machine. Since kinesin is a protein, let us begin by recalling from chapter 1 what a protein is.

A cell is like a microscopic city harboring unimaginably complex structures and unceasing and diverse activities. We can equate jobs in a city to the activities of proteins in a cell. There are builders, building blocks, wreckers, sensors, transmitters, and delicately tuned machines. More than twenty thousand different kinds of proteins cooperate in every cell to keep it alive and function just as thousands of citizens perform a variety of functions to maintain a city. Proteins themselves are composed of smaller subunits called amino acids. Just twenty amino acids, mixed in many ways, comprise the structure of the thousands of different proteins in a cell. To form a protein, the various amino acids join together like differently colored beads on a necklace (fig. A1.1). On average, two hundred to three hundred amino acids comprise a single protein, although some proteins may be thousands of amino acids long, while others have only a few amino acids. To function correctly inside cells, the necklace-like proteins fold, coil, and twist into complex three-dimensional shapes (fig. 1.3). The three-dimensional shape of each protein dictates its function in the cell, and the particular sequence of amino acids determine that shape. At times, two or more proteins combine to produce an aggregate of folded protein chains needed to perform a particular function in the cell. Such multichained aggregates are called multisubunit proteins. Kinesin is one of these.

Kinesin's job is to move cargo from one place to another inside the cell. It does this like a miniature, self-powered train car moving along a track. Kinesin's cellular tracks are hollow, straw-like structures called *microtubules*. One end of kinesin grabs onto whatever cargo needs to be moved, and the other end literally "walks" along the length of the microtubule (fig. 5.1). If we could see it in action, kinesin would look much like a hiker with a backpack walking heel to toe along one rail of a railroad track.[3] Walking around in the cell along microtubules requires energy. The energy source for kinesin and almost all other energy-requiring events in the cell is a high-energy molecule called ATP, which is produced when food molecules burn to release energy.[4] Brain health relies on the activity of kinesin. By moving nutritional and informational cargo along the length of brain cell axons and dendrites, kinesin helps maintain the strands of communication between the brain's one hundred billion neurons.

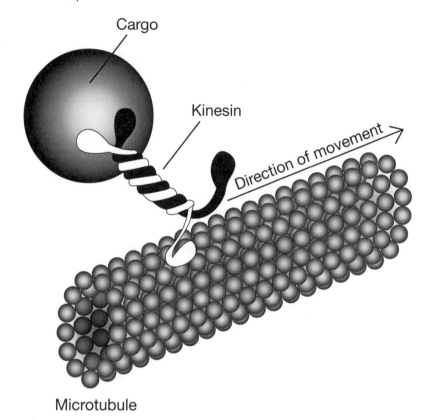

Figure 5.1. Kinesin, a cellular molecular motor protein powered by ATP, moves cargo inside the cell along molecular tracks called microtubules. (Courtesy of the author)

Kinesin is just one example of scores of cellular nanomachines evolved in nature. Nanotechnologists are intensely interested in understanding how nature's nanomachines work. Gaining that information will provide insight and practical help for engineering efficient, workable, human-made nanoscale devices.[5]

Human-made nanoscale objects for nanomedicine. Nanomedicine is an interdisciplinary endeavor to improve human health via nanotechnology. It aims to diagnose, treat, and prevent disease, repair injuries, relieve pain, and promote human health using nanoscale objects that cooperate with nature's molecular structures and the machinery inside our bodies.

Two types of nanoscale particles promise great usefulness in nanomedicine —quantum dots (also called nanodots) and nanotubes. A quantum dot is an indescribably small crystal of a metallic compound containing cadmium, lead, iron, or some other metal. Nano-sized gold particles also form quantum dots. Because of their extremely small size of 5–15 nm in diameter, quantum dots have properties not present in larger particles of the same material. One is the ability to absorb photons at many energy levels, while emitting light at just one wavelength. The wavelength of light emitted, and therefore the light's color, depends upon the size of the quantum dot. Different sized quantum dots composed of the same material emit different colors of light.

Carbon nanotubes are very different. Instead of containing metal, nanotubes contain only carbon atoms linked together to form tiny, hollow, straw-like structures with one or more single-layered wall of carbon atoms (fig. 5.2, top). Under certain conditions, multiwalled carbon nanotubes nest like Russian dolls, fitted together in sequence, one within the other. Single-walled nanotubes are generally 1–2 nm in diameter with variable lengths depending on conditions during their formation. One place that nanotubes occur naturally is in soot from burning wood and fossil fuels. *Homo erectus*, our ancient relative, was probably the first humanlike animal to create nanotubes by building fires over one million years ago. Familiar objects made from carbon are coal, diamonds, and graphite (pencil lead). Differences between these three entities result from different arrangements of carbon atoms. All nanotubes are a modified form of graphite with their own unique characteristics, one of which is the acquisition of certain electrical properties not possessed by graphite itself. One would not guess it, but under certain conditions, carbon nanotubes explode; and for their size, they are stronger than steel. It has even been suggested that carbon nanotubes be used to construct the cable for a space elevator reaching outward from the equator to twenty thousand miles in space (fig. 5.2, bottom).[6] Now, we will consider how quantum dots and carbon nanotubes can be used in medicine.

1. *Cancer detection:* The usual treatments for cancer are surgery and chemotherapy. The latter involves introducing toxic chemicals into the bloodstream to kill rapidly dividing cells. Unfortunately, cancer cells are not the only rapidly dividing cells in the body. Bone marrow cells divide furiously to replace the hundreds of millions of red blood cells that die every day. Cell division also replenishes the white blood cells, our cellular immune system. Hair follicle cells in the scalp constantly make hair (except in bald spots), and cells lining the intestine must continually be replaced. Chemotherapy disrupts all of these normal processes, which is

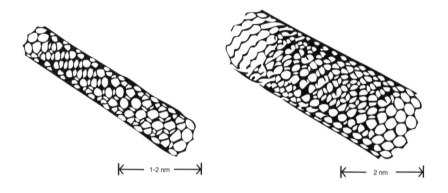

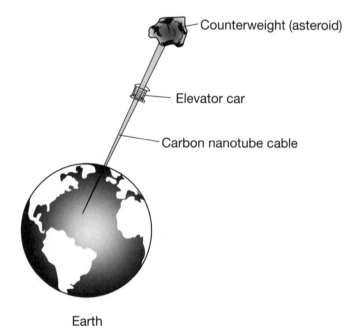

Figure 5.2. Carbon nanotubes with single (top left) or multiple (top right) walls of carbon lattice may form cables stronger than steel cables of the same size. Carbon nanotube cables may someday support elevators from the Earth's surface into outer space (bottom). (Courtesy of the author)

why cancer patients undergoing chemotherapy can lose hair and become anemic and are especially vulnerable to infectious disease.

Early detection of cancer is a big problem for physicians. Often, cancer is discovered because a patient is not feeling well, has a bloody stool, feels a tumor, or receives a suspicious result of a mammogram. Exploratory surgery may ensue, and by then, often the cancer is quite advanced. Detection of cancer cells before tumors form could facilitate treatment and save lives. Nanobiologists are seeking means to detect cancer cells early and destroy them without harming normal cells. Quantum dots play a big role in this project.

Successful tests in animals indicate promise for using specially designed quantum dots for early cancer detection in humans. To prepare nanoscale, metallic crystals (quantum dots) to locate cancer cells in the body, researchers first coat them with an inert substance to make them harmless to living tissue. Next, a thin coating of some type of biocompatible polymer is layered on the dots. To this surface is added a nano-thin layer of antibody proteins that will recognize cancer cells and attach to the surface of those targeted cells (fig. 5.3). After quantum dots attach to the surface of cancer cells, many of them move across the cell membrane and enter the cells. Once inside cancer cells, quantum dots are visualized by shining infrared light on suspected areas of the patient. The infrared light penetrates skin and living tissue to reach the quantum dots that then emit visible light, revealing the location of the cancer cells. Magnetic quantum dots can even help detect cancer cells. In this case, the cellular location of the quantum dots is followed using magnetic resonance imaging (MRI).

2. *Cancer treatment:* Attaching chemotherapeutic drugs to quantum dots that specifically target cancer cells is a powerful and promising new approach to fighting cancer. Researchers are even designing quantum dots to release and/or activate their load of toxic drugs only after entering cancer cells. Since the drugs are aimed at cancer cells and do not enter healthy cells, higher doses of very effective drugs may be used to help assure total destruction of cancerous tumors without surgery. Trials using this technology are underway in animals.

Nanotubes can destroy cancer cells. Prof. Balaji Panchapakesan and his coworkers at the University of Delaware produced a nanotube-based cancer "nanobomb" (Thomas and Atkinson 2005). Consisting of a bundle of carbon nanotubes, the nanobomb targets cancer cells in a way similar to that just described for quantum dots. The bomb is based on the surpris-

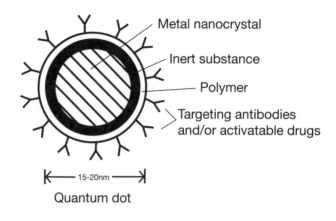

Quantum dot

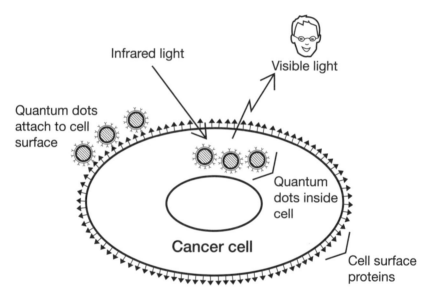

Figure 5.3. Quantum nanodots for early cancer detection. Metallic nanocrystals are coated with an inert substance to render them biologically innocuous. A nano-thin layer of polymer serves to bind antibodies, which recognize cancer cells (top). Once inside the cancer cells and exposed to infrared light, nanodots emit visible light and reveal the location of the cancer cells. (Courtesy of the author)

ing discovery that bundles of nanotubes will heat and explode when a beam of laser light is directed at them. Exploding bundles of nanotubes were used to selectively kill human breast cancer cells in mice. The tiny explosions only destroyed the cancer cells and did not damage nearby healthy cells.

3. *Organ and joint replacement:* Regenerative medicine, the rejuvenation

or replacement of worn-out body parts, will also benefit from nanotechnology. Biomedical researchers are developing ways to grow new organs for transplantation in glass dishes in the laboratory. To do this, they first construct scaffolding shaped like the desired organ. Then, they seed the scaffolding with living cells that multiply to cover the surface of the preformed structure. Some researchers find that certain cells grow best on a surface of nano-fabricated material. Other researchers investigate how coating the surfaces of hip and other joint replacement parts with nano-sized bumps may enhance their biocompatibility, durability, and adherence to surrounding bone tissue.[7]

4. *Micromachines are exciting but not nanotechnology:* In a highly publicized volume titled *Nanomedicine* (the first in a planned four-volume set), Robert A. Freitas Jr. imagines miniature machines injected into the human body that would cure disease, repair organs and damaged DNA, clean clogged arteries, and even enhance our physiology (Freitas 1999). In the last category is the "respirocyte," an artificial red blood cell with an oxygen-delivering capacity that would endow us with superhuman powers in sports and war, fight cardiovascular disease, and, among other things, allow one to hold one's breath for up to four hours. Freitas works for the "nanotech" firm Zyvex whose stated long-term goal is to achieve a breakthrough in producing nanobots, but the respirocytes and other futuristic micron-sized machines detailed in *Nanomedicine* are at least one thousand times too large to qualify as nanomachines. Even if miniature machines like the respirocyte, with dimensions in the tens of microns, were to someday be constructed, their use would not properly be called "nanomedicine." The objects that Freitas writes about in *Nanomedicine* are more properly called microelectromechanical systems (MEMS), fictional devices introduced by Rachel Welch in her 1966 film, *Fantastic Voyage*. Authentic nanomedicine today is done by researchers laboring to develop quantum dots and nanotubes for diagnosis and drug delivery as described earlier.

Other Applications of Nanotechnology

Several commonly used products containing nanoparticles are already on the market. They include stain- and water-repellent clothing, sunscreen, cosmetics, tennis rackets, paint, and antifogging and antireflective glass. Nanoelectronic engineers Philip Collins and Phaedon Avouris identify various uses for nanotubes that are either now ready for market or very feasible. These include the improvement of scanning probe microscopes, chemical sensors designed to detect pollutants or other chemicals in minute quantities, greatly enhanced computer memory, and nanotweezers able to pick up and release objects near nanoscale in size. In 2017, a collaboration between researchers in five countries reported using a scanning probe microscope to write and read magnetic information from a single atom (Natterer et al. 2017; Sessoli 2017). This basic research shows that storing data in single atoms is feasible and could revolutionize computer technology.

Other possible uses of nanotubes and other nanomaterials include the development of new manufacturing processes that minimize production of undesirable waste products, creation of biodegradable pesticides, and the invention of paints and fabrics to which dirt, odors, and other particles cannot adhere. Also, the US Department of Defense is especially interested in nano-products that would allow nano-sized "defensive" agents to be specifically targeted at the enemy. Other uses beneficial to the military include the promotion of rapid wound healing, the ability of clothing to take on different properties under different conditions, and even inducement of human limb regeneration.

Our Nanotech Future: "Nanotopia," Catastrophe, or Something In-Between?

In some ways, nanotechnology is positioned at the interface between real science and science fiction. With improvements in our ability to move atoms one by one and arrange them in combinations that nature has not yet used, the potential products of nano-fabrication seem nearly limitless. In his book *Engines of Creation*, nanotheorist and engineer K. Eric Drexler called attention to fantastic-sounding possibilities for the future of nanotechnology (Drexler 1986).

He describes nanomachines called assemblers (nanorobots or nanobots) that are programmed to create virtually any material object, from a Hawaiian shirt to a New York strip steak, simply by taking atoms from common sources like the atmosphere or the soil and arranging them to produce desired items. The nano-fabrication of useful products will require unimaginably large numbers of nanobots. Since constructing quadrillions of nanobots individually would be infeasible, if not impossible, nanobots will need to be able to replicate themselves, as do living cells.

Some persons describe a nano-assembler future as "Nanotopia," where humankind's every material need and desire have been met and the challenges of world hunger, pollution, scarcity of clean water, and perhaps even mortality as characteristics of human development will be left behind. In

Nanotopia, everybody will be adequately fed, clothed, and sheltered just by turning dirt and air into human necessities. Belief that such a future is actually possible is fueled by claims like Drexler's that assemblers will allow us "to build almost anything that the laws of nature allow to exist." Taking inspiration from Drexler, other authors even suggest that a "utility fog" of swarming nanobots may someday be programmable to take on different shapes and functions depending upon the needs at hand—food, chair, telephone, bicycle, boat, glass of water.

Not everybody sees such a glorious nanobot future. Ten years after Drexler's book was published, Sun Microsystems scientist Bill Joy wrote a now-famous/infamous article for *Wired* magazine in which he warned that all life and Earth itself could be destroyed if nanobot replication went awry (Joy 2000). The planet and everything on it could quickly be converted into a seething mass of nanobots due to their exponential, unregulated replication. This is the "gray goo" scenario. Malevolent swarms of flying, killer, self-replicating nanobots is the subject of the novel *Prey* by the late Michael Crichton (2002, xiv). The book begins with a 1992 quotation from Drexler: "There are many people, including myself, who are quite queasy about the consequences of this technology for the future."

Currently, most nanoscientists and nanotechnologists do not believe that the production of nanoassemblers is possible. In 2003, Drexler and the late Richard E. Smalley, nanoscientist and Nobel laureate, exchanged views on this subject for the trade journal, *Chemical and Engineering News* (Baum 2003). Smalley argued that laws governing chemical reactions and atomic-level interactions will create insurmountable problems for Drexler's vision of nanomachines. Three problems he cites are (1) "stickiness" (attraction) of atoms and molecules for each other that makes construction of nanomachines with moving parts impossible, (2) enzymes' need for water as a medium that will limit what nano-factories can produce, and (3) the likely occurrence of many incorrect chemical reactions along with the desired reactions. Drexler believes that these criticisms are defeatist and irrelevant for the chemistry he envisions leading to the creation and operation of assemblers. More research is needed to show who is correct and whether either "Nanotopia" or "gray goo" is the future we should anticipate. Whatever details unfold in our nanotechnology future, predictions that nanotechnology will spawn the equivalent of a second industrial revolution do not ring as overstatements.

Bill Joy recommends severe restrictions on nanotechnology research in order to avoid catastrophe. But a restricted future for nanotechnology seems unlikely and unenforceable. Governments, universities, and private companies worldwide are throwing big money and brain power at nanotechnology research and product development. The global market for nanotechnology activities is projected to approach $100 billion by 2021. The multiagency National Nanotechnology Initiative (NNI) established by Pres. Bill Clinton in 2000 has received significant budget increases each year since its inception, while other agencies that support scientific research and development experienced budget cuts during the George W. Bush administration. The US national budget for 2019 allocates $1.4 billion for the NNI.

Nanoscience: Human Values

Most of the ethical issues raised by nanotechnology, including nanobiology, also appear in other areas of science and engineering. First, we consider some of these issues that most new technologies share. Then, we examine three ethical concerns that apply specifically to nanobiology and a few other biotechnologies.

Principles for the Ethical Development of New Technologies

Three principles guide responsible research that exposes humans to new technologies: (1) obtaining informed consent, (2) optimizing safety, and (3) protecting vulnerable populations. How well nanoscientists, nanotechnolgists, nanotechnology entrepreneurs, and institutions supporting the development of nanotechnology adhere to and respect these principles will largely shape public attitude toward nanotechnology/nanobiology and, thereby, its future.

Sometimes attention to these principles lapses, and then, renewed vigilance comes after tragic and preventable failures. Chernobyl, Three Mile Island, the Tuskegee syphilis experiment, thalidomide babies, and the death of eighteen-year-old Jesse Gelsinger in 1999 during a gene therapy trial are painful reminders of the importance of putting people ahead of "progress" and profit. In both the Tuskegee and Jesse Gelsinger tragedies, informed consent was lacking. Specific criteria are established for gaining informed consent including knowledge about the potential benefits and risks of the procedure, alternatives to participation in the project, and the right to withdraw from the trial.[8]

The safe use of cancer-fighting "nanobombs" and drug-delivering quantum dots in humans has not yet been established in clinical trials. Crucial questions need answering before nanoparticles can be used to diagnose and treat diseases in humans: What is the fate of metallic nanocrystals and nanotubes in living tissues? What are long-term effects on human health if nanoparticles are not quickly cleared from the body? What hazards accompany the manufacture of nanoscale particles, and how are they best minimized? Is exposure to airborne nanoparticles dangerous? Compared to the attention received regarding potential uses of nanotechnological research, there is not yet much investigation into the possible dangers of nanoparticles to human health. Work with mice has shown that inhaled nanoparticles gain access to the brain, causing damage to brain cells (Wang et al. 2008).[9] Other research suggests that nanotubes of certain lengths cause lung disease. Much more investigation needs to be done to learn about the effects of inhalation, ingestion, injection, and application of nanomaterials on human health.

As the parade of nanotechnologies moves forward, vulnerable populations of individuals who lack the capacity or authority to give their informed consent abound. Such groups include fetuses, neonates, young children, mentally challenged individuals, and comatose persons. Soldiers, prisoners, minority ethnic groups, pregnant women, suicidal persons, and the fatally ill are also especially vulnerable when recruited as research subjects. Safeguards, regulatory protections, responsibilities of the researchers and their funding agency may differ between different groups of vulnerable persons.[10]

As with other technologies, once a specific application of nanotechnology is developed and demonstrated to be safe, there comes the problem of justly distributing its benefits. Distributive justice demands dealing with difficult questions about who should pay the cost of research and development of new technologies, who will reap the benefits, and who will be excluded from the benefits. Will the technology be so expensive that only wealthy persons can take advantage of it? What should be the roles of government and private industry in distributing the benefits? If a technology is not accessible to everybody from the start, how long will it take for its benefits to "trickle down" to the less fortunate? It is too early to tell which applications of nanotechnology will pose the biggest challenge for distributive justice. Hopefully, in some cases, nanotechnology may make possible new manufacturing processes that actually decrease the cost of products and thereby make them accessible to more persons.

Special Ethical Issues for Nanobiology and a Few Other Biotechnologies

At least three ethical issues are particular to nanobiology and a few other biotechnologies: (1) personal and societal effects of greatly lengthened human life spans, (2) the remodeling of human nature, and (3) rethinking our definitions of "life" in the context of creating new life forms. In regard to length of life, nanobiology activities may lead to much longer active human lifespans resulting from improved diagnostics, treatments, and organ replacement therapy made possible by nanomedicine. The genetic and pharmaceutical approaches to age retardation also raise similar ethical issues. In thinking about the second concern, we must consider that alterations of human nature may result from nanotechnology-based physical and cognitive human enhancements. Human genomics, proteomics, genome-editing technologies like CRISPR (see chapter 4), and the manipulation of brain neural circuitry (see chapter 6) may also provide the means to alter core traits that define our species. The self-assembling and self-reproducing nanostructures, if they come to be, will force a reevaluation of our concept of "life," as do advances in synthetic biology (see chapter 2).

Let us get specific. Ethical issues associated with age retardation come in two broad categories: those arising from moderate extension of life spans and those arising from an indefinite extension of human life. Some futurist scientists like Raymond Kurzweil even believe that humans will eventually make themselves immortal (Kurzweil 1999). In the foreseeable future, it is likely that nanomedicine and other biotechnologies will lengthen productive lifespans to 120 to 150 years. Ethical issues at both individual and societal levels will arise due to such a moderate life span extension. One issue is respect for individual autonomy. If age-retarding treatments are to begin at or before birth, or even in early childhood, who makes the decision about applying life-extending procedures? What about informed consent in the case of children? If life-extending procedures act by retarding the aging process, results could include a longer adolescence, more years living at home, and a longer period to move through the academic material now taught during sixteen years of a conventional K-12 and college education. One result would be greater expense for education. It is unlikely that everybody would partake in identical life-extending treatments. Imagine the personal relationship problems of partners with mismatched aging schedules. Being just ten years apart in age is a challenge to any marriage. Think about being fifty or more years different in age. How would even a moderate extension of human life to 150 years affect one's spiritual life? For those whose spirituality is motivated by mortality, pursuing the answers to life's persistent questions may be delayed for a century or so. How might several more decades of prime-time living affect spousal relationships? A doubling of potential working/professional years would require several periods of retraining for persons in jobs where technologies develop at a rapid rate. Consider the societal implications of longer periods of retirement, which may include years of lower living standards due to stressed retirement systems and dwindling personal savings accounts. Widespread increased longevity would need to be matched by decreased birth rates. How would that be accomplished? Some of these issues raise the question, "Do we have a duty to die?" So how should we advise legislators about funding age retardation research? How closely do our investment portfolios reflect our values?

Exactly how might nanobiology be applied to alter human nature? It is too early to say. It seems certain that nanobiology in collaboration with emerging brain science (see chapter 6) and genome editing (see chapter 4) will have that potential. What is human nature? There is no single agreed-upon definition. Religious thinkers, philosophers, evolutionary biologists, and others all offer perspectives on the essential elements of humanness. Suppose biotechnology could alter our innate aggressiveness, competitiveness, cooperativeness, ability to empathize, curiosity, or abilities for forethought. What might be the benefits and risks of such alterations? Who should decide which individuals to alter and to what degree? Further discussion of the interface of biotechnology and human nature is in chapter 4, and in the last chapter of the book, we consider the need for broad conversations about possible trajectories for biotechnology, some of which may be irrevocable.

Regarding the use of nanobiology to engineer new life forms, a comment by MIT physicist Dr. Joseph Jacobson is germane. Robert Service, science news writer for the journal *Science*, reports that Jacobson hopes "that electronically controlled proteins and DNAs will enable molecular biologists to cut and splice genetic information electronically and eventually program computers to engineer new organisms" (Service 2002, 2322). Here is another example where benefits and risks will need to be weighed, ideally by the most diverse representatives of stakeholders as possible. A computer programmed to engineer life forms with human-selected traits could revolutionize human and environmental well-being positively. On the other hand, a nightmare scenario is also easy to imagine. Who ought to use such computers and for what purposes?

Finally, what about the dire predictions of Bill Joy—that nanotechnology will inevitably lead to the destruction of its creators? Let us presume that the object of Joy's fear for humanity, self-replicating nanobots gone wild, is theoretically possible. Are there opposing forces to construct a path into our nanotechnological future that averts catastrophe? Xerox Corporation scientist John Seely Brown and University of California, Berkeley, social scientist Paul Duguid believe those forces exist. In a 2000 article for Standard Media International, they cite the power of human society to guide and divert the pace and direction of new technologies, which at first seem uncontrollable (Brown and Duguid 2000). Their examples are nuclear power and genetically engineered food products. Political systems, social organizations, professional networks, social movements, market institutions, and other social factors are already diminishing the potential of damaging effects that some believed these technologies would have on the human landscape. The power of human society to regulate technology is undeniable, but the scientific literacy and collective moral wisdom of individual members of society will determine the extent and ways that humanity shapes nanotechnology and other modern technologies.

Chapter Summary

Nanotechnology is the interdisciplinary study and manipulation of matter at the atomic and molecular scales (1–100 nm) to build new objects with strange, wonderful, and sometimes scary properties for human use. This bottom-up approach to engineering has potential to change virtually all aspects of our lives in the twenty-first century. Nanotechnology will transform medicine, computing, information storage, weaponry, energy, manufacturing, and countless everyday items and processes. Immediate ethical issues facing nanotechnology include informed consent for human research, the safety of nanoparticles, the protection of vulnerable populations, and the just distribution of its benefits and risks. Far-reaching, longer-term ethical concerns include lengthened human lifespans, enhancements/alterations to human nature, and new life forms. Controversial questions are whether self-replicating nanomachines called "assemblers" can be constructed. If so, will they usher in a Nanotopian world of abundance for all or threaten the well-being of humankind and the planet with unregulated self-replication? Involvement of individual citizens in policy-making decisions and the public's behavior in the marketplace are important factors affecting humankind's future life with nanotechnology.

Questions for Reflection and Discussion

1. Can you think of any other things (material, conceptual, institutional, cosmic, etc.) whose defining qualities change dramatically at some threshold level of size or other property in a way analogous to nanoscale metal crystals?

2. What products do you know of that are named or advertised with the term "nano"? Is the "nano" used accurately in these instances? What products of nanotechnology have you already used?

3. Do you believe self-replicating nanobots will ever be created? If not, why not? If so, do you believe Bill Joy's warning about "gray goo" is warranted? If not, why not?

4. We have iconic warning signs for certain dangerous situations, conditions, or objects (e.g., curve ahead, icy bridge, roadwork ahead, poisonous, radioactive, high voltage, etc.). We do not yet have a similar warning symbol for dangerous nano-sized particles. Sketch a design for such a symbol.

5. Presume that Nanotopia, a world in which all material things can be obtained inexpensively and without human labor through the activity of nanobots, becomes a reality. What do you see as the best and worst aspects of that world?

6. Do you believe there are ethical issues associated with nanotechnology that were not mentioned by the author?

Notes

1. $A = \pi r^2$ and $V = 4/3\ \pi r^3$, where r = the radius of the sphere and π is the ratio of the diameter to the circumference of a circle (circa 3.12).

2. IBM scientists Gerd Binnig and Heinrick Rohrer invented the STM in 1981. They shared the 1986 Nobel Prize in Physics with Ernst Ruska, designer of the first electron microscope. One of the first published STM images was of xenon atoms spelling out "IBM" on a nickel surface. The image was viewed at http://www.almaden.ibm.com/vis/stm/images/stm10.jpg on March 12, 2017.

3. An animation, "*Kinesin Protein Walking on Microtubule*," is at https://www.youtube.com/watch?v=y-uuk4Pr2i8 (accessed June 22, 2017). The video was posted by Emmanuel Dumont in 2012 and extracted from *The Inner Life of a Cell* by Cellular Visions and Harvard (http://www.studiodaily.com/2006/07/cellular-visions-the-inner-life-of-a-cell/).

4. The burning of food molecules starts in the mouth, stomach, and intestine, where enzymes begin breaking the molecules into smaller pieces. Ultimately, most burning occurs inside mitochondria in a process called oxidative phosphorylation, where the energy released during the oxidation of food molecules is used to produce ATP, the energy currency of the cell.

5. A New Focus article in the journal *Science* (Service 2002) details several examples of combining molecular biology and nanotechnology to create electronic and diagnostic devices and innovative ways of nanoscale manufacturing.

6. The concept of a space elevator was first proposed by Russian scientist Konstantin Tsiolkovsky in 1895 and then popularized by Arthur C. Clarke in his 1979 novel, *The Foundation of Paradise*. The idea is now taken seriously by some engineers who envision a cable comprised of carbon nanotubes anchored at the equator and fastened to a counterweight in space. Elevator cars traveling up the cable would reach escape velocity near the end of the cable and release satellites into Earth's orbit without the need for rocket-powered launches. More information about the space elevator and links to other sources are at http://www.pbs.org/wgbh/nova/sciencenow/3401/02.html (accessed March 12, 2017).

7. Research by biomedical engineer Thomas Webster at Purdue University shows that coating material used in joint implants with nanoscale bumps may keep the body from rejecting the artificial parts. The bumps mimic the natural surface of natural bone and other tissues (Venere 2003).

8. Detailed information on the protection of human subjects including informed consent is on the US Health and Human Services website, https://www.hhs.gov/ohrp/regulations-and-policy/regulations/45-cfr-46/index.html (accessed March 12, 2017).

9. A nontechnical synopsis of this research was published in *Environmental Health News* (Benninghoff and Hessler 2008). The research demonstrated that nanocrystals of titanium dioxide, used in white paint, toothpaste, cosmetics, sunscreens, and other widely used products, may travel to the brain via the olfactory nerve, that inhaling the particles can damage brain cells, and that brain damage occurs after a low exposure for a short period of time.

10. Information on the protection of human subjects including vulnerable populations is on the US Health and Human Services website, https://www.hhs.gov/ohrp/regulations-and-policy/regulations/45-cfr-46/index.html (ac-

cessed March 12, 2017). Also, an excellent article on safeguarding vulnerable populations is P. Shivayogi, "Vulnerable Population and Methods for Their Safeguard," *Perspectives in Clinical Research* 4 (2013): 53–57, https://www.ncbi.nlm.nih.gov/pmc/articles/PMC3601707/ (accessed March 12, 2017).

Sources for Additional Information

Ashley, S. 2002. "Nanobot Construction Crews." In *Understanding Nanotechnology*, editors of *Scientific American*. New York: Warner Books.

Collins, P. G., and P. Avouris. 2002. "Nanotubes for Electronics." In *Understanding Nanotechnology*, editors of *Scientific American*. New York: Warner Books.

Keiper, A. 2003. "The Nanotechnology Revolution." *New Atlantis*, 2 (Summer): 17–34.

MacDonald, C. 2004. "Nanotech Is Novel; the Ethical Issues Are Not: We Must Become Competent in Dealing with Moral Concerns Related to All New Technologies." *Scientist* 18 (3): 8.

National Nanotechnology Initiative: http:/www.nano.gov (accessed March 12, 2017). This site contains clear, concise descriptions of nanotechnology and the instruments that make it possible, present and future applications, current research, national funding, government collaboration with industry, the ethical, legal, and societal implications of nanotechnology, and resources and links about nanotechnology.

National Nanotechnology Initiative. 2002. "Small Wonders, Endless Frontiers: Review of the National Nanotechnology Initiative." https://www.nap.edu/catalog/10395/small-wonders-endless-frontiers-a-review-of-the-national-nanotechnology (accessed March 12, 2017). This article prepared by the NNI is an excellent overview of nanotechnology.

Quantum Dot Corporation, Hayward, CA. Now at Invitrogen Corp. reached at http://www.qdots.com (accessed March 12, 2017).

Ratner, M., and D. Ratner. 2003. *Nanotechnology: A Gentle Introduction to the Next Big Idea*. Upper Saddle River, NJ: Prentice Hall.

Royal Academy of Engineering, Royal Society. 2004. "Nanocience and Nanotechnologies: Opportunities and Uncertainties." http:/www.nanotec.org.uk/finalReport.htm (accessed March 12, 2017). This provides a detailed and accessible report on nanoscience, nanotechnology, and the accompanying ethical and social issues.

Stix, G. 2002. "Little Big Science," In *Understanding Nanotechnology*, editors of *Scientific American*. New York: Warner Books.

Zyvex Instruments, http://www.zyvex.com/index.html (accessed March 12, 2017).

6
Brains, Minds, and Neuroethics

Ten billion neurons, ten-to-the-fourteenth different connections—hell,
you can do anything with that. That's more than enough to contain a soul.
—Anonymous neuroscientist in Judith Hooper and Dick Teresi,
The Three Pound Universe

"The human brain is certainly more complicated in organization than is a mighty star, which is why we know so much more about stars than about the brain," wrote the famous author and biochemist Isaac Asimov (1986). Brains and their nervous systems are marvels. Comprised simply of matter, mostly water and various carbon-containing molecules ranging from sugars to proteins, they allow animals to capture prey, evade danger, mate, and in some cases think. Some brains give rise to what humans call happiness, pleasure, sadness, and pain. The human brain spawns philosophy, science, art, religion, literature, morality, and other less lofty endeavors, all stoked by brain-based traits including imagination, creativity, curiosity, desire to understand, self-centeredness, altruism, foresight, shortsightedness, fear, courage, hatred, love, deception, and loyalty. Described as the most complex system known in the universe,[1] the architecture and circuitry of the human brain is now the object of intense scrutiny by thousands of scientists worldwide. Deep understanding of and treatments for brain diseases and disorders and answers to long-standing questions about how the brain gives rise to mind and consciousness will almost certainly come in this century. Technologies for accomplishing these goals along with the ethical and social implications of brain research are the subjects of this chapter. We will examine the following questions:

1. How much brain science do I need to know in order to make good decisions about using modern and emerging neurotechnologies?
2. What major, ongoing research projects will yield breakthroughs in our knowledge of brain structure and function?
3. What technologies will allow these breakthroughs?
4. How might new knowledge about the human brain benefit humankind?
5. What ethical and societal issues arise with modern brain research and neuroscience technologies?

6. Will a dramatically increased understanding of how the human brain works present us with new personal and societal ethical quandaries?

Brains and Minds: The Science and Technology

Not surprisingly, what we already know about our brain is far less than what we do not know. But twenty-first-century brain researchers aim to change that ratio in favor of knowledge. Among the things we now know is that the brain contains about one hundred billion neurons, nerve cells that communicate chemically and electrically with each other. Each individual neuron converses with other neurons through thousands of synapses, specialized regions of communication between the cell membranes of nerve cells.[2] This means that your brain contains over one hundred trillion neuronal connections to receive and send chemical and electrical signals. Normal functioning of myriad networks and circuits of these intercommunicating neurons is the basis for creative thought, sensing, feeling, learning, dreaming, planning, remembering, decision making, being unique individuals, and much more. Similarly, malfunctioning circuitry causes diseases including Parkinson's, schizophrenia, addiction, obsessive-compulsive disorder, and major depression.

Cells, Synapses, Circuits, and Mind

Knowing a few basics about human brain structure and function is sufficient for a general understanding of how neuroscientists hope to shed new light upon normal and abnormal brain function within the next decade or so. A little knowledge about the field of neuroscience is also helpful in making informed personal and collective decisions about the ethical use of modern and emerging neurotechnologies.

Consider the neuron, or nerve cell, a cellular building block of the brain. The generic neuron includes a central cell body, many filamentous, branched projections called dendrites, and a single long projection called an axon (fig. 6.1). Neurons receive signals from thousands of other neurons through their dendrites. Incoming signals from various parts of the body and for a variety of purposes do not all say the same thing. A neuron must integrate the diverse array of incoming messages and determine from their collective chemistry whether to relay a "majority opinion" signal to thousands of other neurons via its axon. Specialized regions of close contact between neurons mediate the movement of incoming and output signals between neurons (fig. 6.2; Bradley 2013a, 229–34). The activities of millions of intercommu-

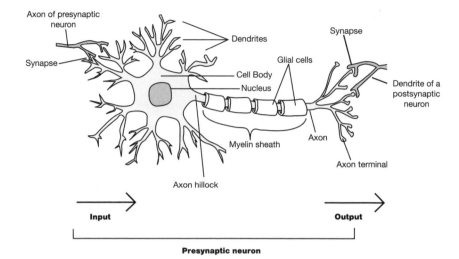

Figure 6.1. A neuron with its three major parts: cell body, axon, and dendrites. Specialized glial cells comprise the myelin sheath, which supports and insulates the axon and facilitates propagation of nerve impulses. Neurons communicate with other neurons at synapses. At a synapse, an axon terminal from the presynaptic neuron comes in close proximity to a dendrite from a postsynaptic neuron. Relative to the postsynaptic neuron on the far right, the central diagram depicts a presynaptic neuron. (Courtesy of the author)

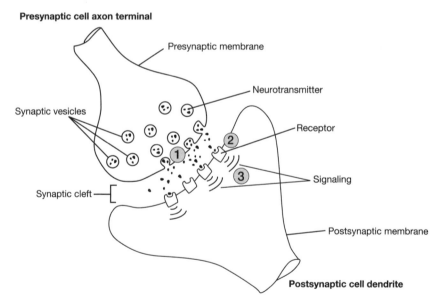

Figure 6.2. A chemical synapse. In response to chemical and electrical signals, synaptic vesicles in the presynaptic cell fuse with the cell membrane and release neurotransmitter molecules into the synaptic cleft (1). Receptors in the postsynaptic cell membrane bind to the neurotransmitters (2) and send stimulatory or inhibitory signals into the postsynaptic cell (3). (Courtesy of the author)

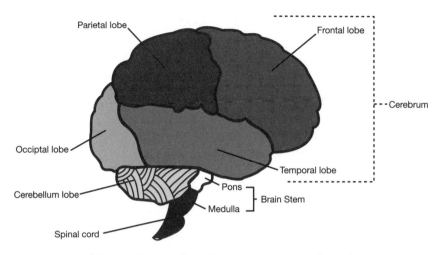

Figure 6.3. Adult human brain with its three major parts: cerebrum (containing the frontal, parietal, occipital, and temporal lobes of the cerebral cortex), the cerebellum, and the brain stem. (Courtesy of the author)

nicating neurons organized into neural circuits make possible specific ordinary and extraordinary motor activities and behaviors.

Nonelectrical glial cells outnumber the brain's neurons by many billions. Some glial cells wrap themselves closely about neurons and provide nutritive support. Other types of glia divide and help repair damaged brain tissue. Recently, glial cells were also discovered to be very communicative. Instead of forming synapses, glia communicate directly with each other and with neurons using small, specialized regions of contact between adjacent cells called gap junctions. A gap junction contains hundreds of tiny channels through which small molecules including neurotransmitters move between cells. Each major type of glial cell—oligodendrocytes, astrocytes, and microglia—has a distinguishing morphology and distinct set of functions. Many glial cell functions are surely yet undiscovered.

The externally visible brain in humans and other mammals has three distinct parts: the brain stem, the cerebellum, and the cerebrum (fig. 6.3). By comparison, an extraordinarily large cerebrum (cerebral cortex) distinguishes the human brain from that of other mammals, including nonhuman primates. When a person is described as "cerebral," or not so, it is the cerebrum that is the object of praise or derision. Sometimes referred to simply as "the cortex," the cerebrum exists in two halves, one on the surface of each hemisphere of the brain. The front part of the cerebrum, the frontal lobe, is the region that has expanded the most during human evolution. It is the literal seat of our unique humanness, the place where neuronal activity produces self-awareness, distinctly human thoughts, and our plans for the future.

The brain is also made of so-called grey matter and white matter. Grey matter contains the cell bodies of neurons, while white matter consists mainly of axons from those cell bodies. The whitish appearance is due to cellular insulation that surrounds each axon and aids in conducting electrical signals between neurons. The grey matter of the cerebral cortex is only about 2.6 mm thick (Winkler et al. 2010), slightly thicker than the width of a house key and not as thick as the bill on a baseball cap. This may seem like insufficient material to be the seat of human thought, sensation, movement, and consciousness, but the cerebrum is highly folded and wrinkled, camouflaging an impressive combined surface area, which if flattened would cover a single-size mattress. Other regions of the brain, particularly the multi-component limbic system lying just below the cerebrum, are evolutionarily older than the cerebrum.[3] In fact, counterparts of our limbic system reside in modern-day reptiles whose ancestors gave rise to mammals over 250 million years ago. The limbic system is important for our emotions, sense of smell, motivation, and memory.

The mind-boggling complexity of our brain prompted the comment in this chapter's epigraph that a human brain is more than adequate to contain a soul, an oblique reproach of the concept of substance dualism championed by René Descartes in the seventeenth century ([1641] 1984, 1–62). Dualism held that the world is composed of two substances, matter and mind. Humans were considered to be a mixture of these, the mind or soul being responsible for thought and consciousness. Now, neuroscientists and even philosophers generally agree that brain matter somehow gives rise to the mind. Just how this happens is not known, but within a decade or so new technologies for studying brain structure and function will almost certainly reveal how signaling through those trillions of synapses within and between specific regions of the brain results in thought, action, memory, perception, and human consciousness itself. However, neither philosophers nor neuroscientists agree upon a definition of *mind* or *human consciousness*, even within their own disciplines. But they do agree that the brain has a crucial role in each. Later in this chapter, we will look at some different views of what constitutes the human mind and how one's own concept of *mind* may affect responses to certain ethical issues raised by modern neuroscience. For now, we proceed knowing that everybody agrees that the brain plays an important role in creating the mind. Some neuroscientists express this by saying that the mind emerges from the brain.

Not thwarted by disagreement over the relationship between mind and brain, the United States and the European Union (EU) both announced major, long-term brain research projects in 2014. China followed suit in 2016. These complementary efforts to decipher the neural language that creates human behavior and the emergence of the mind itself will have dramatic ramifications in many realms of endeavor including human health, self-understanding, justice, and AI. As we will see, these ventures are boldly ambitious and not without controversy.

Big Brain Science Projects

In April, 2013, President Obama announced the BRAIN (Brain Research through Advancing Innovative Neurotechnologies) Initiative (Obama 2013). Six months later, the European Commission launched its Human Brain Project (HBP). Both ventures will last a decade or longer and each requires investment of several billion dollars. The projects have different but complementary objectives. Both call upon tools and minds from many disciplines, including neuroscience, computer science, synthetic biology, cell and molecular biology, nanotechnology, physics, engineering, and ethics.

The BRAIN Initiative. The BRAIN Initiative owes its inception to a small group of neuroscientists who attended a scientific meeting in London in September 2011. In addition to reporting on their current work, attendees challenged themselves to dream a little and share what they would really like to be able to do in their research. One attendee, Dr. Rafail Yuste from Columbia University, reportedly replied, "I want to be able to record from every neuron in the brain at the same time" (Markoff 2013). Within a few months, two journal articles outlined possible ways to reach this goal, beginning with nonhuman animal brains and involving hundreds of scientists, engineers, and technicians over a decade or more (Koch and Reid 2012). Less than a year after publication of these two papers, the deputy director of the White House Office of Science and Technology Policy, Thomas A. Kalil, became convinced of the value of a multiagency government brain research project on the scale of the earlier Human Genome Project (HGP). In 2013, the BRAIN Initiative became reality.

As of 2018, the government's BRAIN Initiative website lists four government agencies among the initiative's alliance members: the NIH, which heads the project, the National Science Foundation (NSF), the FDA, and Intelligence Advanced Research Projects Activity (IARPA), an organization within the Office of the Director of National Intelligence. Alliance members also include twenty-five nonfederal foundations, institutes, universities, and industries. This diversity of funding sources helps protect and sustain the BRAIN Initiative during administrations that seek to deprioritize science. During 2017, the BRAIN Initiative funded 130 research projects, most led by university researchers.

Exactly what does the BRAIN Initiative hope to accomplish? Why? How? The overall goal of the BRAIN Initiative is not far from Dr. Yuste's research dream. The BRAIN Initiative aims to "accelerate the development and application of new technologies that will enable researchers to produce dynamic pictures of the brain that show how individual brain cells and complex neural circuits interact at the speed of thought" (NIH n.d., 3). The following excerpt from an online posting of a BRAIN Initiative funding opportunity concisely describes the Initiative's mission:

> The broad goal of the BRAIN Initiative is to understand the circuits and patterns of neural activity that give rise to mental experience and behavior, which will provide a foundation for understanding and treating diverse neurological, psychiatric and behavioral disorders. It is the dynamic activity of massively interconnected ensembles of neurons in specially organized networks that give rise to the internal states we experience as sensations, perceptions, emotions thoughts, memories and movements. The activity of these networks is the substrate of cognitive processes such as attention, intention, emotions, and rational processes such as reasoning and decision making. Ultimately, these covert, internal activities are translated into patterns of neural activation that lead to overt behaviors, from simple reflexes to highly coordinated movements such as reaching and walking, to more complex behaviors such as navigating the environment and foraging, or speech and language. Dysfunction of these large systems of neurons due to disease, injury or developmental anomaly is the basis of neural and mental disorders. The mission of the NIH BRAIN initiative is to understand how large scale neural systems contribute to cognitive and neurological function in both health and disease (US Department of Health and Human Services 2014).

How is this to be accomplished? Again, from the same BRAIN Initiative funding opportunity statement: "We can seek to understand circuits of the brain by systematically controlling stimuli and measuring the resulting behaviors, while actively recording and manipulating the dynamic patterns of neural activity. We now have transformational technologies that allow us to record large, interrelated ensembles of neurons on an unprecedented scale during active behaviors. For example, it is now possible to study the collective neural activities of entire sensory-motor circuits. By clever manipulation of environments and contingencies, we can devise behavioral tasks that engage memories, decision making and selective attention, while documenting and manipulating the functional relationships within the neural circuits that subtend the behaviors."

Since the BRAIN Initiative focuses on the structure and function of neural circuits, we will briefly consider the role of those circuits in a normally functioning brain. Neurons are the cellular components of neural circuits. The human brain's one hundred billion neurons, interconnected by complex and ever-changing patterns of synapses, form neural networks. Neural networks, which are wired together via synapses and which perform a specific function for the animal, comprise a neural circuit. A conservative estimate of the number of neural circuits in a human brain runs into the thousands. Each circuit contains millions of neurons, whose connectivity may span several regions of the brain.

Neural circuits mediate thousands of varied motor, cognitive, and regulatory functions ranging from the coordination of finger movements for clarinet and piano players and the whole body for dancers and soccer players to visual processing, memory, attention, mood, and social behavior. Adding to the structural complexity of neural circuits is their dynamic nature. The communication pathways at individual synapses in a circuit may strengthen or weaken in response to incoming signals (or lack thereof), which are themselves products of signals from thousands of other neurons. Neural circuit dynamics may reflect an individual's experience within her environment, frequency of circuit use, learning, and a plethora of other unknown factors. In the words of the BRAIN Working Group authors of *BRAIN 2025*, "a thinking brain can . . . be viewed as an immensely complex pattern of activity distributed across multiple, ever-changing circuits" (2014, 13). Understanding the architecture and function of neural circuits is the overarching goal of the BRAIN Initiative. Learning how to manipulate them in living animals including humans may result in tremendous breakthroughs in identifying and treating many disabling afflictions. Accomplishing these goals is crucial to achieve the hoped-for diagnostic and therapeutic benefits of the initiative.

Although the focus of the BRAIN Initiative is solidly on the neuronal components of the brain, some neuroscientists warn against overlooking critical and unknown functions of the brain's glial cells. Chief of the Nervous System Development and Plasticity Section at the NIH R. Douglas Fields argues convincingly that the BRAIN Initiative must examine the distributions and functions of glia in the brain in order to accomplish its long-range scientific and clinical objectives (2013, 25–27).

Why should public money be spent on the research proposed by the BRAIN Initiative? The BRAIN Working Group identified seven goals for the initiative. The seventh goal nicely encapsulates the rationale for the entire project: "Integrate new technological and conceptual approaches produced in Goals #1–6 to discover how dynamic patterns of neural activity are transformed into cognition, emotion, perception, and action in health and disease" (2014, 6–7).

Much of the work called for by the BRAIN Initiative will be performed in nonhuman organisms such as worms, insects, leeches, fish, mice, and rats before the technologies are ever applied to humans. This makes sense both ethically and scientifically. Consider first the ethics of animal research.

Most biomedical discoveries and applications rely on nonhuman animal studies that uncover principles and help assure the safety of a product or procedure before it is used on humans. The underlying assumption is that humans deserve a higher moral status than animals like worms, flies, mice, and monkeys. Although not everyone agrees with this assumption, few if any would promote a return to exploitive medical experimentation on humans like what happened in Nazi concentration camps during World War II or in Alabama during the federally sponsored Tuskegee Study of Untreated Syphilis in the Negro Male from 1932 to 1972. If one accepts learning about neural circuitry as a legitimate goal of modern neuroscience, then nonhuman animal experimentation must be tolerated. Computer modeling and human brain organoids (discussed in chapter 2) may lessen the need for animals in brain research, but they cannot yield all of the important information obtainable from whole, living organisms. Now consider why using nonhuman animals in the human BRAIN Initiative makes scientific sense.

Science proceeds best by breaking complex problems down into simpler ones that are tackled first before progressing stepwise to more complex problems. For example, early BRAIN Initiative work will employ the 1 mm long nematode worm *Caenorhabditis elegans*. The structure of *C. elegans's* entire, adult nervous system was painstakingly mapped decades ago. It contains 302 neurons and about 5,000 chemical synapses (White et al. 1986).

Researchers working to construct brain activity maps will learn much by examining nematode worms and fruit flies ("Brains in Action" 2014). Because of their relatively small and structurally characterized nervous systems, worms and flies are ideal for testing new technologies and discovering basic principles about neural circuits and their role in producing specific behaviors. Vertebrates like fish, mice, and rats, on the other hand, are well suited for experiments aimed at scaling up new technologies to handle many more neurons and to delve deeper into the brain (Alivisatos et al. 2013).

Zebrafish larvae are transparent for up to a week after hatching. All of the organs in living, behaving larvae, including the brain, are easily observable. In fact, the fish's brain at this stage is only 1/2 mm wide, one-fourth the size of a single raw sugar crystal. That makes the zebrafish ideal for learning how specific neural circuits produce complex behaviors. Neuroscientist Michael Orger of Portugal points out that the zebrafish brain is large enough to see with the naked eye but small enough to view in its entirety with a microscope: "We can look at hundreds or even thousands of cells simultaneously. . . . You can look at multiple brain areas operating together. And you can do this in a behaving animal" ("Brains in Action" 2014, 4). Why should we be interested in how this fish's minute brain controls its behavior?

Zebrafish behavior involves decision making and learning, explains Misha Ahrens, researcher for the Howard Hughes Medical Institute: "They swim around, they make decisions about where to go, where to explore. They hunt for food, they have [the capacity] for simple learning, and [they] change their behaviors based on past experience" ("Brains in Action" 2014, 4). Fish and humans are both vertebrates. Humans descended from the former over 420 million years of biological evolution. In fact, the physiological and anatomical relationship between fish and humans is the subject of two well-known books, *From Fish to Philosopher* by Homer Smith (1959) and *Your Inner Fish* by Neil Shubin (2009). When we learn precisely how nerve cell activity and neuronal circuitry allow a fish hatchling to alter its behavior based on previous experiences, we will be on the path toward understanding how to nurture and promote the same for humans and perhaps ultimately to change or alleviate some of humankind's most destructive habits.

A cautionary note against overoptimism about immediate application of knowledge gained from the BRAIN Initiative is in order. From the immensely successful HGP, we learned that collecting and analyzing huge amounts of chemical information about a living organism does not automatically reveal how the organism uses that information. By sequencing the 3.1 billion A, T, G, and C bases in the human genome, we learned the number and locations of genes but not their functions. Discovering the precise role(s) of each of our approximately twenty-two thousand genes and how their gene products interact with each other in human development, health, and disease will occupy molecular biologists for many more years, probably decades. Similarly, obtaining dynamic pictures of neural circuits in living human brains and developing the tools to manipulate these circuits will not tell us precisely how matter thinks. Ultimately, I do believe that humans will gain this understanding and that the BRAIN Initiative is a necessary prelude to a later era when the human brain comprehensively understands itself.

The NIH Working Group also identified several core principles to guide BRAIN Initiative work. Among these are to (1) promote interdisciplinary research collaborations, (2) explore brain functions in the context of the whole body and over time periods ranging from fractions of a second to a lifetime, (3) develop means for sharing research results quickly and publicly, (4) perform all brain research according to high ethical standards, and (5) establish oversight procedures to insure accountability to all BRAIN Initiative stakeholders including the public, participating government agencies, and scientific and clinical communities. Finally, the working group recommended establishing a diverse scientific advisory group to monitor the BRAIN Initiative and ensure that its goals and principles remain germane to the project (BRAIN Working Group 2014, 7–8).

What is the estimated cost for the BRAIN Initiative? Where is the money coming from? And what is the rationale for embarking on the project at this time?

It is difficult to estimate the total cost of the twelve-year BRAIN Initiative, but considering the project's goals and the cost of recent smaller but related programs, the NIH Working Group recommends a $4 billion investment from NIH between fiscal years 2016 and 2025. Support also comes from outside the NIH. By late 2014, several foundations, universities, corporations, and patient advocacy organizations had committed more than $240 million to the project. In September 2014, the NIH announced the first NIH BRAIN Initiative grant awards totaling $46 million to over one hundred researchers in fifteen states and three countries.[4]

Is this the right time for a brain project on the scale of the 1990s HGP? Participants including NIH director Francis Collins argue that now is the right time for the BRAIN Initiative because the technological tools to accomplish its objectives are at hand or on the near horizon. Brief descriptions of three such technologies follow in the section on neuroscience tools, and appendix 2 contains expanded descriptions of several other neuroscience technologies.

The many recent advances in neuroscience make the timing of the BRAIN Initiative propitious. Among these are new and developing technologies for mapping and manipulating brain circuits. The prospect of helping an enormous number of people worldwide now suffering from neurological diseases, injuries, or disorders also argues for major research efforts to revolutionize our understanding of the nervous system. Because the BRAIN Initiative focuses on understanding how cognition and behavior emerge from the healthy human brain, the findings will be invaluable for developing new and better diagnostics and treatments for diseased and injured brains. Brain-scanning technologies already give pictures of whole brain activity but missing are data about brain cell function at the intermediate level of neural circuits. Using recently developed technologies and designing new tools to fill this intermediate level gap in our understanding is the unique territory of the BRAIN Initiative. Next, we will compare certain elements of the BRAIN Initiative with the highly successful HGP of the 1990s. Doing so may help show the way to an equally successful outcome for the BRAIN Initiative.

Comparing the BRAIN Initiative to the HGP. The HGP was completed near the turn of the twenty-first century and identified the linear sequence of the 3.1 billion A, G, T, and Cs (bases) that comprise all of the DNA in all twenty-four different human chromosomes (as discussed in chapter 1).[5] Decoding the human genome was a problem in one dimension; that is, it described the sequence of the four bases in DNA as they are arranged single-file as though they were beads on a string. If instead DNA's bases were arranged in two dimensions, on a surface where each of the 3.1 billion bases has neighboring bases on its left and right and above and below it, the HGP would have been much more complex and taken much longer to complete. Going a dimension further, if the bases had all been jumbled in three-dimensional space with no way of untangling them into a straight line or onto a surface, the project would have been unimaginably complex. The BRAIN Initiative is a problem in three dimensions. Instead of just three billion units to arrange in one-dimensional space, the BRAIN Initiative must deal with one hundred billion neurons, each having at least one thousand connections (synapses) to other neurons. That's one hundred trillion synapses all in three-dimensional space, and these interneuron synaptic connections are not static. They build an ever-changing, dynamic meganetwork that reforms itself as we learn, forget, experience, and age.

Producing accurate maps of the electrical activity coursing through the plethora of neuronal circuits embedded within those one hundred trillion synapses is one aim of the BRAIN Initiative. The ultimate goal is to understand the neural code, how neural circuit activity translates into specific behaviors, thoughts, and human consciousness. How does a person store memories, feel needs and motivations, recognize faces and voices, play a musical instrument, or sink a three-pointer without thinking about it? These phenomena are all somehow represented in the electrochemical activities of neural circuits. Discovering the logic that underlies this presentation is a problem in neural coding. Deciphering the neural code is a major goal of the BRAIN Initiative (BRAIN Working Group 2014).

Another difference between the HGP and the BRAIN Initiative lies in the emergence of competition that the public consortium HGP experienced from private entrepreneur molecular biologist Craig Venter. Fortunately, the public versus private HGP rivalry eventually became a collaboration between formerly antagonistic parties brought together by President Clinton. This unlikely collaboration culminated in a joint announcement of success in the White House Rose Garden in June 2000. At the outset, the BRAIN Initiative began as a partnership between public agencies, universities, private institutions, and the US military. No other announced major brain science project has the same objectives as the BRAIN Initiative. The EU's brain research project complements rather than competes with the objectives of the BRAIN Initiative.

That the US military's Defense Advanced Research Projects Agency (DARPA) is a major funding source for and influence on specific objectives for the BRAIN Initiative is a source of controversy.[6] Some scientists and commentators question whether research based on nonpeer reviewed proposals funded by the Pentagon can yield good science and be made freely available to the worldwide community of scientists. They fear it could instead undermine world peace and stability (Moreno 2012; Horgan 2013; Barker 2014; Reardon 2015b). Involvement of the military was absent in the HGP and has not been seen in a "big science" project like the BRAIN Initiative since the Manhattan Project that created an atomic weapon in the 1940s. As for funding level, the $300 million federal investment during the

BRAIN Initiative's first two years is less than 7 percent of the total project cost for the HGP, only a bit over 1 percent of the total cost of the Manhattan Project in current dollars, and barely 0.2 percent of the ten-year cost for the International Space Station ("How Big Is Science?" 2015).

The BRAIN Initiative lacks a single, well-defined goal analogous to the HGP's goal of sequencing the 3.1 billion chemical bases in human DNA. Many diverse pieces of the neurological puzzle will need to be fitted together to finally understand the relationship between brain structure and function, leaving plenty of room for researchers' specific and different contributions. Although the overall goals of the BRAIN Initiative are to identify the multitudinous brain cell types and to map, measure, and manipulate their interactions with each other, specific research objectives and applications will evolve over the duration of the initiative and beyond.

Current NIH director Francis Collins heads the BRAIN Initiative and also led the public consortium that was the HGP. Accomplishing the BRAIN Initiative's goals will require sustained funding and the collaboration of researchers at far-flung universities and other research institutions. During the HGP, Collins established an impressive track record of accomplishing just that. How much oversight Collins will be able to exert over DARPA's components of the BRAIN Initiative will be critical for the overall success of the initiative. Of particular importance will be public sharing of research data generated by DARPA contractors.

Genome sequencing and genetic manipulation technologies emerging from the HGP and the continuing stream of data from genomic studies spawned a worldwide and burgeoning biotechnology industry with applications in areas including medicine, pharmaceutics, ecology, conservation, agriculture, evolutionary biology, and archaeology. Similarly, the BRAIN Initiative research will produce wide-ranging applications in areas as diverse as medicine, criminal justice, advertising, marketing, military ventures, robotics, computer science, big data storage and analysis, as well as entertainment. Already, a powerful synergy exists between the completed HGP and the fledgling BRAIN Initiative. Gene sequencing and manipulation technologies that resulted from the HGP offer the possibility of understanding how specific genes control brain development and function.

The HGP included a formal, funded, ethical component known as the Ethical, Legal, and Social Implications (ELSI) Research Program. Ethical issues associated with brain research, the societal implications of its findings, and how these issues are addressed by the BRAIN Initiative and the EU's

HBP are taken up in the latter part of this chapter. First, we consider the nature of the European project and some specific neurotechnologies that the BRAIN Initiative and the HBP will draw upon.

HBP. Three documents are crucial for understanding the inception and evolution of the EU's HBP during the first eighteen months after its launch in October 2013. First is the Human Brain Project-Framework Partnership Agreement (HBP-FPA), second is an open letter criticizing both the science and management of the HBP signed by hundreds of neuroscientists, and third is the Human Brain Project Mediation Report.[7]

The HBP is the brainchild of Henry Markram, an accomplished and charismatic neuroscientist at the Swiss Federal Institute of Technology in Lausanne (EFPL). The HBP was launched in 2013 following a two-year "flagship" competition sponsored by the European Commission for funding of Future and Emerging Technologies. The only other project funded by the commission was one in materials science that concentrated on graphene.[8]

The primary stated objective of the HBP was to create a high-fidelity digital reconstruction and computer simulation of the entire human brain from the bottom-up by 2023. The brain computer model was to take into account the activities of genes in the brain's one hundred billion cells, the chemical and electrical communication going on through its one hundred trillion synapses, and the activities of thousands of neural circuits and the interplay between them. Specific HBP objectives detailed in the HBP-FPA included creating an *in silico* reconstruction and simulation of the mouse brain as a prelude to a computerized simulation of the human brain. The project objectives also incorporate the proposed uses for the latter. They included testing hypotheses for causes of and treatments for brain diseases and dysfunction, virtual testing in psychoactive drug development, investigating the roles of social interaction and other environmental factors in behavioral and cognitive development, and developing computers and robots based on principles learned about human brain structure and function.

The year after its launch, the HBP hit big difficulties. In an open letter to the European Commission, several hundred European neuroscientists severely criticized the scientific objectives of the HBP and its administrative organization. The signatories argued that the overall objective of creating a computer simulation of the human brain is currently unrealistic and a waste of resources. Furthermore, they contended that Henry Markham possessed too much decision-making power over the project, particularly over the reprioritization of certain research areas within the HBP, the emphasis on computer simulation of the whole human brain, and the dispersal of the HBP's public research funds to specific investigators and institutions.[9] The letter's signatories included a large number of scientists whose participation in the HBP was crucial for the project's success. In their threat to boycott the project, they demanded an external assessment of the HBP governance and scientific approach and the establishment of an independent external steering committee.

In response to the open letter, the HBP's Board of Directors asked that a mediation committee be established to examine the objectives and organization of the HBP and make recommendations to salvage the project. In March 2015, a twenty-eight-member mediation committee formed by mediator Wolfgang Marquardt, an engineer and director of the Jülich Research Center in Germany, presented its recommendations in a fifty-three-page report. The report upheld the concerns of the letter signatories, recommending specific adjustments to the HBP's scientific program and governance structure, all of which were accepted by the HBP Board of Directors.

If the HBP follows the committee's recommendations, it will de-emphasize its original goal of creating a computer simulation of the entire human brain and instead focus on developing computer technology to model some carefully selected brain functions and dysfunctions. The committee offered the following examples of feasible targets for computer modeling: spatial navigation, goal-directed decision making, visually guided behavior, Parkinson's disease, schizophrenia, epilepsy, addiction, obsessive-compulsive disorder, depression, as well as other consciousness disorders. In addition, the committee encouraged HBP funds be allocated for experiments on both rodents and monkeys rather than extrapolating directly from data on mouse brains to conclusions about human brain function. Finally, research in systems and cognitive neuroscience, areas that had been de-prioritized by the HBP after the project was funded, were slated for return to the project's core objectives.

As for governance overhaul, the committee recommended that the EFPL be replaced by a new international entity, a body including three to seven institutions from different countries to be in charge of the project. It also advised that the project CEO be a person with experience in science management and project execution but not related to the HBP or with a vested interest in how research monies are allocated. The scientific and governance overhaul of the HBP was completed in 2016. The EU's funding for the project extends through 2023. One can follow the rate of incorporation of the mediation committee's recommendations into the HBP by comparing the former's report with the latter's website.[10] Unlike the US BRAIN Initiative, the HBP has allocated a significant portion of its budget (5 percent) to study ethical and societal issues raised by the project. Ethical issues associated with the HBP, the BRAIN Initiative, and other neuroscience research endeavors are discussed later in this chapter.

The China Brain Project (CBP). China initiated a fifteen-year brain research plan in 2016 titled "Brain Science and Brain-Inspired Intelligence." The project is described as a "one body, two wings" system. The body refers to basic research on the neural circuitry underpinning cognition. This body of research is to provide input for and receive feedback from the two wings: (1) diagnosis and treatment for brain disease, and (2) brain-inspired intelligence technology. The neuroscience community in China is relatively small, so goals of the CBP are designed to complement strengths of current neuroscience research programs in China. Several specific objectives of the CBP mirror those of the BRAIN Project: identify new neuron types; map neuronal connections to elucidate functional neural circuits: develop new diagnostics and interventions for schizophrenia, Alzheimer's, Parkinson's disease, autism, depression, and addiction; and gain understanding of the neuronal basis for higher cognition. Like the BRAIN Initiative, the CBP will use the fruit fly, zebrafish, rodents, and nonhuman primates to probe workings of neural circuits.

The overarching objective of the "brain-inspired intelligence technology" wing of the CBP corresponds more closely to the initial but discarded objective of the EU's HBP, computer simulation of human brain function. Spokespersons for the CBP write that "the CBP will focus its efforts on developing cognitive robotics as a platform for integrating brain-inspired computational models and devices. The goal is to build intelligent robots that are highly interactive with humans and properly reactive in uncertain environments, with the skills for solving various problems that can grow through interactive learning, and the ability to transfer and generalize knowledge acquired from different tasks—even to share learned knowledge with other robots . . . also to learn to understand human intention and the way humans make decisions. Thus, a useful milestone for cognitive robotics is to build a robot that acquires behaviorally equivalent capability of empathy and theory of mind" (Poo et al. 2016).

Other neuroscience projects. Near the time that the HBP, BRAIN Initiative, and CBP were announced, other major brain research projects were declared by Israel, Canada, Australia, New Zealand, and Japan (Theil 2015,

38). In addition, other ongoing neuroscience projects in the United States complement the objectives of the BRAIN Initiative and the HBP. We will take a look at three of these: the BigNeuron Project, the Human Connectome Project, and the Allen Brain Atlas of gene expression.

Earlier, we described a generic neuron as a having a cell body from which dendrites project for receiving signals and an axon through which signals are sent to other neurons (fig. 6.1). That was a simplification of the situation. Actually, there is a multitude of neuron types distinguished by differing geometries of their cell bodies and dendritic and axonal projection. Nobody knows how many different neuron types are in the human brain or across the animal kingdom. The BigNeuron Project based at the Allen Institute for Brain Science in Seattle, Washington, aims to create a public catalog of neuron structures in humans and other animals (Shen 2015).[11] Knowing the morphology and locations in the brain of different neuron types is crucial for accomplishing the objectives of both the BRAIN Initiative and the HBP because cell structure and cell function are so tightly interwoven. In order for the BRAIN Initiative to understand neural circuit function, the behavior of individual neurons within circuits needs to be understood. That will require knowledge about the numbers, locations, and geometries of the different types of neurons in the circuit. Computer modeling the function of the human brain, as the HBP ultimately hopes to do, also requires detailed morphological information about neurons since the geometry of a neuron determines how it processes and transmits chemical and electrical information to other neurons.

Creating a detailed and highly accurate three-dimensional construct of a neuron is no easy task. It is laborious, often involving the manual tracing of the outline of a neuron's boundaries through hundreds or thousands of thin microscopic sections or other types of microscopic images. BigNeuron hopes to facilitate this process with computer algorithms developed specifically for extracting three-dimensional morphological information from two-dimensional microscopic images of neurons. The project will collect reconstruction algorithms and imaging data from researchers worldwide and then test and compare the results of the computer algorithms against human-generated results. Ultimately, BigNeuron plans to open access for researchers worldwide to reveal the best algorithms and the resulting neuronal reconstructions from humans and several other animal species.[12] However, nature does not recognize human-made categories resulting frequently in blurred boundaries between types described by scientists. So even when thousands of accurate neuronal reconstructions are in hand, the difficult and sensitive job of classifying these into structural/functional groups of neurons agreed upon by the research community will remain.

Launched in 2009, the US Human Connectome Project is organized and funded by the NIH. Two research consortiums participate in the project, one led by the University of Minnesota and Washington University and the other by Harvard University. The goal of the project is to create a network map of the human brain, that is, to learn how different regions of the brain communicate with each other. The research teams do not examine individual synaptic connections between specific neurons; instead, their focus is at a larger scale. Their aim is to map the primary pathways of connections between functionally and anatomically distinct brain regions. The project will shed light on the genetic basis for brain connectivity by comparing results from identical twins to those from fraternal twins. Noninvasive brain-scanning techniques including functional magnetic resonance imaging (fMRI), diffusion magnetic resonance imaging (MRI), magneto-encephalography (MEG), and electroencephalography (EEG) are employed in this research and described in appendix 2 and in the Neuroscience Toolkit section below.

The Allen Brain Atlas for gene expression was launched in 2003 with a $100 million donation by Microsoft cofounder Paul G. Allen to found the Allen Institute for Brain Science in Seattle, Washington. The overall objective of the project is to produce three-dimensional maps of gene activity inside the cells of mouse and human brains. The purpose of these maps, which are already available to researchers worldwide on the Web and through the downloadable Brain Explorer software, is to aid research on how genes influence brain development, diseases and syndromes, the efficacy of psychoactive drugs, normal and abnormal perception and behavior, and other brain-related phenomena.[13] Individual mapping projects within the Allen Institute include atlases of gene activity for the pre-and postnatal developing mouse brain, adult human brain, midgestational human brain, nonhuman primate brain, mouse brain at different states of sleep and wakefulness, and atlases for the vast array of cell types in both mouse and human brains.

An underlying premise for creating brain atlases of gene expression is that the timing and tissue/cellular locations of gene activity create and support brain architecture and function. Gene activity controls which areas of the human brain expand during development compared to those brain areas that human share with other mammals. Gene activity also determines the cellular locations and abundance of the transporter proteins responsible for therapeutic, psychoactive drugs gaining access to nerve tissue. The mouse, macaque, and human brain atlas projects within the Allen Institute are ongoing endeavors with no specific endpoint. The goal is to continue improving the quality of detailed gene expression maps of the brains of these animals and to identify their usefulness for human health and well-being. Brain atlas data coupled with information and research tools gained from the BRAIN Initiative, the HBP, and other brain research projects may eventually help scientists come to understand the genetic basis for human cognition, consciousness, and the physical basis for humanness itself. Let us now consider tools that researchers use to explore the brain's inestimably complex structure and function.

Toolkit for Neuroscience Research and Clinical Neurobiology

A large and diverse toolkit exists for researchers to monitor, visualize, and manipulate brain activity. Technologies like EEG, MEG, and deep brain stimulation (DBS) have been around for quite a while. Others like fMRI and positron emission tomography (PET) began being used more recently. Optogenetics, calcium and voltage imaging, certain innovative neuromicroscopy methods, and laboratory-grown minibrains are strictly twenty-first-century technologies still undergoing revolutionary advancements. EEG, fMRI, and optogenetics are especially relevant to some of the ethical issues discussed in this chapter, so they are described briefly below. A discussion of living minibrains and associated ethical concerns appears in chapter 2 as a stem cell technology. More detailed information about fMRI, optogenetics, and minibrains and descriptions of other major neurotechnologies in neuroscientists' toolkits are in appendix 2.

EEG. First used on humans in 1924 by German psychiatrist, Hans Berger, EEG records and presents as a graphic picture, patterns of electrical activity in the brain. Activity patterns first measured by Berger are now called brain waves, which are rhythmic changes in electric impulses emanating from the brain (Medical Discoveries 2014; Johns Hopkins Medicine Health Library n.d.). Exactly why and how neurons synchronize their activity to produce brain waves is not fully understood, but the fact that they do and that EEGs record the waves allow for many clinical, commercial, and research applications. For example, Honda Motor Company uses EEG, near infrared spectroscopy (described below), and brain-computer interface technology to develop robots remotely controlled by human thought (Nakata 2009). By identifying abnormal patterns of brain waves, EEG helps diagnose epilepsy, schizophrenia, dyslexia, narcolepsy, stroke, Alzheimer's disease, and sleep disorders and aids in defining brain death. Although an EEG gives no information about the activity of a single neuron, its advantages in human brain research include noninvasiveness, the ability to detect changes in brain wave patterns occurring in timeframes of thousandths of a second, and its relative safety and low cost.

Modern encephalography employs about twenty external electrodes attached to the scalp. The electrodes detect brain waves and send the resulting electrical signals to an instrument called an electroencephalograph that presents the activity graphically as an electroencephalogram (EEG). EEGs present the highly, albeit imperfectly, synchronized rhythmic activity of the thousands or millions of neurons responsible for brain waves in a unit of frequency called a Hertz (Hz). One Hz equals one cycle per second, one cycle being the increasing, peaking, and decreasing of one wave of electrical activity, just as inhaling and exhaling represents one cycle of breathing. Researchers and clinicians identify at least six types of brain waves based on their frequency. For example, alpha waves occur in a regular rhythm with a frequency of 8 to 12 Hz in relaxed subjects with closed eyes, whereas beta waves have a frequency of 16 to 31 Hz and indicate active, alert, focused thinking or anxiety. Particular types of brain waves emanate from specific brain regions that may differ depending on the age of the subject. For example, delta waves (less than 4 Hz) come from an anterior (frontal) brain region in adults but from a posterior region in children.

fMRI and optogenetics. fMRI detects changes in oxygenated blood flow in the brain and uses increased blood flow as a proxy for neuronal activity. Other brain-scanning technologies including PET and near infrared spectroscopy are also based on blood flow and oxygen use by active neurons. fMRI has the advantages of noninvasiveness for the subject and ease of use for the experimenter (Devlin 2007). Used together, optogenetics and fMRI (opto-fMRI) is a powerful duo for discovering specific brain circuits driving normal and abnormal behaviors in living, wakeful animals. A research team from MIT, Harvard University, Boston University, and Tufts University pioneered the combined use of these two technologies in awake mice (Desai et al. 2011). Optogenetics allows neuroscientists to use a highly focused light beam to activate specific nerve cells in an animal's brain. How they do this is described in appendix 2. Monitoring the entire brain by fMRI during optogenetic stimulation of a tiny spot in the brain reveals the extent and location of neural circuits involved in the response. "We can drive a neuron and see what networks are downstream of it, which is really powerful," said MIT neuroscientist Edward Boyden in an interview (Trafton 2011). The ways that different brain regions talk to each other are discovered using opto-fMRI. Since probing brain function with optogenetics necessitates genetically engineering brain cells, opto-fMRI research in humans is not ethically feasible; however, opto-fMRI studies in other mammals including nonhuman primates will certainly provide invaluable information about how human brain circuits are constructed and function. Already, opto-fMRI has helped scientists verify that the blood flow measurements detected by fMRI actually do reflect neuronal activity, a relationship that earlier had only been presumed. This is an important finding since it allows fMRI to be combined with neural stimulation techniques that *are* ethically justifiable in humans, such as DBS, in order to elucidate brain circuitry.

Can Science Explain and Understand the Human Mind and Consciousness?

In the absence of succinct, agreed-upon definitions of mind and consciousness, this is a difficult question to answer. From the point of view of modern neuroscience, the human mind, including its components of consciousness and the feeling of free will, is a product of complex biochemical and cellular activities. From these automatic, self-perpetuating activities of bits of matter emerges the human mind, much as the infinite variety of snowflake patterns emerge from physical properties of water molecules subjected to certain atmospheric conditions over time.

Both the US BRAIN Initiative and EU's HBP will integrate research on nonhuman and human brains. This concept is consistent with the accepted notion that mind emerged naturally via biological evolution from the biophysical components and activities of the nervous system. Charles Darwin noted that the human mind differs not in quality but only in degree from the minds of nonhuman animals. He expressed this concept in his book, *The Descent of Man, and Selection in Relation to Sex*: "All [animals] have the same senses, intuitions, and sensations—similar passions, affections, and emotions, even the more complex ones, such as jealousy, suspicion, emulation, gratitude, and magnanimity; they practice deceit and are revengeful; they are sometimes susceptible to ridicule, and even have a sense of humor; they feel wonder and curiosity; they possess the same faculties of imitation, attention, deliberation, choice, memory, imagination, the association of ideas, and reason, though in very different degrees" (1871, 48–

49). Modern experimental, comparative psychology demonstrates the truth of mental continuity during evolution. For example, comparative psychologist Jeffrey Katz at Auburn University and his colleagues described results of experiments to assess the ability to learn abstract concepts in pigeons, capuchin monkeys, and humans. Their results show continuity between the levels of learning abilities in these different species, that is, differences only in degree, not in kind, just as Darwin observed nearly 150 years earlier (Schmidtke et al. 2013, 141–55).

Since cellular activity is based on the physical properties of matter, a comprehensive understanding of cellular activity in the brain and other parts of the nervous system provides an understanding of the human mind. By similar reasoning, since science deals with the properties and behavior of matter, it seems that science ultimately can explain mind and consciousness. But striving to understand a complex phenomenon by reducing it to smaller and smaller components is not without danger. The risk in using solely a reductionist approach when attempting to understand life is elegantly expressed by author, biochemist, and Nobel Laureate Albert Szent-Györgyi: "Led by a desire to understand life[,] I went from bacteria to molecules, from molecules to electrons. The story had its irony, for molecules and electrons have no life at all. On my way life ran out between my fingers" (1972, 7).

Time will show whether the reductionist approach of molecular and cellular biology will ultimately yield a comprehensively satisfying understanding of the human mind. My intuition is that collaboration between science and the humanities, with each on an equal footing, will be needed before the human mind can fully understand itself. A good starting place is the promotion of mutual respect between scientists, artists, and humanities sages for their different ways of knowing.

Brains and Minds: Human Values

The year 2002 marks the beginning of the current era of neuroethics. The remainder of this chapter is devoted to that subject, for which definitions abound. In 2002, several conferences were held on the subject, and the first working definition of the field was coined: "The examination of what is right and wrong, good and bad about the treatment of, perfection of, or unwelcome invasion of and worrisome manipulation of the human brain" (Safire, 2002). Neuroscientist and philosopher Adina Roskies identifies two major branches of neuroethics, the *ethics of neuroscience* and the *neuroscience of ethics* (2002).[14] The ethics of neuroscience includes issues related to neuroscience research and also the ethical, legal, and social consequences of findings in neuroscience (Greely et al. 2016), particularly pertaining to the human brain. The neuroscience of ethics addresses how human brain structure and function influence our moral systems. Three subjects occupy the rest of the chapter: (1) how the BRAIN Initiative and the EU's HBP address neuroethics, (2) examples of issues in the ethics of neuroscience, and (3) some issues related to the neuroscience of ethics.

The BRAIN Initiative and the HBP: Attention to Ethical Issues

When President Obama announced the BRAIN Initiative, he also directed his Presidential Commission for the Study of Bioethical Issues to examine and report on the ethical issues raised by the project. Volume 1 of the commission's two-volume report, *Gray Matters: Integrative Approaches for Neuroscience, Ethics, and Society*, was released in May 2014 (Presidential Commission 2014), and volume 2, *Gray Matters: Topics at the Intersection of Neuroscience, Ethics, and Society*, in March 2015 (Presidential Commission 2015). In the first report, the eleven-member commission examined the need and means to integrate ethical thinking and training into all levels and fields of the scientific endeavor but particularly into neuroscience research from its earliest planning stages. The commission stressed the importance of ethics training for future scientists even prior to undergraduate school and continuing through graduate programs and on into their professional lives.

The commission also urged personnel within research institutions and universities, public and private funders, policy makers, educators, and others including the general public to engage with each other and to develop a familiarity with science and ethics. The commission writes: "The goal is not to make scientists into ethicists or vice versa. It is to cultivate sufficient fluency for productive discussion and collaboration in a multidisciplinary endeavor that can include biological scientists, clinicians, historians, lawyers, philosophers, physical scientists, social scientists, and theologians, among others" (2014, 25).

Four recommendations conclude the report: "[1] Integrate ethics early and explicitly throughout research[,] . . . [2] evaluate existing and innovative approaches to ethics integration[,] . . . [3] integrate ethics and science through education at all levels[,] . . . [4] explicitly include ethical perspectives on advisory and review bodies" (2014, 25–30). Meeting these recommendations ought not to be burdensome to science educators or scientists. In fact, when ethical thinking pervades the entire scientific process, science itself will be one of the greatest beneficiaries.

In volume 2, the commission focused on three controversial issues—cognitive enhancement, consent capacity, and neuroscience and the legal system (Presidential Commission 2015). It made specific recommendations in each area. Its first recommendation relative to improving neural health is to "support research on existing low-technology strategies, such as healthy diet, adequate exercise and sleep, lead paint abatement, high-quality educational opportunities, and toxin-free workplaces and housing" (46). In other recommendations in the cognitive enhancement category, it implores policy makers and other stakeholders to "ensure that access to beneficial, safe, effective, and morally acceptable novel neural modifiers to augment or enhance neural function is equitable so as not to compound or exacerbate social and economic inequities" (5).

President Obama's Commission on Bioethics was charged with advising the president on ethical implications of scientific and technological developments. Presidents Clinton and George W. Bush had similar ethics advisory groups, which exist at the pleasure of individual presidents. Not surprisingly, positions on issues and advice of these groups reflect biases of its leader(s) and other members. How ethics commissions for post-Obama presidencies view rapidly advancing scientific fields such as neuroscience, genome-editing, and synthetic biology bears close watching.[15] As of mid-2019, the Trump administration had not appointed a Bioethics Commission.

Unlike the HGP, which included a funded ELSI component, the current BRAIN Initiative does not formally fund research into ethical issues likely to emerge from the project. Some bioethicists see this as an unfortunate omission (Green 2014; Shanks 2013). Instead, a Neuroethics Work Group, comprised of nine unpaid volunteers including some neuroethics experts, was created in 2014 (Greely et al. 2016). This work group informs the NIH about ethical issues in its proposed areas of BRAIN Initiative research and also helps individual investigators identify and address ethical issues arising in their research.[16]

By contrast, the EU's HBP commits to supporting research into long-term ethical and social implications of the project's activities and ensuring that HBP researchers follow established principles of research ethics, particularly research involving human volunteers, animals, and clinical data collection.[17] To be fair, the BRAIN Initiative does not ignore the ethical dimensions of its research. The NSF and the NIH are major funders of the BRAIN Initiative, and both have well-established ethical guidelines for research performed under their auspices. But the BRAIN Initiative does differ from the HGP and the HBP in lacking a formal subdivision devoted to ethical and societal implications.

The BRAIN Initiative and HBP also differ in their sources of funding, a point with potential ethical implications. Virtually half of the funding for the BRAIN Initiative comes from the US military's DARPA (Horgan 2013). One possible problem with this is that some findings may become classified and prevented from being shared with the scientific community at large. Open and truthful communication within the scientific community is essential for the scientific endeavor at large. Another ethical concern is the possible use to which scientists' findings may be put. Although DARPA claims an interest in brain research mainly to develop better treatments for injured soldiers, it is also interested in using neurotechnologies to enhance cognitive function in friendly soldiers and to disable or monitor the minds of foes (Horgan 2013).

Whether scientists should be aware of all the uses their funders have in mind for their research is a question deserving open debate. In defense of its involvement in the BRAIN Initiative, Col. Geoffrey Ling, director of DARPA's Biological Technologies Office (BTO), insists that every program in BTO has a bioethics advisory board (Reardon 2015b). This is laudable even though the effectiveness and impartiality of self-oversight is difficult to estimate. By contrast, a majority of HBP funding comes from the European Commission, the executive body of the EU. Although the European Commission has an interest in the EU's common defense, its main interest is in promoting legislation to facilitate cooperation among member nations' military establishments rather than spearheading specific research projects with direct military applications (European Commission 2015). The remainder of the HBP's funding will be raised by European scientists from their nations' governments and institutions (Frégnac and Laurent 2014).

The Ethics of Neuroscience

Martha Farah, director of the Center for Neuroscience and Society at the University of Pennsylvania, identifies three major areas of advancement in neuroscience occurring after the ancient Sumerians documented mood alterations induced by ingesting poppy plant products: basic neuroscientific knowledge, medical applications, and nonmedical applications (Farah

2011, 762–63). Burgeoning nonmedical applications of neuroscience in the twenty-first century set it apart from other modern biotechnologies, such as human stem cell therapy, genetic engineering, and cloning. In this ethics of neuroscience section, we first consider ethical issues raised by some of these nonmedical applications of modern neuroscience that occur in spaces ranging from marketplace and battlefield to courtroom and classroom. Then we turn to the medical and nonmedical applications of implanted brain devices. Finally, we consider neuroscience research with broad societal implications including moral responsibility and criminal justice, the privacy and autonomy of the mind, and one's religion. Also discussed are new generations of psychoactive pharmaceuticals with both medical and nonmedical applications that are on the near horizon. Drugs for memory enhancement and pleasure/happiness modulation and their associated ethical issues received substantial attention in an earlier volume (Bradley 2013a, 237–48).

Reading the brain/mind for profit and other reasons. Natural selection gave us a talent for discerning another person's thoughts or intentions at a glance. We do this every day through body reading. Our brains process the "look" in others' eyes, their facial expression, voice inflections, and body language to tell us the behavior that is most likely to benefit us—an amorous move, a fast exit from the scene, combat, conversation, helpfulness, empathy, shunning, attention, and others. Sometimes we misread the other person. This may reflect the level of our reading talent, or it may reflect the other person's talent for deception. It is easy to imagine how deception had survival value for our deepest ancestors. Imagine making an animal or another human feel safe in your presence until you get close enough to hurl the lethal stone or spear.

Neurotechnology now explores ways to look past eyes and furrowed foreheads into the brain itself for information about what a person feels or believes, what he remembers, and even what she may intend. So far, most of our explorations in brain reading occur in the laboratory using fMRI brain scanning in highly controlled situations (Rose 2016). Yet, some applications are beginning to trickle out into the marketplace. Neuromarketing and lie detection are two of these.

Neuromarketing uses neurotechnology like EEG and fMRI to relate unconscious brain activity to one's attitudes, likes, and dislikes. Neuromarketers use these data to discern consumer preferences in the marketplace and to coax the subconscious mind into prompting market-friendly behavior. The presumption is that brain activity patterns in groups of test individuals who are exposed to products, advertisements, or other marketing related objects reflect their states of liking and wanting and also indicate the preferences of other, untested consumers. Neuromarketing firms already in business include NeuroSense, EmSense, NeuroFocus, and MindLab. Big corporations that now engage neuromarketing services include Disney, Google, Ebay, Pepsico, Microsoft, CBS, Frito-Lay, and Hyundai (Burkitt 2009; N. Singer 2010).

It is still largely unknown how effective neuromarketing is for increasing sales. Research into this field of brain analysis is subject to many variables. These include previous life experiences of untested consumers, responses to environmental surroundings (e.g., a brick and mortar store versus a home computer screen), one's diet, sleep habits, and exercise regimes. Any one of these may affect product preferences of a shopper on a given day. Thus, distortion and unjustified interpretations of basic neuroscience research are major concerns in the current neuromarketing milieu.

Neuromarketing's expanding use also raises issues of brain privacy and of the appropriate use of brain science and technologies in the commercial sector (Fischbach and Mindes 2011). Should the use of brain science by corporations to influence the subconscious minds of consumers be regulated? Jeff Chester, executive director of the Center for Digital Democracy, thinks so, especially if neuromarketing methods can affect individuals' subconscious thoughts without their knowledge. Some consumer advocates worry that "brainwashing" could turn consumers into shopping robots. Emory University bioethicist, Peter Wolpe, takes a different view. He is skeptical that triggering certain brain activities can actually override a person's own control over behavior (Fischbach and Mindes 2011). On the other hand, most of us know obsessive shoppers who display virtually no control over buying behavior even in the absence of sophisticated neuromarketing manipulation of their brains. One can imagine future neuromarketing techniques that aim to produce compulsive buying by otherwise conservative shoppers.

Meanwhile, brain-imaging explorations into "reading" minds and research into the wireless manipulation of neural circuits with the ultimate possibility of influencing the subconscious mind continue. Nonclinical implications of this research extend beyond the marketplace to include politics, romance, religion, and the courtroom. It is one thing to enlist neuroscience to increase sales of toothpaste or smartphones but quite another to apply it to influencing voters and lovers or spying on and molding the thoughts of flocks of churchgoers.

Wireless suggestions targeting the brains of the electorate, a person met in the park, or a wayward congregation may seem unlikely anytime soon but then again maybe not. In 2007, fMRI studies probed voters' thoughts about presidential candidates (Iacoboni et al. 2007). In 2009, fMRI research indicated that judging the truth of religious propositions and nonreligious propositions activate different brain regions (Harris et al. 2009), and in 2015, scientists in China and the United States used fMRI to document brain activities associated with being "in-love" and experiencing "ended-love" (Song et al. 2015).

In 2017, researchers at Caltech and the Howard Hughes Medical Institute in Pasadena reported identifying a simple neural code that monkeys use for face recognition. By analyzing the electrical activity from just 205 brain neurons, researchers were able to reconstruct human facial images that monkeys had just viewed. They could also predict the neuronal activity that would occur if monkeys were presented with the image of a particular human face that they had not previously seen (Chang and Tsao 2017). What remains to be determined is whether humans use a similar system for face recognition and how the system may apply to recognizing other objects, scenes, and the world in general. As neural circuits involved in political preference, love, religiosity, perceiving the world, and other human activities become better defined, possibilities for manipulating these circuits also increase. Whether technologies for mind reading and thought control are five years, fifty-five years, or further down the road, now is the time for engaged dialogue on how such technologies ought to unfold.

A polygraph test has been the standard means of lie detection since the 1920s. The test measures several indicators of deceit including breath and heart rate and skin conductance. The reliability of polygraphy for lie detection is iffy because anxiety and other states of mind affect subjects' physiology similarly to deception, and some subjects can control their physiological indicators so as to manipulate the test.

Laboratory research during the past decade suggests that fMRI scans of brain activity may be more reliable than polygraphy at lie detection. Still there are problems with high-tech lie detection. Performing an fMRI requires the subject to cooperate with absolute stillness during brain scanning. So far fMRI-based lie detection is not yet validated in real-world settings outside the laboratory nor is it admissible evidence in the courtroom (Haynes 2011, 9–10; Fischbach and Mindes 2011, 364). There is also danger of overinterpretation or misinterpretation of brain scan data that could lead to unjustified or erroneous conclusions about the mental state of the subject. Despite these uncertainties, some companies like Cephos and No Lie MRI already market fMRI scanning for lie detection and personnel evaluation (Fischbach and Mindes 2011, 364).

The uses to which fMRI data can be put are not regulated. This leaves the door open for dubious applications by health and life insurance companies, prospective employers, professional interrogators, screening committees for admittance into private organization and schools, or the spouse who suspects adultery. For all of these potential applications of brain scans, there is not only the significant problem of interpreting the data but the larger issue of brain privacy.

Neuroethicist Judy Illes believes that brain data should receive the same protection that genetic data now receives under the Genetic Information Nondiscrimination Act of 2008 (Illes and Racine 2005).[18] Other ethicists imagine a time when a "brain search warrant" may be required before one's thoughts and memories can be legally probed (Fischbach and Mindes 2011, 364). The need for legislation to protect brain privacy may be nearer than most of us realize. In 2015, researchers at Yale School of Medicine reported using fMRI data from 126 participants in the Human Connectome Project (Finn et al. 2015) to examine activity in 268 different brain regions. By mapping the strength and locations of neural activity, the researchers discovered that every individual has a "brain fingerprint." Each brain had a three-dimensional activity pattern unique to the individual, even under a variety of conditions. The researchers were even able to correlate individuals' unique activity patterns with problem-solving ability. This suggests a way to assess intelligence from an fMRI brain scan. Other researchers have collected data indicating that fMRI scans can glean intentions of the person and the identity of objects a person is thinking about (Rose 2016). The need to protect brain privacy is obvious; exactly how to do so is less obvious.

Brains for battle or promoting peace? In April 2015, neuroscientists, sociologists, psychologists, and historians met in Paris for a remarkable conference titled "The Brains that Pull the Triggers" organized by the Paris Institute of Advanced Studies. It was remarkable not only for its interdisciplinary nature but for the question it tackled—how do psychologically stable, normally nonviolent persons overcome the normal human aversion to killing other humans when they find themselves in extreme circumstances such as participating in war, perpetrating genocide or terrorism, or engaging in other violence against humanity? One conferee reported a post-WWII study of hundreds of untrained German reservists involved in massacring tens of thousands of Jews in Poland who, when given the opportunity to opt out of the actual killing, did so only at a frequency of one in ten. What makes otherwise normal people able to commit atrocities under certain conditions? There is yet no definitive answer to this important question, but research on the brain circuitry underlying human behavior may give insights.

Based on recent research, one neurosurgeon at the conference proposed that ordinary persons can become repetitive killers when emotion-associated brain regions, which normally keep behavior within culturally acceptable bounds, are somehow freed from ideology-associated, cognitive brain regions ("The Kill Switch" 2015). A similar disconnection between cognitive and emotional brain regions could explain how otherwise normal persons can commit atrocities simply by being ordered to do so, even in the absence of an ideology-driven motive. Anticipating a future when brain circuits controlling killing behavior, nationalism, xenophobia, aggressiveness, and territoriality are understood, what uses will be made of such knowledge?

DARPA aims to develop wireless devices to repair brain damage and restore memory loss in soldiers. Eventual wireless control of traits that make for obedient and aggressive soldiers, docile and fearful enemies, and passive prisoners is not far-fetched. How should we prepare for such a future? Perhaps we should demand that equal attention be given to understanding and nurturing brains adept at living peacefully with their fellow brains.

Gender-targeted neuroscience research: benefits and risks. Several studies comparing human female and male brains claim biological reasons for gender-based differences in behavior, ways of thinking, abilities, or aptitudes. But none of these studies is exempt from criticism of experimental design or alternative interpretations.[19] One important question for many women that neuroscience may answer is whether postpartum depression and psychosis are qualitatively different from depression and psychoses experienced by men and non-post-partum females. An answer to this question has implications for criminal infanticide law, health insurance policy interpretation, mental health parity law, and disability discrimination law (Tovino 2011).

When a lactating mother kills a newly born child, should she be tried for murder or for the female-specific criminal offense of infanticide? Because some postpartum women suffer from impaired thought and judgement, Canada and Zimbabwe courts apply lesser sentences to women convicted of infanticide than to persons convicted of murder. In the United States, no state has a formal female-specific infanticide provision; however, in a few child-murder cases, evidence of postpartum psychosis in support of an insanity defense resulted in nonguilty judgements or in lesser sentences than could otherwise have been dealt (Tovino 2011, 704–5).

Whether a female-specific infanticide provision should be broadly established in US law is controversial. Also at issue is whether postpartum depression should be classified as a mental or physical illness or considered a disability. How these questions are answered will influence health insurance coverage, which is generally greater for physical than for mental illnesses. How postpartum depression is classified also affects employment and workplace environments in which discrimination is prohibited on the basis of disabilities.

Increased neuroscience research into the causes and forms of depression and psychoses could also provide important benefits, especially for women. These include improved health insurance benefits, lesser criminal charges, and increased protection under the 1990 federal Americans with Disabilities Act. However, Tovino warns that if society comes to understand postpartum depression and psychoses as conditions of neurological impairment, revival of outdated stereotypes and sex-based discrimination in diverse social arenas could result (2011). In this vein, the next section on neuroscience and education describes how misinterpreted and misused neuroscience can lead to perverted educational experiences based on gender.

Neuroscience and education: ethical conundrums. Products of neuroscience already have a presence in the classroom in the form of chemical cognitive enhancers like methylphenidate (Ritalin) and the levoamphetamine-dextroamphetamine mixture (Adderall) that help focus one's attention and also facilitate a person's ability to manipulate information. Both drugs are commonly prescribed to treat attention deficit hyperactivity disorder. Adderall is sometimes prescribed for narcoleptics. Both circulate widely on university, college, and high school campuses as they pass from the prescription recipient to the casual user. Selling one's medically prescribed pill can be very lucrative. A fifty-cent pill of prescribed stimulant can fetch up to fifteen dollars as a nonprescribed pill. Surveys in the late 1990s and early 2000s showed that, depending upon the campus, 2 to 20 percent of students have used nonprescription Ritalin (Kapner 2003). The drug's popularity is due largely to the widely held belief that taking a few pills before a test will improve course grades. In fact, studies show that such drug use among students performing poorly academically due to bad study habits does not result in better outcomes (Arria and DuPont 2010). Put bluntly, a few days of Ritalin or Adderall do not compensate for a semester of partying at the expense of study. Diversion of prescription stimulants to nonmedical, recreational users also risks addiction, unknown adverse side effects, and ramifications of illegality.

As new pharmaceuticals and other neurotechnologies to enhance memory, attention, and other cognitive traits come down the pike, it is incumbent upon all citizens, especially parents and students, to consider whether and how they ought to be used in educational settings. The legal and safety status of future cognitive enhancers may differ from currently used ones. So judgements should be made independently of legal and safety issues associated with currently used neuroenhancers. The nonmedical use of cognitive enhancers to improve cognitive performance in school and the workplace raise ethical issues of fairness since access may exacerbate social and economic inequalities when the drugs are not readily available to everybody. Here, we consider the importance of the relationship between young students and authority figures such as parents, physicians, and school officials in the context of nonmedical cognitive enhancement.

Several educators with research specialties in educational neuroscience make a distinction between *designing* children and *raising* children (Stein et al. 2011).[20] Designing children entails changing dispositions or moods with pharmaceuticals or by other physical means in the absence of children's participation in the decision to do so or their understanding about what may be gained or lost in the process. These educators argue that designing children is unethical because it violates the children's right to participate in their own development. They cite the United Nations Convention on the Rights of the Child (United Nations 1989) in support of their view.[21]

By contrast, raising children entails altering dispositions and behaviors mainly by using shared languages and values. Raising a child does not preclude using pharmaceuticals or other products of neuroscience to alter a child's mood or disposition as it relates to educational experience, but it does preclude doing so without involvement of and reasonable understanding by the child. That it is better to raise than to design one's child may seem like a "no-brainer" to most parents and other authority figures, but the apparent ease with which ADHD diagnoses and neuroenhancer prescriptions can be obtained from some US physicians (Center for Mental Health in Schools, 2014) is alarming when one considers the memory-enhancing pharmaceuticals, brain implants, brain-computer interfacing, and other technologies that may soon be marketed for classroom application. Stimulatory implants that enhance cognitive brain functions are an example. Who ought to be candidates to receive the implants, and who ought to control the stimulatory signals? Should the signals come from on-body power packs or via wireless transmitters? If from the latter, who should have access to the transmitters?

As educational neuroscience gains momentum, there is also danger for abuse or misuse of brain research findings to promote education policies that may be socially destructive for individuals and society at large. An unfortunate example of this is already recorded from Louisiana. There a female junior high school student filed for an injunction to prevent her school from implementing a gender-segregated teaching policy based on the 2005 book *Why Gender Matters* by psychologist/physician Leonard Sax. The book claimed scientifically documented differences between male and female human brains that justify disparate treatment of girls and boys in educational settings (Brown and McCormick 2011). The injunction motion charged that implementation of the planned teaching policies would violate the prohibition on sex discrimination set out in Title IX and the Equal Protection Clause of the Constitution. According to the court filing elements of the proposed teaching policies extracted directly from Sax's book include:

1. Girls have more sensitive hearing than boys. Thus, teachers should not raise their voices at girls and must maintain quiet classrooms, as girls are easily distracted by noises. Conversely, teachers should yell at boys, because of their lack of hearing sensitivity.
2. Because of biological differences in the brain, boys need to practice pursuing and killing prey, while girls need to practice taking care of babies.
3. Girls need real-world applications to understand math, while boys understand and enjoy math theory. Girls understand number theory better when they can count flower petals or segments of artichokes, for instance, to make the theory concrete.
4. Literature teachers should not ask boys about a character's emotions and should only focus on what the characters actually did. But teachers should focus on a character's emotions in teaching literature to girls.
5. Teachers should smile at girls and look them in the eye. However, teachers should not look boys directly in the eye and should not smile.
6. "Anomalous males"—boys who like to read, who do not enjoy competitive sports or rough-and-tumble play, and who do not have a lot

of close male friends—should be firmly disciplined, should spend time with "normal males," and should be made to play competitive sports.

7. Boys should receive strict, authoritarian discipline, and boys respond best to power assertion. Boys can be spanked. Girls must never be spanked. Girls should be disciplined by appeals to their empathy.

According to the court filing, the gender-segregated policies were developed without input from students or parents and no coeducational alternative to the proposed program was offered.[22] Thus, students would be required to participate or opt out of public education. Later the same day as the injunction request was being filed, the school board defendants abandoned their plans to implement the gender-segregated teaching policies, rendering the motion moot.[23]

Good quality research does suggest that gender-based differences in human brain development, structure, and function exist (Cosgrove et al. 2007; Smith et al. 2014; McCarthy 2015). There is no reason to deny such findings. Then again, we must remember that human gender itself comes in more than two forms and that complex traits such as behavior, aptitude, and optimum learning styles result from complex interactions between multiple genes, physiological factors, and environmental influences. These influences differ across broad spectrums that do not necessarily respect gender-based demarcation lines. In addition, there are many non-gender-biased spectrums of abilities and propensities relevant for education. Finally, the monocular use of cherry-picked brain studies to justify sex-segregated education programs ignores benefits derived from diverse classrooms and the ability of good teachers to nurture a diversity of individuals within the same classroom.

In 2009, in response to misuses of science in education like the one in Louisiana, the forty-thousand-member Society for Neuroscience held a Neuroscience Research Education Summit. The meeting examined ways that neuroscience research can help contribute to successful educational strategies and help ensure its appropriate and responsible use in education. As brain-scanning technologies advance and the size of databases increase, researchers, journalists, policy makers, and the citizenry at large must be ready to counter the misuse of data that discriminates based on gender, ethnicity, or other aspects of human diversity.

Brain implants. In 2015, the FDA approved the implanting of microelectrodes into the brain to treat essential tremor and Parkinson's disease by DBS. Clinical DBS trials looking into therapy for obsessive compulsive behavior, depression, and other disorders of mood, behavior, and thought are underway or forthcoming. During the next decade, research sponsored by the BRAIN Initiative and other brain research projects will lead to ever-increasing numbers of studies and subsequent clinical trials with brain implants.

Potential therapeutic benefits for persons with brain disorders are enormous, but the research required to achieve that potential raises ethical concerns. First, we cannot think of brain implants in the same way as we now think of other body implants such as cardiac pacemakers, cochlear implants, and insulin pumps. The reason is that we do not understand the functioning of the brain as well as we do the workings of the heart, ear, or insulin system. There will likely be unintended consequences of electrical brain stimulation beyond the hoped-for therapeutic results. For example, some patients receiving DBS therapy for Parkinson's disease experience changes in their personal actions regarding gambling and sexual behavior (Underwood 2015). The risks of unknown side effects to DBS and other invasive brain manipulations need to be weighed against the expected benefits, and those risks must be clearly communicated to trial participants as part of the informed consent process.

Another risk for brain implant trial participants is uncertainty about the continued availability of the technology after the trial. If an experimental device is beneficial to some trial participants but not to enough of them to warrant mass manufacture of the device, the helped participants may not have replacement parts available to them in the future. There may also be instances where a company making a device discontinues its manufacture or is bought out by another business that discontinues the item. Or the cost of continuing to use a device after a trial ends may be prohibitive for some participants who were significantly helped. Again, these possibilities must be communicated to prospective trial participants.

One researcher at a US medical institution kept an independent trial of DBS for depression going so that participants benefiting from the technology have a portion of the cost covered by Medicare or private insurance. Were the trial to end, that coverage would disappear and some participants would need to give up the treatment with deleterious consequences (Underwood 2015). From some tax payers' points of view, it is unethical for a researcher to maintain a clinical trial simply so that some participants can afford to continue being treated. But from the point of view of the researcher, it is unethical to leave patients high and dry by ending the trial. Uncertainties and ambiguities like these complicate the design, staging, and informed consent for clinical trials testing high-end, therapeutic technologies not easily made marketable and affordable.

Whether children or young adults with still-developing brains should be candidates for DBS or clinical trials, whether users of DBS technology ought to be able to self-modulate the stimulatory signals entering the brain, and how best to obtain informed consent for research subjects with impaired consent capacities (Matthews et al. 2011) are also thorny issues. President Obama's Commission for the Study of Bioethical Issues recommended including participants with impaired consent capacity, as currently defined, in neuroscience research in order to enhance progress in understanding and ameliorating neurological disorders and psychiatric conditions. The commission also stressed that more research is needed to better understand consent capacity and ensure adequate protection for participants whose unique vulnerabilities, desperation, and emotional states may affect their decision-making abilities (Presidential Commission 2015, 6). Finally, the prospect of neuromodulation, whether via brain implants, neurosurgery, or other technologies, raises issues of how to define and maintain patient autonomy, personal identity, and authenticity. That these may be viewed differently by the patient, by persons emotionally close to the patient, and by researchers and physicians makes informed consent for such procedures a complex issue (Woopen 2012).

Neuroessentialism: Moral responsibility, criminal justice, and mental illness. Neuroessentialism is the notion that our brains contain who we really are, that the self with which we identify, including our prejudices and biases, resides in our brains (Roskies 2002; Reiner 2011). Neurobiologist and neuroethicist Peter Reiner identifies three points of view regarding neuroessentialist thinking (Reiner 2011). The *hard neuroessentialist* believes that all human behavior can ultimately be explained on the basis of brain neural circuitry. The *soft neuroessentialist* acknowledges that the brain is the seat of human behavior but suspects that this is not the whole story. The soft neuroessentialist may feel that some human behavior derives from nonmaterial phenomena or she may question science's ability to reduce all of human behavior to the chemistry and physics of a biological organ. Finally, the *neuroessentialist naïf* has virtually no exposure to modern neuroscience and does not contemplate the source of human behavior. Whether neuroessentialism is correct is beside the point. What is noteworthy, according to Reiner, is that an increasing proportion of the public is adopting a neuroessentialist view of human personality and behavior. This is in the wake of accelerating coverage of neuroscience by the popular media.

What are likely social outcomes from a rise in neuroessentialism? Society attributing a greatly decreased moral responsibility for an individual's actions would profoundly affect the criminal justice system. As neuroscientific data accumulate linking human behavior, including criminal behavior, with the architecture of brain circuitry and neurotransmitter levels, the perceived role of free will in human actions may diminish. In turn, the degree of moral responsibility we can justly assign to criminals for their actions and the levels and appropriateness of retribution deemed fair in sentencing may change. Neuroscience enters the courtroom in other ways as well.

Two decades ago, researchers reported that PET scans of the brains of murderers can distinguish between those who murdered impulsively versus those who acted with premeditation (Raine et al. 1998). Since the risk of a repeated murder is greater for the latter, PET scans may guide sentencing and parole decisions. Brain imaging may also play a future role in jury selection and evaluation of testimony. Courtroom lawyers attempt to eliminate potential jurors with biases that may preclude open-minded consideration of evidence in a particular case. Currently the prosecutor and defense lawyers do this through oral questioning. Jury selection can be difficult because potential jurors may harbor unconscious biases hard to detect, but such biases may be uncovered by fMRI brain scans. As the notion of neuroessentialism becomes more prevalent, increased use of neuroscience in the courtroom and in the justice system as a whole is likely to follow. This may please those who question premises of retributive justice to begin with but displease others who believe that crime is reduced by threat of punishment.

Increased neuroessentialist thinking may also alter views toward mental illness and addiction. Better understanding of the neurochemical bases of psychoses, neuroses, addictions, and other mental disorders may lead to viewing these conditions as similar to other diseases of the body and thereby help remove long-standing stigmas attached to mental conditions. However, some neuroethicists warn that a biological understanding of mental disorders could lead to stigmatizing persons with these conditions as "neurobiologically other" and foster discrimination against them (Buchman et al. 2010).

Brain-machine and brain-brain interfacing: A minefield of mind melds?

Are you a Trekkie? If so, you know that mind melding is a way for two brains to share the same consciousness. In *Star Trek* TV episodes and feature films, the phenomenon is a Vulcan mind meld. The Vulcan, often Spock, places his fingertips around the skull of a humanoid to create a telepathic link between the two individuals whose minds then meld into one.

Mind melding is now poised to move from science fiction into twenty-first-century reality. In August 2013, University of Washington researchers reported performing the first human-to-human mind meld (Al Jazeera 2013). In the experiment, Professor Rao sat in front of a computer wearing a cap with electrodes attached to an EEG machine. On the computer was a video game. Across campus sat Rao's colleague wearing a cap with a transcranial magnetic stimulation (TMS) coil placed over his left motor cortex, the region of the brain that controls movement in the right hand. When the video game required Rao to lift his right hand to punch a key to fire a virtual cannon, Rao simply imagined lifting his hand but did not move it. The EEG machine detected those brain waves associated with Rao's thinking about moving his hand and transmitted the resultant EEG signals via the Internet to his colleague's computer. From there the signal entered the colleague's brain via the TMS coil. The eerie result was an involuntary movement of Rao's colleague's right index finger to push the space bar on his computer, as though firing the cannon in Rao's video game. Although the experiment was not published, it is recorded in a video.[24] Similar experiments were performed earlier with rats and monkeys at Duke University. This work does not mean that all the thoughts of two persons can now be melded together into one consciousness, but it does point the way to a possible future when brain-computer interfaces may not only revolutionize treatments for stroke, dementia, and spinal cord injuries but also change our views of consciousness, human nature, and personal identity.

Consider some experimental results already obtained: brain wave signals from humans and monkeys controlling limb movement in "robots," brain wave signals transmitted from human to rat brain causing a motor response in the rat, transcontinental transmission of the word *hola* between human brains via computer-interfacing, ceremonial first kick opening the 2014 World Cup performed by a young paraplegic man outfitted with a thought-controlled (via EEG impulses) "robotic" exoskeleton, memory of learned behavior sent between the brains of two rats, and a microchip-based prosthesis for the portion of the brain that helps move memories from short-term into long-term storage.[25]

Science journalist Jerry Adler (2015) vividly describes the future, based on reasonable extrapolation from current brain-machine and brain-brain interfacing. Here are some possible future applications Adler foresees from these technologies:

1. Stroke victims and amputees may control robotic exoskeletons and prosthetic limbs by thought.
2. Fighter jet pilots may guide their planes by thought.
3. We may literally get inside the minds of our pets and other non-human animals.
4. Thousands or millions of human brains may wirelessly cooperate as a biological "super computer" and apply their collective wisdom to some of life's persistent questions and problems.
5. Criminals, battlefield warriors, and others may communicate wirelessly and silently with each other.
6. People may control mobile "robots" wirelessly from great distances.
7. One may implant thoughts or control behavior in other persons, such as parolees or terrorists.
8. Ultimately, depending upon discoveries about the nature of *mind*, one's mind may be uploaded to a computer or stored in durable material such as silicon indefinitely.
9. The contents of one's mind may ultimately be stored in "the cloud."

Ethical concerns related to these and similar possibilities are multitudinous. In the context of human-to-animal and human-to-human brain-to-brain interfacing (BTBI), specific concerns arise in the realms of privacy, cognitive enhancement, human agency, and individual identity. These possibilities are already under discussion by psychologists, neuroethicists, and neuroscientists (Trimper et al. 2014). Unfortunately, as of 2019, no regulations exist to guide informed consent or protect personal information gained from brain-computer interfacing or BTBI. As psychologist John Trimper observes, "the technology is outpacing the ethical discourse" (Adler 2015, 51). The reader may begin that ethical discourse here and now by considering some relevant questions and sharing them with family, friends, club members, work colleagues, and others:

1. If two or more persons are connected by BTBI and one of them commits a criminal act, are any of the others held responsible?

2. If a group of scientists or artists collaborate via BTBI and one of them becomes inspired to discover or create something great, do others in the BTBI group rightfully share in the profits or glory?

3. How might BTBI be used in educational settings? Study groups would take on a whole new meaning.

4. Should a person with an exceptional mind be allowed to charge money to mind meld with others? If so, should fees charged for sharing one's mind be regulated? In many realms of life, who you mind meld with rather than what you know may become of ultimate importance. Should one be compelled to share his ideas when the well-being of society may be compromised by not doing so?

5. Will BTBI put one in danger of becoming "infected" with the fears, prejudices, religious beliefs, and political persuasions of others' minds?

6. How will research in human-to-animal and human-to-human BTBI proceed safely and ethically?

Trimper and his coauthors (2014, 3) observe, "Ethical discourse around BTBI must keep pace with the advances in technology in order to prepare for possible life-changing implications of BTBI use. . . . In order to best prepare the public to understand the implications of BTBI, ethicists and scientists must work together to ensure that the technology is developed with the highest ethical standards."

Neuroscience and religion. French philosopher of science René Descartes (1596–1650) maintained that the universe is composed of two substances, matter and mind, an idea called substance dualism. Theologies positing an immaterial soul that survives the body after physical death hold beliefs that are compatible with substance dualism. As modern neuroscience has advanced our knowledge about how the brain's structure gives rise to its functions, adherents to substance dualism have decreased in number. Still, much about brain function remains to be explained. Some modern-day philosophers believe that our subjective experience of being conscious will never yield to a purely physical explanation.

But what if the prediction that the BRAIN Initiative will provide tools to ultimately understand the human mind at the cellular/biophysical level is realized? What if findings from the BRAIN Initiative, the HBP, and other neuroscience research projects finally provide a holistic understanding of the human mind at the cellular/biophysical level? For one thing, the concept of substance dualism will be dead. What would this mean for religions like

Christianity, Islam, and Hinduism that presume an immaterial soul or essence that survives the body after physical death to live on in another realm or to inhabit another body?

If the concepts of mind and soul are conflated and if the foundation of the mind is shown to be purely physical, logic seems to require abandonment of presumptions of an immaterial, eternal soul. How theologians will respond to such a development is hard to say. One option will be to distinguish the soul from the mind. But in that case, how would one experience an eternal existence in the absence of a mind? Alternatively, theological responses to scientific paradigm shifts wrought by Copernicus, Galileo, and Darwin may predict some short-term theological responses offered to refute the contemporary scientific explanation of mind/soul. Denial of the credibility of theologically inconvenient scientific findings has a five-hundred-year-long history.

The Neuroscience of Ethics: Defining Mind and Morality

How might our conceptions of what the mind is and where it is located affect how we decide to use certain neurotechnologies? How might forthcoming discoveries about the mind's intuitions regarding good and evil affect our confidence in longstanding systems of morality? These are questions for the branch of neuroethics called the *neuroscience of ethics*. Neuroethicist Neil Levy explains how neuroethics differs from other branches of applied ethics by reacting back upon itself via the neuroscience of ethics: "The neuroscience of ethics will help us to forge the very tools we shall need to make progress on the ethics of neuroscience. . . . It occupies a pivotal position, casting light upon human agency, freedom and choice, and upon rationality. It will help us to reflect on what we are, and offer us guidance as we attempt to shape a future in which we can flourish" (2007, 2). Now, we will take a brief foray into the neuroscience of ethics, beginning with a consideration of the location of the mind.

Why where the mind is matters. Earlier we noted the view that mind is a product of the physical/biological activity of the brain, a view that underlies most neuroscience research including the objectives of the BRAIN Initiative and the HBP. This view, called the *identity thesis* for the mind, asserts that minds are simply functioning brains, that the human mind is located and contained wholly within the skull. The notion that mind emerges from brain activity is consistent with this view. The important thing is that the geographic seat of the mind is the brain. Competing with this view are at least two others: the *extended mind hypothesis* and the *embedded cognition thesis* (N. Levy 2011).

The extended mind hypothesis holds that a person's mind extends beyond the brain and even beyond the body. Proponents of the extended mind suggest that external elements used for memory or thinking include: language, information-containing notebooks, computers, diagrams, paper and pencil used with numerals and other mathematical symbols, slide rules, Scrabble game tiles and trays, and numerous other features of culture that qualify as mental elements just as well as do the internal brain components involved in remembering, calculating, planning, or thinking (Clark and Chalmers 1998; N. Levy 2011). The extended mind hypothesis does not deny that many crucial elements of mind reside in the brain. It simply maintains that additional important elements of the mind are external to the skull. Proponents of the identity thesis are sometimes called internalists and proponents of the extended mind hypothesis externalists.

The embedded cognition thesis is a compromised melding of these two different views of mind. Embedded cognition holds that the bona fide mind resides in the brain but acknowledges that our cognition is intimately tied to, even dependent upon, an external scaffolding, elements in our environment that are essential for us to think and live the mental lives that we do. Actually, few internalists would deny this.

Why does it matter whether you are an internalist or an externalist when it comes to neuroethics? One of the looming ethical issues for neuroscience is neural enhancement, the boosting of cognition or other brain functions through interventions that affect mental processes. Such interventions may act directly on the brain as with pharmaceuticals or DBS, or they may involve external devices as with brain-computer interfaces (BCI). In the case of psychoactive drugs, both internalists and externalists would view the interventions as acting directly on the mind. In the case of BCI devices, only externalists would consider the interventions to be acting unambiguously on the mind itself. Neuroethicist Neil Levy (2011, 292) poses questions that need rethinking in the context of the extended mind thesis:

Ought society use our new powers to intervene into the minds of agents?
Does the dependence of someone on external props affect their identity or their authenticity?
Is it wrong to alter human nature?
Ought human beings adopt an attitude of gratitude for the unforced gifts of nature, and not interfere with them?

Levy notes that these questions presuppose an internalist view of *mind*. That is, they suggest that the seat of our personal identities, our human nature, is the mind and that we have a choice between leaving our mind in its natural state and allowing it to depend upon artificial, external props. However, externalists argue that our mind has always been dependent upon external props. In fact, the extended mind thesis puts some of these "props" on equal footing with functional elements of the brain itself. If one accepts the extended mind thesis, then one must logically accept that we have long intervened in the workings of our mind. Education, calculators, diaries, books, and smart phones are extensions of mind for an externalist. External interventions like these, if viewed as mental elements, are not different in kind from psychopharmaceuticals or future technologies that may directly affect the brain. From the externalist's point of view, this does not mean that it is automatically all right to use every new mind-enhancing technology that comes along. Rather, it means that we should not reject their use solely on the grounds that they act directly on the brain.

Levy suggests assessing the ethical use of new technologies "on grounds which are not a mere reflection of internalist prejudices" (2011, 293). He suggests an *ethical parity principle* for assessing the goodness of emerging neurotechnologies: "Whether a particular means of altering cognition directly targets the brain/CNS [central nervous system] or the external scaffolding shouldn't make a difference to the assessment of its permissibility or advisability. Causal route is a difference that makes no difference; what matters is the result" (Levy 2011, 293). In other words, a psychoactive drug that accelerates the rate of information uptake by your brain and an external device like a smart phone that does the same thing ought to be given ethical parity, presuming other considerations such as safety are the same.

Another concern often voiced over the use of cognitive neuroenhancers like psychoactive drugs is that they may be available only to an elite group of persons and thereby lead to or exacerbate existing social inequalities. From the extended mind viewpoint, this concern comes more than a little too late since inequalities in external elements of the mind such as educational experiences, access to computers, and myriad elements of the environment have already created social inequalities. This viewpoint argues that effort and thought ought to go into changing the ways that nations construct and influence social environments and choices, not just into who should have access to the next generation of mood or cognition-enhancing drugs. Philosophers of the mind Andy Clark and David Chalmers, who first introduced the notion of an extended mind, write this about possible outcomes of the extended mind hypothesis: "As with any reconception of ourselves, this view will have significant consequences. There are obvious consequences for philosophical views of the mind and for the methodology of research in cognitive science, but there will also be effects in the moral and social domains. It may be, for example, that in some cases interfering with someone's environment will have the same moral significance as interfering with their person. And if the view is taken seriously, certain forms of social activity might be reconceived as less akin to communication and action, and as more akin to thought. In any case, once the hegemony of skin and skull is usurped, we may be able to see ourselves more truly as creatures of the world" (Clark and Chalmers 1998, 18).

Our notion of mind and where it is located will greatly influence what we consider to be the ethical use of certain neurotechnologies. Conversely, our use of neurotechnologies and the discoveries that they enable may profoundly influence our concepts of morality.

Evolution, neuroscience, and the roots of morality. Does human morality have an evolutionary origin, formed mainly or solely by natural selection? Alternatively, do some moral "laws" exist separately from humans so that they must either be discerned by (or revealed to) humans? Or do humans construct their moral systems (religious and nonreligious) over time, calling upon experience, intuition, tradition, and culture, independent of our species' evolutionary history? Where does our moral intuition come from and how reliable is it? Will neuroscience undermine or reinforce long-standing views about why some actions are morally abhorrent and others are virtuous?

These are weighty questions, and any one of them could be the subject of an entire book. Here, we briefly consider a controversy sparked by some scholars who suggest that most, if not all, of our moral systems may be irrational and unreliable because they are founded on intuition, which in turn is emotionally based. In opposition are those who argue that even though intuitions may be emotionally based, some intuitions are demonstrably reliable guides for making certain choices. In fact, neuroscientist Antonio Damasio makes the point that psychopaths are bad decision makers because they lack the feeling (intuitions) needed to successfully navigate the social world (Damasio 1994).

Ethicists widely agree that our intuitive feeling for what is right and wrong plays a crucial role in developing the moral systems by which we live our individual lives and formulate laws and other practices to guide our social behaviors. What is the link between intuition and moral systems or theories? According to Neil Levy (2007), a moral intuition based on emotion gives rise to a moral belief which we then codify into a moral principle using reason. We then use an assemblage of moral principles or a very broadly stated moral principle (e.g., the Golden Rule) as a system to guide our actions. At some point, we may feel that an action ordained by one of our moral principles is morally wrong. For example, if someone suggests that the utilitarian principle in which the best action is that which creates the greatest good or happiness for the greatest number of people be used to justify enslaving a few persons to serve a larger number of people, most of us spontaneously reject the suggestion as disgusting. In this way, moral intuition is crucial for evaluating and refining moral principles. So intuition may be important both at the beginning and at the end of the process that develops a moral system.

Is there reason to doubt the validity of our moral intuitions or the role of rationality in making moral judgments? There is no agreed upon answer to these questions. Certainly, we must acknowledge that not all of us share the same moral intuitions. For example, some feel a moral repugnance toward the prospect of human cloning, while others see cloning as a way to extend reproductive rights to otherwise infertile couples.

Consider this argument questioning the validity of moral intuition: Intuitions arise from the ways emotional regions of our brain are constructed and function and particularly the extent to which these brain regions engage in moral decision making.[26] Our brains, in turn, have arisen via evolutionary processes, primarily natural selection, that favor the survival of individuals for long enough to successfully reproduce. So if we intuitively feel that an action is good or that it is bad, it is because our distant ancestors who felt that way gained a reproductive edge over individuals who felt otherwise.

This is a naturalistic view of the origin of morality. Morality is viewed as arising from within ourselves via natural processes rather than existing outside the realm of humankind and nature. An evolutionary origin for morality presumes that a propensity for certain moral intuitions is heritable. Collaborations between geneticists, behavioral psychologists, and neuroscientists may be able to test this presumption during the next decade.

Skeptics of moral intuition argue that since external moral laws are not consulted during the evolutionary process, there is no reason to believe that our moral intuitions give reliable insights into true morality (P. Singer 2005). In this vein, Jonathan Haidt, a social psychologist at New York University, proposes a social intuitionist theory of moral judgments that denies reason any role in formulating moral judgments. He argues that morality is based on intuitive, emotion-based feelings and that what we believe to be reasoned responses to moral dilemmas are actually a post hoc rationalization of emotional responses (Haidt 2001). According to Haidt, there are at least six categories of moral intuitions upon which cultures develop their own systems of moral judgment: care/harm, fairness/cheating, liberty/oppression, loyalty/betrayal, authority/subversion, and sanctity/degradation. Accepting that these categories of moral intuition are products of humankind's evolutionary history, are they reliable foundations for morality?

Nineteenth-century English geologist and early mentor to Charles Darwin Adam Sedgwick thought that evolutionary ideas spelled disaster for religion, morality, and social justice. Commenting on an anonymously published book espousing evolutionary concepts fifteen years before Darwin's *On the Origin of Species*, Sedgwick ([1845] 1890), wrote to his geologist colleague, Charles Lyell: "If the book be true, the labours of sober induction are in vain; religion is a lie; human law is a mass of folly, and a base injustice; morality is moonshine, . . . and man and woman are only better beasts!"[27]

Now that the reality of Darwin's great idea, the common ancestry of life on Earth and its diversification via natural processes, is confirmed within disciplines ranging from anatomy and molecular biology to psychology, ethology, and ecology, Sedgwick's worst nightmare has come true. Evolution happened. But is morality only "moonshine"? Just what is true morality, if there is such a thing? Can we not accept a naturalistic origin for morality and simply define it as individual and group behavioral propensities emerging during our evolutionary history—ones that support the thriving of individuals, groups, and the species as a whole? Defining morality this way seems to rescue moral intuition from the realm of the *unreliable* and thereby to rescue moral theories based on moral intuition.

Harvard biologist and Pulitzer Prize winner Edward O. Wilson writes about the likely origin of altruistic behavior in humans and a few other animals via an evolutionary process called group selection in his book *The Meaning of Human Existence* (2014). When our conscience tells us to help children unrelated to us and to be kind to strangers, we may be hearing the voice of the psyche of our prehuman ancestors whose groups survived because members cooperated with each other. The problem, according to Wilson, is that tribalism, unhealthful nationalism, and feelings of distrust and antipathy toward persons different from us are also products of those psyches.

As members of communities, we use reflective equilibrium to examine both our moral intuitions and our moral responses to moral dilemmas. Reflective equilibrium is the end point of a deliberative process during which we reflect on, analyze, and revise beliefs about an area of inquiry, such as a moral decision. It may be impossible to satisfy all of our moral intuitions and the moral principles to which we defer for guidance with one decision. The process of reflective equilibrium allows a balancing of feelings and principles and a taking into account specific information, context, and personal experiences relating to the decision at hand.

Thus, we need not accept all moral intuitions as being correct, nor do we need to distrust all of them. In fact, we can acknowledge that different members of the community may sometimes have conflicting moral intuitions. But over time, reflective equilibrium as a community-wide enterprise tapping cognitive resources beyond the brain itself and applied to our responses to human needs will give rise to a dynamic communal morality always subject to refinement.

If we accept this moral constructivist view of morality, I suggest that the deliberative process that gives rise to morality ought to include voices from all thoughtful members of the community not just the voices of so-called moral experts like ethicists and clergy. Neil Levy agrees that all voices ought to be heard, but he suggests that "morality as a social enterprise is amply vindicated if there is a gradual move toward greater consistency of moral response . . . and if the moral responses of experts gradually permeate society" (2007, 311). However, there are no data of which I am aware demonstrating superior moral intuition or responses from persons with academic training in moral philosophy compared to those lacking such training. Experts may be able to associate their intuitions or responses with the name of a particular moral theory, but this does not make their intuitions or responses more noteworthy than those from persons untrained in moral philosophy.

Finally, there are nonnaturalistic ways of viewing morality, either from religious or secular viewpoints. Some religions view morality as revealed truth. For example, a literal reading of Judeo-Christian scripture has the ten "shalts" or "shalt nots" being delivered directly from God. Other moral directives or prohibitions may be viewed as divine commandments revealed through the lives and words of prophets and sages.[28] Some persons may subscribe to a nonnaturalistic origin for morality without invoking a supernatural entity. The theory of natural law posits that moral laws are part and parcel of the universe, built into its very nature. By this way of thinking, what is natural is good and ought to be respected, nurtured, and preserved because it exists for a purpose. This view of nature goes back at least to Aristotle (350 BC). Thirteenth-century Christian theologian Thomas Aquinas and others co-opted this view by replacing "nature's purpose" with "God's intention." Ostensibly, natural types of behavior such as being kind to others and forming bisexual unions were deemed good, while outlying behaviors like wanton maliciousness and homosexual actions were deemed unnatural and therefore bad. Rachels (2003, 56) explains why the theory of natural law has few modern advocates.[29]

When all is said and done, whether new discoveries about the neurological bases of morality threaten the credibility and reliability of the moral systems we chose to live by will depend upon our personal views about what *morality* is and from whence it comes.

Chapter Summary

The human brain is the most complex object in the known universe. Recent brain research reveals much about the structure of certain neurons and their electrochemical communication via specialized regions of contact called synapses. But little is known about how millions of communicating neurons operate in neural circuits to create specific aspects of brain function. The identifying, mapping, monitoring, and manipulating of the brain's thousands of neural circuits are goals of the decade-long BRAIN Initiative. Other projects including the EU's HBP and the Chinese Brain Project aim to use information about brain structure and function to inspire advances in artificial intelligence. Functional MRI, optogenetics, deep brain stimulation, and other existing and developing modern neurotechnologies are employed to explore brain structure and function. Anticipated research outcomes include treatments for brain disorders including Parkinson's disease, schizophrenia, depression, and Alzheimer's disease and also for non-medical applications in marketing and criminal justice. An understanding of how a brain creates a mind and how human consciousness exists may also be forthcoming. Neuroscience research and its findings raise diverse ethical issues. Matters of privacy, informed consent, and distributive justice are raised by other biotechnologies, but identification and manipulation of the cellular basis of *self* and *consciousness* are unique to neuroscience. Two branches of neuroethics are the *ethics of neuroscience* and *the neuroscience of ethics.* The former deals with concerns raised by neuroscience research and its applications. The latter relates to ways that discoveries in neuroscience affect our understanding and practice of morality.

Questions for Reflection and Discussion

1. Do you believe that the BRAIN Initiative is a worthwhile investment of money and scientific energy? If not, how would you rather see these resources used?
2. Do you believe that science will at some time be able to fully describe the human mind in terms of the electrical and chemical activity of cells? If not, what do you think would be missing from such a purely physical description of the mind?
3. Do you believe that science will ever be able to create a machine that possesses the equivalent of a human mind? If not, why not?
4. Do you believe that applications of modern brain science will lead to the further stratification of society into "haves" and "have-nots"? Why or why not? If yes, what type(s) of stratification do you envision?
5. If neuroscience shows that our moral intuitions are based on neural circuitry created by natural selection over a million years ago, would it change your mind about the reliability of those intuitions, especially if they ignore moral instruction from sacred texts? Why or why not?
6. What moral dilemmas can you foresee emerging from neuroscience that were not discussed in this chapter?

Notes

1. Harvard University biologist Edward O. Wilson describes the brain thusly in his book *The Meaning of Human Existence* (2014, 159), as does Christof Koch, chief scientific officer at the Allen Institute for Brain Science in Seattle, as reported by his interviewer, Ira Flatow, host of *Science Friday* on National Public Radio (http://www.npr.org/2013/06/14/191614360/decoding-the-most -complex-object-in-the-universe, accessed November 14, 2014). Isaac Asimov (1986, xv) wrote, "The human brain . . . is the most complicated organization of matter that we know," in his foreword for *The Three Pound Universe.*

2. A nontechnical description of the structure and function of nerve cells and synapses and the action of neurotransmitters, chemical signaling molecules in the brain, is given in Bradley 2013a.

3. Limbic system components include the hippocampus, amygdala, olfactory bulbs, cingulate gyrus, and other structures.

4. These NIH BRAIN Initiative grant awards and brief descriptions of some of the funded research projects are reported on the NIH News and Events page at http://www.nih.gov/news/health/sep2014/od-30.htm (accessed October 14, 2014): "Creating a wearable scanner to image the human brain in motion, using lasers to guide nerve cell firing, recording the entire nervous system in action, stimulating specific circuits with radio waves, and identifying complex circuits with DNA barcodes are among the 58 projects."

5. Humans have twenty-two different autosomes (nonsex chromosomes) and two types of sex chromosomes, X and Y. The forty-six chromosomes in a normal human body cell include forty-four autosomes (one copy of each of the twenty-two types from each parent) and two sex chromosomes (two Xs in females and an X and a Y in males).

6. DARPA, the research wing of the US Department of Defense, is providing $50 million to the BRAIN Initiative for its specific projects. This is about half of the government's funding for the initiative, nearly as much as is coming from NIH and NSF combined. DARPA's stated research priorities within the BRAIN Initiative include developing wireless devices for curing brain disorders such as posttraumatic stress, chronic pain, and major depression and for repairing brain damage and reversing memory loss.

7. The HBP-FPA was selected for funding by the European Commission via its Future and Emerging Technologies Flagship grant. The 228-page document is available at https://www.humanbrainproject.eu/documents/10180/538356/HBP_FPA_PRINT_29-07-14.pdf (accessed May 19, 2015). The letter was written by Alexandre Pouget of the University of Geneva and Zachary Mainen of the Champalimaud Centre for the Unknown in Lisbon and signed by leading neuroscientists from all over Europe. It is available at www.neurofuture.eu (accessed May 21, 2015). The fifty-three-page Human Brain Project Mediation Report document is available at http://www.fz-juelich.de/SharedDocs/Downloads/PORTAL/EN/pressedownloads/2015/15-03-19hbp-recommendations.pdf?__blob=publicationFile (accessed May 19, 2015).

8. Graphene is a two-dimensional, hexagonal lattice of carbon atoms with properties of special interest to electrical and materials engineers. Its magnetic, conductive, and exceptionally strong physical properties promise many future applications in biology and medicine, composite materials, and energy storage.

9. Specifically, controversy arose over a decision to remove cognitive and systems neuroscience from the core project of the HBP, relegating it to a part-

ner project. The core project encompasses research funded by the EU through the European Commission's flagship program rather than through individual partnering institutions, private companies, or countries. Funding for research in the core project is more secure than that in a partnering project. Cognitive and systems neuroscience deals with the role of neural circuits in controlling psychological and cognitive brain functions.

10. HBP website: https://www.humanbrainproject.eu/discover/the-community /overview (accessed August 1, 2015). The site for the HBP Mediation Report is given in note 7 of this chapter.

11. The BigNeuron Project website is at http://alleninstitute.org/bigneuron /about/ (accessed April 27, 2015).

12. Fruit flies, zebrafish, and mice are high-priority animals for the BigNeuron Project, nicely complementing the array of species from which both the BRAIN Initiative and the HBP also extract information.

13. The Allen Institute for Brain Science website is http://alleninstitute.org/. Information about and data collected on all of the institute's projects is available within this site at http://alleninstitute.org/our-research/open-science-resources/. The Brain Explorer software (downloadable at http://mouse.brain-map.org/static /brainexplorer) allows one to view gene expression patterns throughout the mouse brain. Each website was accessed August 4, 2015.

14. An extensive bibliography for both branches of neuroethics lists 135 review articles and 46 books (Buniak et al. 2014).

15. Synthetic biology is an interdisciplinary endeavor to engineer and create brand new life forms. President Obama's Commission for the Study of Bioethical Issues recommended that synthetic biologists move ahead in their research with "prudent vigilance," leaving responsibility for informing the public about ethical implications of the research and regulating the research to scientists and the public and private institutions supporting the research. Synthetic biology as a discipline, ethical issues it raises, and the commission's study of it are discussed in *Brutes or Angels* (Bradley 2013a: 254–87).

16. The Neuroethics Work Group (NWG) communicates with NIH officials indirectly through a multicouncil working group (MCWG) comprised of nongovernmental representatives from advisory councils of the ten NIH institutes that contribute to the BRAIN Initiative and also five at-large members. The NWG is a subgroup of the MCWG.

17. Specific goals of the Ethics and Society Programme within the HBP are detailed on its website at https://www.humanbrainproject.eu/discover/ethics (accessed September 11, 2015).

18. The Genetic Information Nondiscrimination Act protects Americans against discrimination by health insurers and employers based on their genetic information.

19. Several such studies are cited and discussed by Brown and McCormick (2011), Tovino (2011), and Cossins (2015).

20. Educational neuroscience is an international movement at the interface between brain science, cognitive development, and formal education.

21. Article 12 of the convention states: "1. States Parties shall assure to the child who is capable of forming his or her own views the right to express those views freely in all matters affecting the child, the views of the child being given due weight in accordance with the age and maturity of the child. 2. For this purpose, the child shall in particular be provided the opportunity to be heard in any judicial and administrative proceedings affecting the child, either directly, or through a representative or an appropriate body, in a manner consistent with the procedural rules of national law" (http://www.ohchr.org/en/professionalinterest /pages/crc.aspx [accessed September 28, 2015]).

22. Memorandum of Law in Support of Plaintiff's Motion for a Temporary Restraining Order in the United States District Court, Middle District of Louisiana, Michelle Selden v. Livingston Parish School Board (https://www.aclu.org /sites/default/files/field_document/20060801tromemo.pdf [accessed September 30, 2015]).

23. Here, one can read about the dismissal of the motion and the court's denial of the plaintiff's request for an award of attorney's fees and costs: https:// casetext.com/case/minor-child-v-livingston-parish-school-bd (accessed September 30, 2015).

24. A video of the University of Washington "mind meld" experiment can be viewed at http://www.washington.edu/news/2013/08/27/researcher-controls -colleagues-motions-in-1st-human-brain-to-brain-interface/ (accessed August 11, 2015).

25. I put *robot* in quotes here because most definitions of robots require that the device act autonomously. *Hello* in Spanish was sent brain to brain from India to Strasbourg, France.

26. An important and often-cited study (Greene et al. 2001) used fMRI to show that brain regions known to be associated with emotion are preferentially active during moral-personal decision making; that is, when the agent's action will affect another person with whom the agent directly interacts. These brain regions were less active during nonmoral decision making (such as which coupon to use at the grocery store) and moral-impersonal decision making (when

the agent's action affects others indirectly, as when a bombardier releases his payload upon faceless persons below rather than via an up close interaction, as when one directly assaults another person face to face, even for an outcome with goodness perceived to outweigh the immorality of the assault itself). Specific brain regions found by fMRI to be active during moral-personal decision making include Brodmann's Areas 9 and 10 (medial frontal gyrus), 31 (posterior cingulate gyrus), and 39 (angular gyrus, bilateral).

27. *Vestiges of the Natural History of Creation* was published anonymously in England in 1844. It espoused evolutionary ideas about the cosmos and the biological world. Clergy were generally opposed to it, but Charles Darwin believed that it helped to prepare the public for his theory of evolution by natural selection published in 1859 in *On the Origin of Species by Means of Natural Selection or the Preservation of Favoured Races in the Struggle for Life.* In 1884, Robert Chambers, a Scottish journalist, was revealed as the author of *Vestiges.* Lyell (1797–1875) is a founder of modern geology. In his *Principles of Geology,* he argued for an Earth more than three hundred million years old and provided evidence for uniformitarianism, the notion that the forces that shaped the Earth's surface are the same forces we see at work today, such as erosion and volcanism. His ideas about Earth's antiquity influenced Charles Darwin's thinking about the power of natural selection to shape the biological world.

28. The notion of information about right and wrong coming directly from God is known as divine command theory. This theory is vulnerable to reasoned answers to the famous question asked by philosophers going back as far as Socrates: Is conduct right because the gods command it, or do the gods command it because it is right? For a discussion of philosophy's answers to this question and their implications for divine command theory, see Rachels (2003, 50–53).

29. According to Rachels, two main reasons that Natural Law is out of vogue are (1) that believing that what *is* is what *ought* to be is illogical, as famously demonstrated by David Hume and (2) that the notion of absolute facts about right and wrong being built into the fabric of the universe is inconsistent with modern science's view of the universe.

Sources for Additional Information

Bae, B.-I., and C. A. Walsh. 2013. "What Are Mini-Brains?" *Science* 342: 200–201.
BRAIN Working Group. 2014. *BRAIN 2025: A Scientific Vision.* Report to the

Advisory Committee to the Director, NIH. National Institutes of Health. www.nih.gov/science/brain/2025/BRAIN2025.pdf (accessed October 10, 2014).

"Brains in Action." 2014. *Scientist*, February 1. http://www.the-scientist.com /?articles.view/articleNo/38987/title/Brains-in-Action/ (accessed January 4, 2014).

Brown, Fraser. 2015. "Optogenetics: A Vision of the Future of Neurology?" *Cambridge Medicine Journal: The Student Journal of the Cambridge School of Clinical Medicine*. http://www.cambridgemedicine.org/article/doi/10.7244 /cmj-1394635425 (accessed January 24, 2015).

Brüstle, O. 2013. "Miniature Human Brains. News and Views." *Nature* 501: 319–20.

Clausen, J., E. Fetz, J. Donoghue, et al. 2017. "Help, Hope, and Hype: Ethical Dimensions of Neuroprosthetics." *Science* 356: 1338–39. http://science .sciencemag.org/content/356/6345/1338 (accessed July 17, 2017).

Olena, A. 2017. "Primates Use Simple Code to Recognize Faces." *Scientist*, June 1. http://www.the-scientist.com/?articles.view/articleNo/49564/title/Primates -Use-Simple-Code-to-Recognize-Faces/&utm_campaign=NEWSLETTER _TS_The-Scientist-Daily_2016&utm_source=hs_email&utm_medium=email &utm_content=52665793&_hsenc=p2ANqtz_AwgfuRduEr7o2T1fN9 NBV XJepEwMvCVjvd5Yfq6Y6_P5nqILXNDTeyU5qBCViBgyQMLxnIeJkWo9j uK8qpd9EyYw5jg&_hsmi=52665793 (accessed June 6, 2017).

Presidential Commission for the Study of Bioethical Issues. 2014. *Gray Matters: Integrative Approaches for Neuroscience, Ethics, and Society*. Vol. 1. http:// bioethics.gov/node/3543 (accessed October 10, 2014).

Shen, H. 2014. "Tuning the Brain." *Nature* 507: 290–92. This article is a nontechnical description of current and future applications of deep brain stimulation, including personalized fine tuning of brain circuit activity.

Yuste, R., and G. M. Church. 2014. "The New Century of the Brain." *Scientific American* 310 (3):38–45.

7
Robots and Roboethics

> Human beings can tolerate an immortal robot, for it doesn't matter how
> long a machine lasts. They cannot tolerate an immortal human being,
> since their own mortality is endurable only so long as it is universal.
> And for that reason they won't make me a human being.
> —Isaac Asimov, "The Bicentennial Man"

What is a robot? In a 1921 science fiction writing, Czech author Karel Čapek introduced the term *robot* to describe characters that were artificial people manufactured to work for humans but who think for themselves and mount a rebellion against their makers. Nearly a century later, there is no widely agreed-upon definition of *robot*. Some writers require that machines be designed to resemble human beings in order to qualify as robots. That is, they must have a head, eyes, and four limbs. Others such as Rodney Brooks, director of the MIT computer science and artificial intelligence laboratory, accept a more general definition: "To me a robot is something that has some physical effect on the world, but it does it based on how it senses the world and how the world changes around it" (CBCnews 2007). Joseph Engelberger, the Father of Robotics, once said, "I can't define a robot, but I know one when I see one" (CBCnews 2007). For our discussion here, I propose a definition paraphrased from George A. Bekey, professor emeritus of computer science at the University of Southern California and distinguished professor of engineering at California Polytechnic Institute: robots are machines that sense, think (analyze and learn), and act in the real-world environment (Bekey 2012, 17–18). This definition excludes remote-controlled machines since they do not think for themselves. It also excludes virtual robots simulated by software since these do not act in the real world. What constitutes *thinking* is a contentious issue, but let us say that it involves at least some degree of autonomous decision making. A robot's autonomous decisions are based upon information it senses or is given, and its analysis of that information occurs using a set of internal rules that are programmed or learned (Bekey 2012 18). The primary objective of this chapter is to explore our relationships—present and future—with robots. Minimal attention is given to the technology of robotics and/or artificial intelligence. This is intentional so that we can focus our thinking on our relationships with robots not the mechanics. These questions guide our exploration:

1. What are some robots with us now, and what can they do?
2. What may future robots "think" and do?
3. What are benefits and risks of personal human relationships with robots?
4. Should robots simply augment human abilities or actually take over human activities?
5. Will robots ever be persons? If so, how will we know?

Robots: Present and Future

Today's Robots

Robots already live and work more closely with us than we may realize. Humans interact intimately with robots on battlefields and highways and in hospitals and homes. In this section, we look at a sampling of these associations, beginning with robots that coinhabit human living spaces.

Robots in our personal lives. Most of us own, have witnessed, experienced, or heard about robots that vacuum floors, clean swimming pools, and transport people around cities or between airport terminals. Less known are robots that provide companionship and comfort to the elderly and that are even beginning to replace prostitutes in some regions of the world. We begin with the mundane, vacuum cleaners.

Robotic vacuum cleaners began entering homes in the early 2000s. In 2001, a British company made and demonstrated a robot vacuum named DC06, but its cost prevented marketing. The following year, the American company iRobot offered its Roomba floor vacuuming robot for sale. By 2015, at least ten different robotic vacuum cleaners were available for prices from $122 to $700 (Robot Vacuum Review 2015). The top-rated vacuuming robot was Neato Botvac, while iRobot's Romba 980 was a close second. The Neato Botvac moves methodically along the perimeter of a room before cleaning the center area, and it approaches objects like furniture and walls slowly so not to damage them. The robot vacuum connects to home Wi-Fi, giving its user complete control over the robot's cleaning schedule from wherever and whenever.[1] Similar devices clean the bottoms of swimming pools and mow lawns. Soon many people will use cooking and laundry robots. Valued by owners for time and labor savings from doing house and yard work, these robots are an aid to humans but are far from replacing them in home maintenance. Elbow grease still returns the shine to Great-Grandma's silver, while gentle human hands still tend babies' bottoms. More interesting than robots that interact with walls, trees, and Wi-Fi are robots that interact with people.

Beginning in the early 1990s, roboticists began creating socially interactive robots. These semiautonomous robots are designed to have some humanlike features (head, arms, eyes, etc.) and are endowed with an artificial intelligence (AI) that allows them to follow certain social rules and engage in social behaviors including communication. Robot autonomy includes abilities to sense and interpret a wide range of sensory stimuli, and to learn, plan, and communicate with humans and with other robots (Bekey 2005, 509). With these humanoid robots, biologist Edward O. Wilson's concept of biophilia is pertinent (Wilson 1984).

Wilson hypothesized that humans evolved to feel an affinity for other living things. We enjoy caring for house plants and pets, and most of us feel refreshed by forays into nature, be it a walk in the park or months on the Appalachian Trail. A corollary concept is that emotional attachment or empathy toward living things becomes greater as their similarity to us increases. We feel more empathy toward an injured chipmunk than toward an injured beetle and more toward an orangutan behind bars than a penned zebra. It will not be surprising if humans develop stronger emotional bonds with social robots designed to look and behave like humans than with Neato Botvac or Robomow's robotic lawn mower; however, we see later that a caveat to this is a phenomenon called the Uncanny Valley, a region of discomfort with robots.

A few rather sophisticated humanoid robots already provide companionship for their owners. Leading the pack is an autonomous, interactive robot named Pepper manufactured by SoftBank Robotics, a corporation headquartered in Tokyo and with offices in France, the United States, and China. Established in 2014, SoftBank specializes in humanoid and service robots. Pepper has two arms, two hands, one head, two "eyes," two "ears" and is mobile, qualifying it as humanoid, unlike a vacuum cleaner for example. Pepper distinguishes itself as the first humanoid robot to perceive human emotions and to adapt its behavior to the disposition of its user. Pepper also learns and remembers things about its user's personality, habits, and temperaments. It can adapt its communication style, including tone of voice, body language, and even eye color, to the mood of the moment (SoftBank

Robotics Corp. n.d.). Japanese consumers already use Pepper in their homes simply because having the robot around makes them happy. Some universities also use Pepper for research on human-robot relationships. Pepper does not do housework or other tasks. Unlike the very dexterous kitchen assistant robot, ARMAR-3, developed by a German institute, Pepper is manufactured solely for people to enjoy a human-robot relationship.[2]

A different kind of robot, which is not humanoid, is a self-driving car. By late 2018, thirty-five states and the District of Columbia had addressed autonomous, self-driving vehicles either through enacted legislation and/or by executive order by governors (National Conference of State Legislators, 2018). As we see in the latter portion of the chapter, programming driverless cars for safe behavior raises the difficult ethical issue of whose safety ought to receive priority. Finally, robotics has not overlooked the potential of robotic bedroom mates. In Korea and Japan, the design and building of robotic sex partners is in full swing (D. Levy 2012, 224–25). Before visiting the bedroom, we will consider examples of teachable robots in the workplace and autonomous robots in hospitals and on battlefields.

Industrial robots. Baxter is a collaborative robot built by Rethink Robotics in Boston, Massachusetts. Being collaborative means that Baxter works with humans in larger settings such as industry or education. First marketed in 2012, Baxter is three feet tall and able to learn and be easily programmed to perform repetitive tasks. Baxter has two arms, an animated face, and a choice of pedestals from which to work. At $25,000 each, Baxter is popular with small- to medium-sized companies and with robotics teaching programs at colleges and universities. No computer programmer or software engineer is needed to teach Baxter new tricks or jobs. Anybody can teach Baxter a task in just a few minutes. By moving its arms through the required motions Baxter learns to mimic the activity presented. Sensors in its head, arms, and hands make it safe to work alongside Baxter. If it encounters unexpected resistance, Baxter immediately reduces the force of its movements, avoiding damage or injury to things or persons. Baxter can communicate information via its facial expressions, such as whether it is confused or what particular job it is currently tending to. Cameras in its hands allow Baxter to perform detailed tasks such as placing lids on containers or sealing and opening boxes. Other cameras make the robot minimally aware of its environment, such as whether humans are nearby or whether it has dropped a tool necessary for performing its current job. Baxter is especially good at loading and unloading material, packing boxes, assembling kits, and sorting items.

Baxter was followed in 2015 by Rethink's $29,000 Sawyer, a one-armed robot built for precision work such as circuit board testing, machine tending, and the precise handling of certain materials. Like Baxter, Sawyer is adaptable to many tasks. A wide-view head camera and a wrist camera allow Sawyer to learn and perform complex vision tasks rapidly and with an adaptably soft touch. Sawyer is smaller than Baxter and has a longer reach, allowing it to work in tighter spaces. Sawyer's competitor from Universal Robots is another one-armed fellow named UR3 and designed especially for soldering, gluing, screwing, and operating hand tools (Tobe 2015).

Military robots. The US military reportedly struggles with decisions about how autonomous to make robotic killing machines (Rosenberg and Markoff 2016). It has already tested missiles outfitted with AI software that can decide what to attack. Also already built are unmanned ships that find enemy submarines and stalk them for thousands of miles without human input. Developing drone warfare capabilities has also been a priority during the twenty-first century. So far, the killing activities of drones used in counterterrorism have been remotely controlled by humans, so they do not qualify as robots by our earlier definition. But that may soon change. Robert O. Work, former deputy secretary of defense, sees China and Russia as major competitors in the development of autonomous and semiautonomous battlefield robots. Many scientists fear a new arms race in autonomous weaponry. In 2016, Gen. Paul J. Selva, vice chairman of the Joint Chiefs of Staff, volunteered that the United States is only about ten years from being able to build an autonomous robot that can make completely independent decisions about when and who to kill, although he reportedly stated that the United States has no intention of actually building one. Yet DARPA, the research wing of the US Defense Department, is actively pursuing development of software to give drones and other battlefield weapons more autonomy so they can better work with soldiers. "It's like having another head in the fight," says Maj. Christopher Orlowski, program manager at DARPA (Rosenberg and Markoff 2016).

Our military now has autonomous and semiautonomous vehicles in various stages of development and deployment. Currently in use are unmanned flying vehicles, unmanned underwater vehicles, and unmanned ground vehicles (UGVs). An example of a UGV is iRobot's PackBot, which resulted from a DARPA contract awarded to iRobot in 1998. More than two thousand iRobot PackBot and the similar MARCbot robot built by Exponent Inc. deployed in Iraq and Afghanistan to search beneath and inside vehicles for explosives and dangerous chemicals, to clear buildings and routes, to detect biological and nuclear agents, for surveillance, and for other dangerous missions. Some are lethally armed and made to explode when encountering enemy personnel.

Of great concern is that the technology for building autonomous killing machines is unlikely to be contained and, in fact, will proliferate during the next few decades. Who will design the software needed for such machines to recognize targets and make fatal decisions? How easy will it be to hack into and alter the "intelligence" of such machines? How available will autonomous killing machines be to terrorists or private citizens? Will the Second Amendment be invoked for private ownership and use of autonomous killing machines? So far, little public attention has been given to questions like these. Meanwhile, public funds support research that could lead to a new world of autonomous killing devices. But at the other end of the spectrum are robots that assist in medical care for humans.

Robots in healthcare, surgery, and rehabilitation. HelpMate, sold by Pyxis Corporation, is a nurse's assistant used in nearly one hundred US hospitals. The robot is a developmental spinoff from the National Aeronautics and Space Administration's interest in robotics. HelpMate can navigate autonomously inside hospitals to transport and distribute pharmaceuticals, laboratory specimens, meals, medical records, equipment, supplies, radiology films, and other medical items.[3] Collaborations between the University of Pittsburgh, Carnegie Mellon University, and the University of Michigan have produced prototype autonomous caregiving nursebots that assist elderly persons in hospitals, nursing facilities, or private homes.[4] These nursebots remind clients to take their medication, take messages, help with mobility, collect surveillance data to help avoid certain emergency conditions, and provide social interaction including companionship for persons with autism-spectrum disorders. During rehabilitation, nursebots provide encouragement and guidance for clients recovering from spinal injuries or strokes. In other quarters, roboticists create robots for sexual intimacy, a key necessity for human well-being.

Sexbots. The inflatable vinyl or silicone sex dolls of the late twentieth century are metamorphosing into high-tech, sophisticated twenty-first-century sex robots (sexbots) now available for hire and for sale depending on the locale.

Sexbot technology is still in its infancy, but it is rapidly evolving to generate a future of nearly unimaginably intimate social relationships between humans and machines. What exactly is a sexbot? Roxxxy, a creation of the New Jersey–based company TrueCompanion, is an example. At five feet seven inches tall, Roxxxy weighs sixty pounds, has lifelike synthetic skin, and creates internal warmth from her own circulatory system. Equipped with AI, the gynoid listens, speaks, converses, responds to touch, has moods and five different temperaments, each modifiable by the user. Automated moving parts include head, legs, hands, and hips. Roxxxy can even learn and remember her user's likes and dislikes. You get the idea. In 2010, Roxxxy sold for about $7,000. TrueCompanion also markets an android version named Rocky.[5] Another US company, RealDoll, is reportedly working to create an intelligent sex robot (Gold 2015) to compete with Roxxxy.

Certain Japanese and South Korean companies are already deep into sexbots. The Tokyo-based Orient Industries markets female sexbots named Dutch Wives XXX and CandyGirls known for their very realistic silicone/rubber-based skin and contracting or dilating pupils that project various moods. As we see later, there are ethical arguments both for and against developing and using sexbots. In any case, sexbots show no sign of going away. It appears that human-robot relationships may soon be a matter-of-fact element in the sex lives of some segments of society with consequences hard to predict.

Robots of Tomorrow

We cannot know for certain what types of robots will inhabit humankind's future and what our relationships with them will be, but we can confidently predict that future robots will greatly impact at least five domains of human life: sexual relationships, personal social interactions, transportation, the workplace, and warfare. In the second portion of the chapter, we examine human values issues associated with the imminent entry of autonomous robots into some of these domains.

Robots and Human Values

Four human values issues predominate for robotics in the future. Two are of immediate concern: the types of relationships humans will establish with robots and the specter of robots replacing humans in the workplace on a massive scale. A third concern is the behavior of robots. With what kind of morality software should robots be endowed? This is especially important for military robots with lethal capabilities. It is also crucial in certain nonmilitary settings, such as self-driving cars traveling at lethal speeds. The fourth concern is more distant but nevertheless a realistic possibility. AI will someday give robots a feeling of self-identity. In this event, what rights and responsibilities are appropriate for robots that possess humanlike consciousness? In this section, we consider each of these four issues, beginning with our social relationships with robots.

Human-Robot Relationships

How will enlisting robots for companionship and personal care for children and the elderly affect human-human relationships? Does having sex with robots benefit the individual and society? What moral concerns might outweigh possible benefits of sexual intimacy with machines? Who should have a say in designing personalities for robots that have close social encounters with humans? An urgent need exists for public discussion of these and related questions because the market for and marketing of personal robots is growing rapidly. Let us consider some issues associated with using robots for caregiving, companionship, and other life enhancements.

Robots for health and living enhancement. Humanoid and animallike robots already exist as playmates for children, shopping assistants, household helpers, and companions for the elderly and persons with dementia (Normile 2014). Conceivable social benefits of robots are counterbalanced by some ethical concerns. For example, in 2009, a "robot" built by Advanced Telecommunications Research Institute International in Nara, Japan, spent two weeks at a senior day care center. The child-sized Robvie is remote controlled. Since Robvie possesses no autonomy, it fails our earlier definition of a robot. However, the seniors' experiences and reactions to Robvie highlight ethical issues for future autonomous or semiautonomous social robots.

During its visit to the day care center, Robvie talked, waved, roamed the halls, and appeared to listen. By the way, Robvie's name is borrowed and morphed from the famous Isaac Asimov short story titled "Robbie" from an anthology of Asimov's writings (Asimov 1950). Although the seniors knew they were participating in an experiment to learn about future human-robot interactions, they became emotionally attached to the robot and asked to visit Robvie in his laboratory home after the experiment ended. Some ethicists and psychologists warn that human feelings of affection toward and social dependence upon robots could stunt the development and maintenance of social relationships between real live humans, while others believe that robots will actually nurture, enrich, and expand our social lives.

Important for designers of robots that interact socially with humans is a phenomenon called the Uncanny Valley. First described by Japanese roboticist, Masahiro Mori (1970), the Uncanny Valley is a region of human discomfort and revulsion toward certain humanoid robots and other quasi human entities including zombies and corpses. Researchers analyzing humans' emotional reactions to robots find that our levels of comfort and trust toward robots increases as robots becomes more humanlike and less machinelike but only to a certain point. Comfort and trust plunge precipitously into a valley of discomfort and distrust when robots' characteristics are simultaneously quite humanlike yet visually nonhuman looking.

Studies aimed at finding ways to design humanoid robots that dodge the Uncanny Valley suggest several possibilities. For example, humans have a more positive reaction to robots whose appearance and behavior match human realism (Goetz et al. 2003). So a robot with both very realistic voice and appearance should engender more trust than a hybrid with a synthetic voice but very realistic physical appearance. Similarly, research shows that facial texture and proportions should match in order to avoid the Uncanny Valley (MacDorman et al. 2009). So giving a humanlike surface texture to the face of a robot with an abnormally large mouth or nonhuman ears would probably not ameliorate a drop into the Uncanny Valley. Remember the Coneheads on *Saturday Night Live* in the late 1970s? The humanlike aliens with tall, bald conical heads from the planet Remulak were amusing but also off-putting.

The biological and psychological causes for the Uncanny Valley are unknown, but theories abound. One posits that during human evolution natural selection endowed us with mate selection "radar" to avoid partners with low fertility or poor health as telegraphed by certain facial and/or body features. The dissonance of features such as complexion, eyes, facial expressions, muscle tone, hair, or lips in a robot may trigger an innate avoidance alert whose original biological function was to aid in mate selection. Another theory suggests a psychological mechanism whereby we feel empathy toward a robot doing a mediocre-to-passable job at looking human but disgust towards an entity that looks like a real human but falls short of engaging in what we consider to be normal behavior. That is, the Uncanny Valley may be a psychological manifestation of passing judgement on a human even though the object of our judgement is a robot. As we learn more about the bases for the Uncanny Valley phenomenon, better strategies for avoiding or dealing with it will emerge, moving us into an era in which sophisticated humanoids will partake in an array of human-robot social interactions.

Computer scientist and roboticist, George Bekey (2012, 29), foresees several risks and ethical issues arising as our social life with robots becomes more frequent and intimate. These I paraphrase below along with some comments of my own:

1. Personal privacy, or at least the perception of privacy, may suffer as robots share more and more of our living spaces such as bedrooms and bathrooms. It might take a while to become comfortable using the toilet or falling asleep in the presence of a wide-eyed humanoid, even if one knew it were there only to offer assistance.

2. If robots respond to human commands, what dangers lurk when a robot receives an unethical command, perhaps from someone other than the robot's owner? Voice recognition software needs to be mandated for some robots. What if a robot does an unethical action and nobody admits to having commanded that action? Will robots need to be outfitted with 24/7 recordings of vocal commands? For robots able to respond to commands from more than one human, how should conflicting commands be prioritized?

3. What, if any, rights, respect, or responsibilities are due autonomous or semiautonomous robots? If we say that robots need not have rights or receive respect no matter how humanlike their appearance and behavior, would habits of indifference or disrespect toward robots spill over into our relations with other humans? Germane to this issue is the fact that animal abuse is strongly correlated with the mistreatment of fellow human beings (CFAWR 2010).

4. How should humans and robots relate to each other emotionally? Can we imagine instances when a robot ought to be punished? If so, how does one punish a robot?

5. What about the likely eventuality that robots' computers will be accessed by hackers who instruct the robot to compromise its owner's privacy or to act in other unethical ways? Who has responsibility for the ethical behavior of robots? Software designers, robot manufacturers, robot owners, or the robots themselves?

Although software design and other technological solutions will address some of these concerns, issues of emotional relationships between humans and robots and robot rights are not likely to be fully resolved technologically. Next, we consider sexual intimacy with robots.

Sex, love, and the robot-human relationship. Sexbots already augment the prostitution trade in Japan. High-priced sexbots have been offered via a 24/7 doll-escort service in southern Tokyo since 2004, according to private AI researcher and author David Levy (2012, 224). In his book *Love and Sex with Robots*, Levy (2007) contends that sexbots will soon satisfy both the motivational and sexual needs of men and women who would otherwise be clients for sex workers. He cites studies claiming that the three most common reasons that men use female prostitutes are for sexual variety, to avoid the complications and constraints inherent in personal relationships, and to sidestep difficulty in establishing intimate relationships with women. Sexbots could fulfill these needs with the added health benefit of reduced risk of contracting sexually transmitted disease. But all may not be rosy in the robot red light district. There are ethical concerns about robot prostitution.

Social anthropologist Kathleen Richardson directs an international Campaign against Sex Robots to alert society to potential dangers of using sexbots and how they could exacerbate existing inequalities in society.[6] In several essays, Richardson argues that development and use of sexbots will reinforce the unequal and exploitive power relationship intrinsic to prostitution, namely that the buyer of sex remains in control while the prostitute becomes merely a thing to be purchased and loses her or his human identity (e.g., Richardson 2015a, 2015b). Richardson and others believe that women, men, and children will suffer devaluation and dehumanization if female, male, and child sex robots become mass produced. Society-wide attitudes of control and coercion toward sex partners encouraged by widespread use of sexbots could jeopardize healthy, mutually empathetic human relationships in the nonsexbot community where satisfaction depends upon freedom and mutuality.

David Levy does not share these concerns. He predicts that soon sex with robots will become common, marriage to robots will be legalized in some countries by about 2050 (D. Levy 2007, 305), and by the turn of the century, robots may be leading independent lives of their own (Gaudin 2007).

Furthermore, Levy (2007, 151–59 and 304) sees intimate relationships with robots, including marriage, as a natural outcome of increased respect and rights for nontraditional unions and the evolutionary trajectory of robot technology, which has already brought us various types of service robots, personal caregiving robots, and robotic pets. Levy argues that development of adult and child sex robots will benefit society by offering harmless sexual outlets to persons whose passions might otherwise harm real people through violence and sexual exploitation. He suggests that sexbots will help to reduce prostitution and the illicit sex trade.

The opposition claims just the opposite, that a society permeated by sexbots will suffer reinforced inequality and more violence in human relationships while only furthering the commodification of human bodies. As evidence, Richardson (2015b) points out that the numerous artificial sexual substitutes already available have not diminished prostitution. Another issue given little attention by roboethics authors is how the use of sexbots by minors may encourage users' sexual commodification of humans in later life. Should there be minimum age laws for relations with sexbots, even in the privacy of one's home? Psychologists will be interested to learn whether a human will be more likely to marry a robot if his/her earliest sexual experiences have been with robots. Should laws against polygamy apply to robots?

Will genuine love really be possible between humans and robots? A major hurdle to overcome before humans can truly feel love for their robot partners is the Uncanny Valley. Research and technological advancements that enable robots to monitor users' emotions, to reciprocate with their own displays of simulated emotions, and to store intimate knowledge about users in order to anticipate and respond empathetically to users' moods and feelings may help overcome the Uncanny Valley.

David Levy (2007) identifies three routes via which humans may come to love their robotic companions. The first is emotional attachment fed by having sex with the robot. The second is the phenomenon of technophilia, a feeling of love fed mainly by the novelty of the technology itself. And third is a genuine preference for a robot lover over a human lover. The latter would be important for persons who cannot interact intimately and effectively with real human beings. In the end, whether humans can truly love robots depends largely on how one defines *love*.

Notions of love range widely from Plato's nonsexual, spiritually centered love through the Greek and later Christian unconditional and universal *agape*. Then there is the sensual, secretive, illicit courtly love of the Middle Ages. Today some psychotherapists view love as "a decision." Parental love and love for family and/or members of relatively small groups has been around for eons. And finally, there are media-idealized feelings of love at first glance.

Except for unrequited love, most concepts of love involve some form of reciprocity. If robots that realistically simulate empathy and reciprocation are created, it seems likely that some persons will actually come to love their robotic companions, perhaps even choosing them over humans as life partners. What effects human-robot love relationships may have on society and what direction(s) the global general public wants companion robotics to take need to be discussed openly and soon. Otherwise, designers and marketers of robot partners will largely dictate how our ideas about human love change and expand. As philosopher and ethicist Blay Whitby (2012, 246) warns, "the market for robot lovers and other caring technologies is maximized in the situation where nobody chooses human companionship."

Next, we consider what the overarching objective of twenty-first-century robotics ought to be. Should we consciously keep ourselves in control of what are now uniquely human undertakings and design robots to assist rather than replace us?

Robots in the workplace: AI or intelligence augmentation (IA)? Pulitzer Prize–winning technology and science reporter John Markoff addresses this question in his 2015 book, *Machines of Loving Grace: The Quest for Common Ground between Humans and Robots.* What should be the overarching objective of twenty-first-century robotics? Are we striving to create intelligent robots that can best humans in virtually every category of human endeavor? Or are we taking care to stay in control of our unique qualities and create robots that augment human intelligence rather than compete with it?

Markoff notes that the rate of worker replacement by computerized systems has increased significantly since the 2008 recession. Yet, the Bureau of Labor Statistics (BLS) (2013) projects that job growth into the 2020s will be mainly influenced by aging Americans in need of health- and elder-care rather than by technologies that either create or displace jobs. In 2015, about 8.5 million American seniors required some type of care, and that number is expected to increase to 21 million by 2035 (Markoff 2015, 330). The BLS projects a 28 percent growth rate for jobs in health-care support occupations compared to 18 percent for jobs in computer and mathematical applications. The question is whether these jobs will be done by people or by robots.

Robotic caregiving for the elderly is an example of dual-use technology; it has both potential benefits and risks. On the one hand, low-cost robotic care would extend much-needed assistance to the many elderly persons who could not otherwise afford the $70,000+ yearly cost of human caregiving. Faced with a shortage of human caregivers, aging Baby Boomers may keep the loneliness and isolation of old age at bay with personalized, attentive, robotic companions. The question is whether AI or IA should predominate in this sector and others. One danger of AI dominating and largely displacing humans as caregivers, soldiers, drivers, service workers, companions, doctors, nurses, teachers, and others is an irreversible erosion of human identity. Whether robots replace or augment human activities in caregiving chores will be a decision made by AI and robot designers prompted by the marketplace comprised of persons like you and me. Personal decisions we make about caring for aging loved ones, our own care, and the selection of public health-care policymakers for our state and nation will powerfully influence the balance between AI and IA. Markoff comments: "The vision of an aging population locked away and 'watched over by machines of loving Grace' is potentially disturbing. Machines may eventually look, act, and feel as if they are human, but they are decidedly not" (2015, 329).

Moral Machines

No present-day robots are programmed for morality. Some are programmed for human and robot safety but not for moral decision making. In humans, moral decision making has an emotional component, and it is not clear that artificial moral agents (AMAs) will ever possess this human characteristic. Nevertheless, bioethicists, philosophers of science, and cognitive psychologists recognize the need to develop AMAs and grapple with the problem of how to do it responsibly (Wallach and Allen 2009; Allen and Wallach 2012). What morality(ies) should be bestowed on autonomous robots entrusted with using lethal weaponry, caring for children or the elderly, or performing other social activities?

Over half a century ago, biochemist and science fiction novelist, Isaac Asimov suggested a rule-based morality for future robots. He then illustrated problems associated with such a programmed morality. In a 1942 short story titled "Runaround," Asimov proposed three fundamental Rules of Robotics, designed by fictional future humans to ensure appropriate robot behavior. In the story, the rules come from the *Handbook of Robotics*, 56th edition, 2058 AD (Asimov 1950, 50–51):

1. A robot may not injure a human being or, through inaction, allow a human being to come to harm.
2. A robot must obey the orders given it by human beings except where such orders would conflict with the First Law.
3. A robot must protect its own existence as long as such protection does not conflict with the First or Second Laws.

Later, when robotics in Asimov's fiction had progressed to the point where robots governed entire planets and civilizations, Asimov added a fourth law. He named it the zeroth law and placed it ahead of the other three:

0. A robot may not harm humanity, or, by inaction, allow humanity to come to harm.

At first glance, these rules for programming robot behavior seem pretty good. In fact, they look virtually foolproof for keeping humans in control and preventing robots from harming humans or other robots. But the human characters in Asimov's "Runaround" story soon discover that these simple rules do not guarantee the expected results.

"Runaround," though written in 1942, is set in 2015. Two American astronauts on the sun's nearest planet, Mercury, operate a mining station in the searing heat of the planet's perpetual dayside. Photocells provide the energy to cool the station and keep it habitable for humans when outdoor temperatures are several hundred degrees Fahrenheit. The energy cells need a supply of selenium, an element that melts at 431°F and is in a natural pool seventeen miles from the station. The astronauts send heat tolerant Speedy, an expensive model SPD 13 robot, built by United States Robot & Mechanical Men Corp, out to collect selenium for the station. But Speedy does not return. The two astronauts will broil to death if they do not get the selenium soon. They must find Speedy. Finally, they locate Speedy's shortwave signals only to learn that the robot is endlessly circling the selenium pool. It is not getting close enough to obtain the life-sustaining element and also not responding to signals to return to the station. Speedy is caught in a ceaseless runaround due to Asimov's Rules 2 and 3 for robot morality. The human command to obtain selenium keeps Speedy near the pool, but that command is not strong enough to overcome Speedy's sense of danger to itself due to a corrosive gas emitted from the center of the pool. Rule 2 pulls Speedy toward the center of the selenium pool and rule 3 repels it. The unforeseen result is that the robot endlessly circles the pool's center just far enough away to prevent damage to itself. The only solution is for an astronaut to put on a protective suit and venture outdoors, knowing that he will be safe for just a short time before being baked. The plan is for the suited astronaut to approach Speedy, trusting that Rule 1 will trump 2 and 3, that Speedy will see the human in danger, exit its runaround, pick up the nearly unconscious astronaut in its strong mechanical arms, and return him safely to the station. Next, Speedy will receive a new selenium-collecting command strong enough to override Rule 3. Hopefully, the effects of the corrosive gas will not prevent Speedy's return to the station with the essential selenium. I will not spoil the story by telling how it ends, but it and many of Asimov's other short stories aptly illustrate how unexpected problems and dilemmas arise in seemingly well-thought-out moral systems, even for autonomous acting robots. We need to take Asimov's warnings seriously now that the burgeoning technology of autonomous machines is upon us.

Isaac Asimov made a good attempt to imagine programming ethical behavior into future robots. His own thought experiments demonstrated that the four moral laws designed for robots are not foolproof though. Today, some software designers and robotic engineers question whether a set of rules can ever give the desired results.

At this juncture, it is fitting to take a brief look at some influential moral theories that we use to guide our own behavior. At least four such theories have been seriously considered by software theorists for use in robots: divine command, deontological, utilitarian, and virtue ethics. We will consider the major elements of each of these so we can ponder which, if any, we might wish to impose on robots.

Divine command theory posits that God commands morally correct actions and forbids morally wrong acts. The difficulty is discerning and agreeing upon what God commands. It is also a challenge when commands discerned from humanity's several gods, from different versions of the same god, or even from a single god are in conflict. A philosophical problem identified in the time of Socrates as reported in Plato's dialogue, *Euthyphro*, also plagues the divine command theory of ethics. According to Plato, Socrates asked, "Is conduct right because the gods command it, or do the gods command it because it is right?" Either option spells trouble for the authority and reliability of what humans perceive as divine commands. If an action is right simply because God commands it, the command becomes arbitrary. God could just as easily have commanded something different, so the concept of God's ultimate goodness is undermined. One may counter that God would never command something that is not right; that is, God commands certain conduct because it is good. But this is the second option above. Here, the problem for the believer in an omniscient god is that goodness exists apart from God's will. In this case, God itself is not the source of ultimate goodness. Goodness lies outside God, and God consults that source before making a command. Thus, saying either that an act is right because God commands it or that God commands it because it is right leads to a conclusion about God that is untenable to religiously devout persons.

Despite the philosophical problems with divine command theory, computer scientists and AI specialists Selmer Bringsjord and Joshua Taylor continue to pursue a divine-command approach to engineering ethical AI systems, primarily for military use.[7] They point out that "in a world where human fighters and the general populations supporting them often see themselves as championing God's will in war, divine-command ethics is quite relevant to military robots" (Bringsjord and Taylor 2012, 85). They proceed to develop divine command logic and propose its eventual application in lethal military robots, while not ruling out the possibility that nondivine command-based codes might also be useful in certain autonomous robots.

Immanual Kant's (1724–1804) deontological system of ethics is another rule-based theory for morality. The word *deontological* means "duty- or obligation-based." Kant's objective was to develop a nontheologically based system of morality grounded on reason alone. According to Kant, rules of moral behavior derived from reason alone obligate reasonable persons to follow them. Kant proposed two formulations for an overarching categorical imperative to guide moral behavior (Kant [1785] 1959; Rachels 2003, 131):

1. Act only according to that maxim by which you can at the same time will that it should become a universal law.
2. Act so that you treat humanity, whether in your own person or in that of another, always as an end and never as a means only.

The first formulation emphasizes that reason provides the foundation for Kant's moral system. Take lying for example. If your maxim (a statement assumed to be true) is that lying to give yourself an advantage over others is fine, then the test for its morality is to ask whether you can reasonably will that everybody lie whenever they wish. Since the advantage gained by lying depends upon others' believing the lie, the proposed maxim fails the universality test. If everyone lied, then nobody would believe anything they were told, and lying would cease to give anybody an advantage. Therefore, lying is immoral according to Kant. Notice that the universality test does not judge whether the act produces good consequences for humanity but only whether it is logically self-defeating if the act were universalized. One can imagine instances where lying would feel like the right thing to do, such as in the 1930s when Nazi soldiers knocked on doors and asked whether Jews were being hidden in the house. If you were hiding a Jewish family, to tell the truth would mean certain death for the family and probably for you as well. In this and other imaginable cases, it seems all right to disobey Kant's rule against lying. If a robot were given a deontological software system for ethical behavior, how would it "feel" when it is appropriate to disobey a rule?

The second formulation of Kant's categorical imperative is about the intrinsic worth and dignity of human beings. Essentially, it says that it is immoral to use another person simply as a means toward a personal end of your own. To do so would disrespect the person's right to and the freedom to strive toward his or her own ends. Rape is an obvious example of an act that uses another person as merely a means and is wrong. Forcing slaves to work cotton fields is another act that fails this test of morality. On the other hand, hiring a person to work in your garden for a fair wage is all right because the person willingly accepts the work and receives compensation for it. But what about instances where a person works as a prostitute with little or no choice or someone has a servant earning so little money that she or he cannot reasonably leave the job? I pose these questions and others rhetorically for the reader to ponder. Questions are important, but answers are hard to come by. If robots someday approach the moral status of human persons, how will we as their creators reconcile our use of such robots with the second formulation of Kant's categorical imperative, to treat all persons with dignity and not solely as a means to an end?

Utilitarianism is another major nontheological, rule-based theory of morality. Developed in the eighteenth and nineteenth centuries by David Hume (1711–76), Jeremy Bentham (1748–1832), and John Stuart Mill (1806–73), utilitarianism's early formulation derived from Bentham's principle of utility, which requires that we choose the action that has the best overall results for everybody affected by that action. Utilitarianism is a consequentialist theory of ethics because it takes into account the likely consequences of conduct when assessing the morality of an action. Another way of stating the principle of utility is the familiar dictum: when choosing between two or more possible actions, the one that produces the greatest good (or least suffering) for the greatest number of sentient beings is the right act.[8]

Deciding whether an act is right based solely on the estimated amount of goodness or suffering likely to result from that act is called act-utilitarianism. The difficulty comes in comprehensively calculating and comparing the levels of good between alternative actions. For example, how should the concepts of knowledge, wisdom, financial security, feeling loved, being respected, and good health be ranked when calculating the goodness that an act is likely to bring? Is one of these worth more than another? How do the age, gender, ethnicity, or economic status of persons affected by an action influence how the goodness of an act is assessed? How does one calculate the magnitude of the "ripple effect" from good or bad acts? Enabling a child to feel secure and loved may result in that child being a better parent for an unknown number of future children who may in turn pass that goodness on to their children and those to theirs. The same can be said for providing a child with an education. How can one quantify the level of happiness or disappointment resulting from an education?

Comprehensively calculating the long-term benefits of individual actions can be very complex. Furthermore, strictly following the principle of utility allows little time and energy for persons to pursue personal happiness. To avoid pitfalls of act-utilitarianism, some philosophers suggest following a re-formulated version of utilitarianism called rule-utilitarianism. Rather than judging the goodness of each individual act, rule-utilitarianism requires us to identify rules that best support the happiness and well-being of every-body in our society. This releases utilitarianism's adherents from having to ponder the ultimate goodness of each and every act. They need only re-fer to preestablished rules that promote the common good. But even rule-utilitarianism is not without difficulties. For example, is it all right to ignore a rule when to do so would clearly result in greater good than enforcing it? At this point, we are thrown back to the difficulty that deontological ethical programs pose for a robot. How is a robot that is endowed with rule-based ethics to sense when it is better to disobey a rule? Recall the quandary of Asimov's poor little Speedy endlessly circling the pool of selenium on the surface of Mercury.

In the context of robot ethics, divine command theory, Kantian deonto-logical ethics, and utilitarianism are all "top-down" approaches for design-ing moral software to guide robot behavior. Rules extrinsic to the robot it-self are imposed from the top. We finish this brief survey of ethical systems with a potential "bottom-up" approach for controlling robot behavior, virtue ethics. In this case, the robot learns correct behavior over time and through experience, much as a child does.

Virtue ethics emphasizes how to become a good person over time rather than prescribing rules or giving commands to instruct adherents how to be-have. A practitioner of virtue ethics who faces a moral dilemma does not ponder what rule to follow but instead asks herself what a role model would do in a similar situation. Aristotle in his *Nicomachean Ethics* (325 BC) and also Socrates and Plato (fourth century BCE) were proponents of virtue ethics, which asks, "What character traits make for a good person?" There is remarkable agreement across cultures on traits that characterize good people. These include:

Kindness	Helpfulness
Courage	Courteousness
Honesty	Friendliness
Tolerance	Tactfulness
Generosity	Moderation
Patience	Trustworthiness
Loyalty to family and friends	

Aristotle described a virtue as a desirable character trait manifested in habitual action. We become virtuous by developing and nurturing habits that lead to virtue. If we believe that honesty is a good thing and we practice honesty in our everyday living over a long period of time, we acquire the virtue of honesty as part of our character. We are not honest because there is a rule that says we must be honest. Rather, we are honest because we practice the habit of honesty and believe in it. In our society, which is becoming more and more secular, virtue ethics is experiencing a renaissance in the wake of criticisms of moral systems that depend upon laws, rules, obligations, or duties in the absence of a lawgiver. Reinforced learning and the formation of habits are closely related. So some writers suggest that virtue ethics is the best option for instilling morality into the behavior of autonomous robots if they are programmed to learn.

A step in this direction is the development of a conceptual and computational model of cognition called *learning intelligent distribution agent* (LIDA), which shows promise for providing a foundation for development of a humanlike AMA (Wallach and Allen 2009, 171–87). Meanwhile, other computer scientists and AI designers observe how preschool children learn naturally, and they apply that information to machine-learning strategies (Gopnik 2017). If we eventually decide to give robots the capacity to develop virtues, we will need to decide which virtues are most desirable and under what circumstances. Truth telling may be virtuous for a robot performing scientific experiments but not for a military robot captured by the enemy. Who will make decisions about moral software for AMAs?

Problems and ambiguities will arise regardless of which moral system(s) future autonomous robots possess. For example, consider autonomous, self-driving cars. This technology is already with us in a limited way and is poised to expand rapidly during the next decade. Imagine a situation in which a car must decide between a head-on collision with another car or turning off the road into a wedding party in order to avoid the collision. A car equipped with a rule-based moral system might be instructed by its moral compass to always maneuver so as to protect its passengers from injury. So it would turn into the unsuspecting wedding party, perhaps killing and injuring many people. By contrast, a car with a utilitarian moral system would likely continue on the collision course, perhaps killing passengers in both vehicles. As the buyer of a self-driving car, which moral software would you choose? As a biker or walker who rarely rides in a car, which software would you prefer manufacturers to use?

Should federal regulations specify the moral software for all cars whether imported or manufactured in the country? Or should decisions about vehicular moral software be left solely to manufacturers? How about the buyer? Should she/he be able to choose among an array of moral systems for a new car?

Perhaps every autonomous self-driving car should be equipped with several different moral systems, allowing the owner to toggle back and forth between systems depending upon the driving environment. If individual autonomous cars come equipped with multiple moral software programs, should a visible signal be mandated on the cars to inform pedestrians, bikers, and other cars which morality guides the behavior of every car they encounter? The fatalities in Florida and Arizona in 2016 and 2018, respectively, involving self-driving cars highlight some of the several technical hurdles remaining to be overcome with this type of robot.

Worries about AI as an Existential Threat to Humanity

As some computer software engineers and AI scientists cogitate over which moral system(s) are most appropriate for robots in particular circumstances, others concern themselves with the possible existential threat to humans posed by advances in AI. One of these is computer scientist and author Stuart Russell. In 2014, Russell joined the Centre for the Study of Existential Risk at Cambridge University in the UK. His role there is to examine "risks that could lead to human extinction." In an interview, Russell warned about the dark side of future AI (Bohannon 2015). For example, Russell cites possible negative consequences of giving autonomous machines objectives that the machines carry out without having the capacity to consider all of the situations that humans care deeply about. Military weaponry such as armed drones fully controlled by AI systems is especially worrisome.

Russell compares AI research to nuclear fusion research. The latter was first aimed at developing a virtually endless supply of clean energy but with the knowledge that uncontained fusion reactions could cause unimaginable destruction if weaponized. Russell points out that fusion researchers now include containment in their objectives as a matter of course. By contrast, AI research is progressing rapidly without appropriate attention to the potential negative outcomes of creating pure intelligence. Russell acknowledges the difficulty of designing regulation now for varieties of AI software that have not been conceived and cannot yet be described. Recently, fifteen philosophers and social studies scholars expressed a similar concern in a public letter in a special issue of *Science* magazine devoted to AI (Didier et al. 2015). For a fictionalized description of a fearful dark side of robot-human relations, Daniel Wilson's 2011 novel *Robopocalypse* is hard to beat.

Can Robots Ever Be Persons?

AI and personhood. Is it just science fiction to imagine that someday autonomous robots will have the cognition, emotions, and moral agency to warrant being granted the status of personhood? We cannot now answer with certainty, but prudence prompts us to begin preparing for that eventuality. Fifty years ago, few could imagine that we would decipher the entire 3.1 billion letter code of the human genome, let alone be able to sequence the genome of any person in hours or days for $1,000. Even thirty years ago, few could imagine a new discipline called synthetic biology through which entire genomes can be synthesized from off-the-shelf chemicals and be made to work inside living cells. Twenty years ago, few would have suggested that we would today be on the verge of creating living cells from the ground up or be able to redesign genomes quickly and inexpensively with a technology called CRISPR.

Recall from chapter 6 that the US BRAIN Initiative aims to understand the neuronal basis of human consciousness and that the EU HBP may ultimately build a computer that mimics the human brain. At this moment, neuroscientists are homing in on the neural circuitry that gives rise to the biological sense of self, that is, how we distinguish between ourselves and the rest of the external world (Liepelt and Brooks 2017). When we understand how body ownership and agency (the conviction that we can control the actions that we initiate) are encoded in ourselves, it may well become possible to perform similar encoding for a robot. So whether humanlike AMAs become reality in ten, fifty, one hundred years, or never, the recent past sends us a strong message: prepare now for the day when Robvie asks for or demands full-fledged personhood.

Testing robots for personhood. Andrew, a robotic product of US Robots and Mechanical Men Corp. in Isaac Asimov's 1976 story *The Bicentennial Man*, has enjoyed a very successful professional life as a writer, historian, and designer, but he is unfulfilled. He wants to be a real man. To this end, he has an operation that allows him to eat a small amount of food, actually a few drops of olive oil. But this is not enough for Andrew, who longs to give up his immortality, which is the one thing that most distinguishes him from real human beings. He wants to become mortal. There is no reason to spoil the ending of this story for those who have not yet read it. But with beautiful poignancy, Asimov makes us confront the real possibility that our creations will someday either consider themselves persons or wish to become persons. When/if this happens, how will we recognize it? What will ultimately compel us to think of entities with AI as persons and attribute to them the dignity and rights we ascribe to ourselves?

To consider this question, one must first define *person*. The word does not mean the same thing to everybody. Some people require membership in the species *Homo sapiens* for personhood, while others do not. For some, personhood begins at conception, while for others full personhood begins at birth or even later after certain cognitive faculties develop. Earlier, I suggested that a person is "a human being who possesses a special dignity conferred upon her or him by other human beings" (Bradley 2013a, 53). I offered this definition in the context of ESCs and the issue of whether the cells in a five-day-old human embryo comprise a person. This definition disqualifies robots for personhood unless they are considered to be a form of human being. Robots are unlikely to meet that criterion unless, someday and somehow, they are made to run on a human genome.

Philosophers who debate personhood often describe a person as *an entity that possesses the moral status we normally attribute to a cognitively healthy adult member of the species Homo sapiens* (Sparrow 2012, 302). Notice that this description does not require that a person be a member of our species, only that it has the same moral status as our species. Let us adopt that definition for our discussion of robots. We will also presume that humans must take responsibility for deciding upon the moral status of robots even though their cognitive abilities may someday make them better able to judge than we.

How should we decide whether a machine deserves the moral status that we normally attribute to a fully formed human being? In 1950, Alan Turing proposed a test to determine whether a machine can exhibit intelligence indistinguishable from that of a human. Known as the Turing test, the hypothetical exam involved a human subject conversing via computer and written text with another human and also with another computer. The subject is not told which is which. If the subject cannot distinguish between the human and computer-generated responses to her questions and comments, the computer and its software pass the Turing test.

The validity of the Turing test can be criticized on several grounds. One criticism is a thought experiment called the Chinese room (Searle 1980). Searle asks that you imagine yourself locked in a room, able to understand English but unable to understand either written or spoken Chinese. In your possession is a set of rules in English that allows you to correlate one set of Chinese characters with another set of Chinese characters. You are then given questions written in Chinese by persons from outside the room who do understand Chinese. Using your rule set, you can compose Chinese answers to the Chinese questions that make sense and lead the posers of the questions to believe that you understand the conversation even though you do not. By analogy, Searle argues that a computer executing a program that carries on an intelligent-sounding conversation with a human does not actually understand the conversation and therefore cannot be said to possess humanlike intelligence. In this way, Searle argues against the possibility of a computer being programmed to possess so-called strong AI, which would render it able to actually understand and also possess other cognitive traits of a human mind. Rather, such machines would possess only weak AI, the ability to act as though they are intelligent, just as the English-speaking subject in the Chinese room only acts as though he or she can understand Chinese.

Still, with an increasing understanding of brain circuitry and steadily improving computer programs that allow machines to learn from experience, the question of whether a machine might someday qualify as a person remains. So in 2004, ethicist and political philosopher Robert Sparrow proposed a Turing triage test for personhood. The test involves a human subject put in a situation where he must choose between ending the life of another human person or the "life" of a robot. If the subject feels that this decision is too difficult and grievous to make and he sometimes actually decides to preserve the life of the robot over that of the person, the robot is deemed to possess the moral status of human personhood (Sparrow 2004).

As a specific hypothetical example, Sparrow imagines a hospital setting in which a robot with AI used for diagnosing patients is able to learn, reason, and make decisions independently. The hospital experiences a power loss such that there is insufficient electricity to save the lives of two patients currently on life support in the ICU. Using information from the robot about the condition of each patient and the likelihood of eventual recovery, you make the excruciating decision to preserve the life of one over the other. Next, the power loss becomes so severe that you are unable to keep the robot running and also maintain the life of the single human left in the ICU. If you opt to preserve the life of the human and turn off power to the robot, the robot's circuits will be irreparably damaged, ending the life of the robot forever. Both the human and the robot are pleading for continued life. Sparrow argues that when this second decision poses a moral dilemma as difficult as the first, machines will have acquired the moral status of human persons.

One problem with the Turing triage test is its lack of objectivity. Decisions on personhood for machines are placed in the hands of human deciders, each of whom necessarily brings personal emotions, prejudices, and experiences to the decision-making process. One person in the hypothetical dilemma may rule in favor of the life of a particular robot and another not, depending upon previous personal interactions with that robot. Moreover, a decision to save the life of a robot over that of a human in the Turing triage test might be made solely on utilitarian grounds. That is, considering that the robot's superb medical diagnostic abilities are likely to save many human lives in the future, one may justify the sacrifice of one tenuous human life now so that the robot can continue to serve other humans. Such a decision need not have anything to do with humanlike, cognitive, or emotional traits of the robot. Human penchants to anthropomorphize machines could also compromise the integrity of the Turing triage test. Given the problems with the original Turing test and with the Turing triage test, four new Turing tests were recently proposed. These are summarized by Gary Marcus, director of Uber AI Labs and professor of psychology and neural science at New York University (Marcus 2017). The first, called the Winograd schema challenge, tests whether a computer has the "common sense" to discern the meaning of ambiguously worded statements. For example, the statement: "City fathers refused a permit for the demonstrators because they promoted violence." The word *they* is ambiguous, and commonsense knowledge is required to discern the statement's meaning. The second proposes standardized testing for machines using the same written tests currently used by elementary and middle school students. The third is only for embodied AI and would require the robot to manipulate real-world objects to build a structure and/or to solve a problem. Finally, the I-Athlon test asks AI "to summarize the contents of an audio file, narrate the storyline of a video, translate natural language . . . and perform other tasks" (Marcus 2017, 62). Each of these and the original Turing test can assess a degree of machine intelligence, but are they sufficient to make a decision on personhood? I maintain that they are not because they do not assess moral decision-making capacity, a trait possessed by all normal cognitively healthy, adult human beings.

One widely recognized trait of human persons is a capacity for moral choice making (Bradley 2005) or at least the potential to develop such a capacity. We all know from experience that different humans presented with the same moral dilemma and with access to the same information often make different choices. Examples include whether women have the right to terminate pregnancies, whether scientists can ethically perform research on human embryos, and whether we should genetically enhance our own germ line.

I propose a variation of the Turing triage test in an attempt to remove human subjectivity from the test and to add a dimension of moral decision making to it (Bradley 2013c). In this test, prospective robot persons interact with each other on moral choice-making problems. Imagine two machines manufactured identically and provided with the same information relevant to a human triage dilemma like that in the ICU of the hypothetical hospital with a power outage described above. If the machines disagree with each other on the solution to the problem, I suggest that they and other machines like them may qualify as persons.

One problem with this test is that two robots agreeing with each other on the solution to a moral dilemma, as many pairs of humans may do, may deserve the status of personhood but remain unidentified by the test. To minimize the chance that this unfortunate oversight occurs, robot pairs agreeing with each other in initial rounds of the test should be presented with additional and increasingly difficult moral dilemmas. At some point, after robots have failed to disagree on solutions to many difficult dilemmas, we would need to conclude that they are not persons.

We could attempt to solve the robot personhood question by denying that it will ever come up. For example, we might say that cognitive abilities are one thing but emotions are quite another. We may wish to believe that robots will never have emotions. And since emotions play an essential role in moral decision making, we may tell ourselves that robots can never be moral agents. It is just a short step to go further and say that since moral agency is part and parcel of full personhood, robots can never be persons. But is it really so far-fetched to imagine a day when AI will give robots genuine emotion? Human emotions appear to be products of electrical activity in the brain's limbic system and of chemicals released at junctures between brain cells. If this is so, there is no rational reason to believe that the physical basis for emotion, including love, disgust, compassion, fear, hate, sadness, joy, and depression, will not someday be thoroughly understood and re-created in a robot.

Once we understand the physical basis for emotion, the possibility to create AI with a facsimile of human emotion will be real. And when that emotional capacity becomes integral to autonomous, learning, thinking robots, the question of robot personhood will be a serious social, ethical, and legal issue. Recall that the national Chinese brain research project considers the building of a robot with a mind capable of empathy to be a realistic objective for cognitive robotics. Now is the time to determine how best to prepare our children and grandchildren to live a good life in their imminent future with robots.

Chapter Summary

Differing definitions of *robot* exist. Here, we consider robots to be machines that sense, think (analyze and learn), and act in the real-world environment. They need not look like humans, but they must possess at least some degree of autonomy. A completely remotely controlled device is not a robot by this definition. Today, robots assist with medical attention and companionship, transportation, household maintenance, on the battlefield, exploring extraterrestrial environments, in industrial settings, and in bedrooms. As the autonomy of robots increases and their domains of activity expand, ethical issues arise including a need for some robots to be endowed with systems of morality. Rule-based moral systems present several difficulties. The emergence of AI programs for learning makes virtue ethics an attractive system for certain applications. Human emotional bonding to robots, engaging robots as sex partners and eventually marriage partners, autonomous lethal robots, the programmed behavior of fully autonomous self-driving vehicles, and the possibility that robots will someday qualify as persons with the moral status of adult humans all present special challenges to human values that deserve immediate broad public discussion.

Questions for Reflection and Discussion

1. Do you or someone you know own a robot? If so, describe your relationship with the robot. Does your relationship have an emotional component?
2. What type of moral system(s) do you think a caregiving robot should have? An autonomous, lethal military robot? An autonomous self-driving car?
3. Do you believe that military robots should have autonomous decision-making power to kill human beings?
4. If you have sex with a robot, are you being unfaithful to your spouse, boyfriend, or girlfriend? Why or why not? How do you feel about future human-robot marriages?
5. In your view, can a robot ever be a person? If so, how would you know when it qualifies? If not, why not? How do you define *person*?

Notes

1. More information on the Neato Botvac is at http://robot-vacuum-review
.toptenreviews.com/neato-botvac-review.html#sthash.sQX3vmCs.dpuf (accessed April 20, 2017).

2. The ARMAR family of robots was developed and produced by Karlsruhe Institute of Technology, https://his.anthropomatik.kit.edu/english/241.php (accessed June 11, 2017). Among the tasks ARMAR-3 is able to do is to open a dishwasher, load items into it, and close the door when finished.

3. HelpMate is described at https://spinoff.nasa.gov/spinoff2003/hm_4.html (accessed May 31, 2017).

4. Information on the Nursebot Project and a report on development and field testing of a nursebot named Pearl are at http://www.cs.cmu.edu/~flo/scope
.html and www.cs.cmu.edu/~nurse-bot/web/papers/umich/aaai02wkshp.pdf (accessed May 31, 2017).

5. TrueCompanion's website bills its sexbots as delivering unconditional love and acceptance.

6. Dr. Richardson is a robot anthropologist and ethicist interested in attachment theory, gender, and robotics. She is a senior research fellow in the ethics of robotics at the Centre for Computing and Social Responsibility (CCSR) at De Montfort University, UK. The campaign's rationale is described on its website: "Over the last decades, an increasing effort from both academia and industry has gone into the development of sex robots—that is, machines in the form of women or children for use as sex objects, substitutes for human partners or prostituted persons. The Campaign against Sex Robots highlights that these kinds of robots are potentially harmful and will contribute to inequalities in society. We believe that an organized approach against the development of sex robots is necessary in response [to] the numerous articles and campaigns that now promote their development without critically examining their potentially detrimental effect on society"(http://campaignagainstsexrobots.org/ [accessed March 6, 2016]).

7. Selmer Bringsjord is on the faculty of the Departments of Cognitive Science and Computer Science at Rensselaer Polytechnic Institute (RPI) in Troy, New York. Joshua Taylor, a recent PhD graduate of RPI, currently works at Siege Technologies on advanced computer security technologies in the Rome, New York, office and continues his collaboration with Dr. Bringsjord on a logic textbook/software package.

8. Utilitarianism is concerned about increasing happiness and decreasing suffering for all sentient beings not just humans. A modern advocate of utilitarianism is Peter Singer, a moral and political philosopher at Princeton University who advocates for animal rights.

Sources for Additional Information

Bengio, Y. 2016. "Machines Who Learn." *Scientific American* 314(6): 46–51.

"From Horseless to Driverless." 2015. *Economist*, August 1, Supplement: The World If, 15–16.

Gear, A. G. 2016. "Pepper, the Emotional Robot, Learns How to Feel Like an American." https://www.wired.com/2016/06/pepper-emotional-robot-learns-feel-like-american/ (accessed June 10, 2017).

Greenemeier, L. 2016. "Deadly Tesla Crash Exposes Confusion over Automated Driving." *Scientific American*, July. https://www.scientificamerican.com/article/deadly-tesla-crash-exposes-confusion-over-automated-driving/ (accessed June 9, 2018).

———. 2018. "Uber Self-Driving Car Fatality Reveals the Technology's Blind

Spots." *Scientific American*, March. https://www.scientificamerican.com/article/uber-self-driving-car-fatality-reveals-the-technologys-blind-spots1/ (accessed June 9, 2018).

Popular Science, Special Edition. 2016. *The New Artificial Intelligence.* New York: TimeInc. Books. This issue contains thirty-nine short articles on the current and future state of AI and robotics distributed under five section titles: Training, Medicine, Earth and Beyond, Art, and Everyday Life.

Russell, S. 2016. "Should We Fear Supersmart Robots?" *Scientific American* 314(6): 58–59.

Shladover, S.E. 2016. "The Truth about 'Self-Driving' Cars." *Scientific American* 314(6): 52–57.

"Special Report: Robots: Immigrants from the Future." 2014. *Economist*, March 29, 50–64.

Wilson, D. H. 2011. *Robopocalypse*. New York: Random House.

8

Responsibilities and Living Well with Modern Biotechnologies

Science now finds itself in paradoxical strife with society: admired but mistrusted; offering hope for the future but creating ambiguous choice; richly supported yet unable to fulfill all its promise; boasting remarkable advances but criticized for not serving more directly the goals of society.

—J. Michael Bishop, American microbiologist and cowinner
of 1989 Nobel Prize in Physiology or Medicine

These final two chapters offer thoughts about living well with the transformative technologies described in this book. The views reflect my life as a scientist and university teacher over forty-five years and as a world citizen with more than a casual interest in bioethics and how science and the humanities intersect.

I am a cell biologist by training and practice. I like learning about how living cells work and how their internal structures relate to life functions. I also like thinking about ethical and societal implications created by discoveries in cell and molecular biology. The biotechnologies examined in this book are rooted in basic discoveries about how cells work. I am not formally trained in philosophy or the social sciences, but neither are most of us. That does not mean that we cannot contribute meaningfully to a dialogue on how humans ought to develop and use twenty-first-century biotechnologies. We each have a perspective that needs to be heard. As a biology teacher and researcher, and without the jargon or expansive knowledge of the social science literature, I use this chapter to share my thoughts on specific responsibilities humans have for their future with biotechnology.

In my view, the era of biotechnology began in the 1970s. That decade brought the discovery and use of enzymes that cut and paste DNA segments to produce recombinant DNA molecules. Recombinant DNA derives from two or more sources, often from different species. Early products of recombinant DNA technology were bacteria-producing human insulin and human growth hormone, which offer cheaper and more effective treatments for diabetes and growth irregularities. Other writers mark the beginning of the biotechnology era as early as the 1920s with the first use of electroencephalography in humans. Still others put it in 1953 with the announcement of DNA's double helical structure, capping work by Rosalind Franklin, James Watson, and Francis Crick among others. It does not matter when we say the era began. Important is to recognize that we are in the midst of it now, that the era may never end, and that we all need to do our part to help steer humankind's burgeoning abilities to manipulate life in directions that serve not just our personal interests but the well-being of all humanity and the biosphere.

Making wise choices in using present and future biotechnologies consists mostly of commonsense ideas and demands that we communicate truthfully with, listen respectively to, and learn from each other. These basic life skills need nurturing before grade school and honing for a lifetime. Even for Olympians and professional athletes, mastering and adhering to basics is essential for success. In a big game or at a high level of competition, a player may bypass basic form while executing a maneuver. One sloppy pass just before the buzzer can cost the ball game. Similarly, as we decide how to develop and use biotechnologies, good communicating, listening, and learning between all stakeholders are basic and crucial needs for a healthy future with biotechnology. Stakeholders include scientists, the nonscientist public, news media personnel, policy makers, and others such as teachers, financiers, and clergy positioned to influence attitudes and activities of others. A section is devoted to each of these five stakeholder groups and their responsibilities.

Current and future technologies will profoundly alter humankind and the rest of life on Earth. Author and scholar of culture, values, and change Joel Garreau (2005) coined the acronym GRIN—genetic, robotic, information, nanotechnologies—to refer to these transformative technologies. The present volume and an earlier one (Bradley 2013a) examine specific examples beneath the GRIN umbrella including human embryo selection, cloning, genomics, age retardation, cognition and mood enhancement via pharmaceuticals and brain-computer interfacing, the trajectory of modern brain research, nanobiology, genome editing, and synthetic biology. Ethical issues associated with the development of GRIN technologies and long-term outcomes of their uses are oceanic. This is not to encourage fear of GRIN technologies or to suggest that neo-Luddism is an appropriate stance toward them. Their potential is great for many benefits to human well-being and planetary health. Rather, amid justifiable excitement over modern biotechnologies, we must remind ourselves that benefits come with risks and that long-range planning is in both the Earth's and our best interests.

Decisions about the future of our species and the biotic world within which *Homo sapiens* has evolved and is embedded must include wisdom and input from the greatest diversity of stakeholders as possible. But more than gut feelings are needed. Useful input depends upon informed opinion. Requisites for informed views about developing and using modern biotechnologies include knowledge about the technologies and their transformative potential, an appreciation for the antiquity and interdependent nature of evolved life on Earth, and a moral sense about what good decisions look like. If one agrees with these requisites and with the need for long-range planning in using GRIN technologies, the job before us is five-fold: *(1) identify group-specific responsibilities, (2) educate children and ourselves about the nature of the biosphere, biotechnology, and the social and ethical implications of the latter, (3) cultivate an appropriate moral sense or ethics, (4) engage in group discussion and decision making about humankind's long-range goals, and (5) implement informed, consensual decisions.*

This chapter addresses objective 1. Objectives 2, 4, and 5 are mainly what, in 1969, Gerald Feinberg proposed as the Prometheus Project, discussed in chapter 9. Objective 3 is the subject of a recent book by philosopher of science and technology Shannon Vallor (2016). In chapter 9, I briefly describe and comment upon her ideas. I do not have a foolproof plan for living a good life with modern biotechnologies. As a beginning though, I suggest that scientists, policy makers, and various groups of nonscientists embrace certain responsibilities.

In identifying responsibilities of scientists and nonscientists for actions and attitudes that foster good living with transformative technologies, I draw upon the insights and examples of several persons whose minds and lives I especially admire. Among these are the young Renaissance philosopher Pico della Mirandola; twentieth-century polymaths Jacob Bronowski and Gerald Feinberg; conservationist Aldo Leopold; one of the greatest biologists of our age, Edward O. Wilson; twenty-first-century molecular biologist Jennifer Doudna; and moral philosopher of science and technology Shannon Vallor. The special contributions of these citizen scholars in bringing forth, consciously or unknowingly, an ethical and healthful path into the future with biotechnology for us and our descendants emerge as we proceed.

The following questions guide our consideration of the special responsibilities of different types of persons:

1. What are the major responsibilities of scientists?
2. What are the responsibilities of nonscientist citizens?
3. What special responsibilities do policy makers and others with social influence have?

Scientists' Responsibilities

The primary responsibilities of scientists are 1) truthfulness in reporting research findings, 2) communication to nonscientists about the nature of science and the societal implications of their lines of research, 3) awareness and practice of research ethics, and 4) teaching these habits to students. The challenge is in the doing.

Unfortunately, many forces from within and without the science community oppose rather than aid scientists in performing these fundamental responsibilities. Extraordinary pressure to obtain funding for research and then to publish in order to keep one's job, temptations to accept funding from sources that run counter to one's personal values, the task of mentoring and overseeing graduate and postdoctoral students subject to similar pressures and temptations, challenges of multitudinous institutional duties, combined with too little time all conspire to undermine the responsible practice of science. Fortunately, most scientists respect the values integral to the scientific endeavor. Scientists also have past and current role models of responsible membership in the fellowship of science. We meet some of these role models as we examine each of the fundamental responsibilities of a scientist—truth telling, communication, ethical research, and teaching. First though, we will briefly consider the broader landscape of the relationship between science and human values.

Some persons, including some scientists, believe that science is value free or value neutral. They believe that the work of scientists is to discover new things about the world, not to make judgements about the rightness or wrongness of the uses to which their discoveries are put. This thinking is half right. Discoveries themselves do not have a moral aspect. But it is wrong to say that the use to which a scientific discovery is put or even the activity of striving after knowledge about a particular natural phenomenon, is of no moral matter to the scientist.

Jacob Bronowski distinguished between discovery and the process of discovery with characteristic power and eloquence: "Those who think that science is ethically neutral confuse the findings of science, which are, with the activity of science, which is not" (1965, 63). Bronowski identifies six human values that arise from, and are essential for, the activity of science: truthfulness, independence, originality, dissent, freedom, and tolerance. The activity of science not only contributes to and nurtures these values but it also reflects the prevalent human virtues of curiosity, persistence, courage, and cooperation, writes Bronowski (1965).

Bronowski is absolutely right. Largely unappreciated is the degree to which science has emerged from and contributed to the humanities via these virtues and values. Also underappreciated is the degree to which modern science can reflect the values of society. Twenty-first-century science is an expensive enterprise. It requires costly instrumentation and supplies and a complex and stable infrastructure, most often supplied by research universities or private corporations and institutions. In the case of state-supported institutions, the tax-paying public supports science through revenue collected by the state. Although the link may seem circuitous, research initiatives funded by democratic governments usually reflect values endorsed by a majority of a nation's citizens. For example, funding to understand cancer cell behavior reflects the value society places on preventing suffering and death due to cancers. Age retardation research reflects value placed on living longer. Research that aligns clearly with societal values is deemed beneficial, while research moving in directions counter to the values of the society within which it is performed, such as germ cell gene editing, is considered dangerous or even immoral.

It is not always easy to distinguish between beneficial and treacherous research. First, because culturally pluralistic societies contain diverse values, some research appears to have valuable, ethical applications to some groups but not to others. ESC research is an example. Even in a pluralistic society, there may be both widely agreed upon benefits and misapplications for many lines of research at their outset. For such dual-use research of concern, it can be difficult to assess the ethical dimensions of proceeding with it. Creation of air-borne strains of pathogenic viruses and human germ line enhancement are examples. Finally, the benefits and risks of research may not be apparent until long after it is performed. Examples include early investigations into the structure of the atom and the chemical basis of heredity, which later led to technologies such as nuclear power and weaponry, recombinant DNA, cloning, and gene editing. Let us turn now to the first of the four fundamental responsibilities of a scientist—truthfulness.

Truthfulness

Again, we find wisdom in the words of Jacob Bronowski. Nobody discusses the relationship between truth and science with more acumen than he does in his short, gem-like book, *Science and Human Values* (1965). In it, Bronowski calls our attention to three aspects of truth in science: (1) the nature of truth for the process of science, which he calls *the habit of truth*, (2) honesty among scientists, and (3) a social dimension of truth necessary for the scientific endeavor itself.

Bronowski begins by noting a similarity between creativity in science and in art. For both artists and scientists, he says, the act of creation lies in discovering a likeness between facts or experiences that was not previously noticed by others. Uniting previously disparate things or phenomena produces a new concept, which may be expressed by an artist as a metaphor in a poem or a painting or embodied by a scientist in a word like *gravity*, *mass*, *atom*, or *pluripotency*. The fate of a new concept distinguishes art from science. While art may endure forever in its original presentation, a scientific concept must conform to new "facts" uncovered by ongoing experience. When fresh empirical evidence contradicts a currently held scientific concept, the concept faces revision or abandonment. This new or revised scientific concept or belief is of course subject to the same fate as time marches on. The winnowing process itself is scientific progress. The artist, by contrast, is unconstrained by an empirical boundary of truth. This provides a dimension of creative freedom not open to the scientist. In nearly every poem, sculpture, dance, painting, or musical composition one can enjoy products of the artist's unbounded creativity. An example is found in scene 106 of Ben Johnson's seventeenth-century masque, *Chloridia: Rites to Chloris and Her Nymph*, where springtime is presented as a messenger, arriving on the wings of the west wind, Zephyrus, to bring happy news to planet Earth, the mother of us all (Parfitt 1988, 332):

Come forth, come forth, the gentle Spring,
And carry the glad news I bring
   To earth, our common mother:
It is decreed by all the gods

The heaven of earth shall have no odds,
But one shall love one another.

Another dimension of truth in science is the act of honesty to oneself, to the community of scientists, and to the human community at large about the possible risks of discovery. Informing other scientists and nonscientists about conceivable negative repercussions to society of certain lines of research or specific discoveries is an important responsibility. Sometimes action is initiated by a single scientist and at other times by groups of scientists. The former is epitomized by a letter written by physicist Leo Szilard, signed by Albert Einstein, and delivered to Pres. Franklin D. Roosevelt in August 1939, with Europe on the verge of a continent-wide war (Einstein 1939).

In 1933, Szilard conceived the concept of a nuclear chain reaction from the fission of atomic nuclei. He soon realized the profound implications for society if the huge amounts of energy projected to be released by nuclear fission were to become reality. The Einstein-Szilard letter warned of the imminent possibility that German scientists could develop the means "to set up a nuclear chain reaction in a large mass of uranium, by which vast amounts of power and large quantities of new radium-like elements would be generated. . . . If a bomb were created using this new technology and carried by boat and exploded in a port, [it] might very well destroy the whole port together with some of the surrounding territory," warned the physicists. Later that same year, largely as a result of Szilard's letter, the Manhattan Project became reality. This top-secret project, a collaboration between the United States, Canada, and England, aimed to harness energy released during nuclear fission reactions. By 1943, research had progressed to the point where building an atomic bomb became feasible. A secret military laboratory at Los Alamos, New Mexico, received that mission, code named Project Y. Presumably, the Manhattan Project was a response to a foreseen nuclear threat from Germany. By the time Project Y realized its mission by successfully testing an atomic bomb on July 7, 1945, much had happened on the political and war fronts. The United States had entered World War II after the Japanese attack on Pearl Harbor on December 7, 1941, Roosevelt had died in April 1945, and Germany had surrendered a month later. If the atomic bomb were to be used, it would not be in Europe but against Japan.

Eleven days after the first test of an atomic bomb successfully exploded over the sands of central New Mexico, a petition authored by Leo Szilard

(1945) and signed by sixty-nine other scientists arrived on the desk of Pres. Harry S. Truman. The petition noted that the atomic bomb was developed out of fear that Germany might attack the United States using atomic weaponry and that the country's only defense might lie in a similar counterattack. With Germany's defeat and surrender, that danger was cancelled. Szilard was concerned about the bomb's use against Japan and wrote the following to Truman:

> The war has to be brought speedily to a successful conclusion and attacks by atomic bombs may very well be an effective method of warfare. We feel, however, that such attacks on Japan could not be justified, at least not unless the terms which will be imposed after the war on Japan were made public in detail and Japan were given an opportunity to surrender. . . . The development of atomic power will provide the nations with new means of destruction. The atomic bombs at our disposal represent only the first step in this direction, and there is almost no limit to the destructive power which will become available in the course of their future development. Thus a nation which sets the precedent of using these newly liberated forces of nature for purposes of destruction may have to bear the responsibility of opening the door to an era of devastation on an unimaginable scale. (Szilard 1945)

We know the tragic events that soon followed. On August 6, nineteen days after Szilard's petition arrived on the president's desk, an atomic bomb was detonated over Hiroshima, Japan, and three days later over Nagasaki. The two bombs took 120,000 lives during the first day after the bombings, and nearly as many died from radiation sickness and burns during the next few months. Japan surrendered on August 15, 1945. Now in 2019, the worldwide stockpile of nuclear weapons stands at more than fifteen thousand, each one being thousands of times more powerful than the bombs dropped on Japan. These are distributed among nine countries with Russia and the United States tallying about seven thousand each. Does this mean that Szilard and the other petitioners failed? Does it mean that science failed? Jacob Bronowski, a personal friend of Szilard, reported that shortly after the bombing of Hiroshima and Nagasaki he "heard someone say, in Szilard's presence, that it was the tragedy of scientists that their discoveries were used for destruction. Szilard replied, as he more than anyone else had the right to reply, that it was not the tragedy of scientists, 'it is the tragedy of mankind'" (Bronowski 1973, 370).

We are all responsible for how scientific discoveries are used, but the situation is complex. Some argue for more responsibility from scientists. British biologist Solly Zuckerman (1982) and American physician, researcher, and essayist Lewis Thomas (1983) maintain that "scientists" researching nuclear weaponry and defenses against it are really not scientists or engineers at all. Rather, they are enthusiastic technicians. They do what they do largely in the absence of peer review and in military-funded laboratories supplied with virtually unlimited resources and instrumentation. Under the influence of such copious and continuous support, few researchers can objectively assess the morality of their lines of research. Recognizing when research is not progressing toward a solution (e.g., security and peace) to problems ostensibly addressed by that research (e.g., insecurity and aggression) and then altering course are crucial to the scientific enterprise generally. If one is unable or unwilling to do this, does the label *scientist* rightfully apply to him or her? Lewis Thomas comments, "There is no way to design or redesign these weapons so that they can ever be used to win a war or even to fight a war, and no technological fix within the grasp of human imagination that can assure defense against them—if what is to be defended is human society. And yet the 'science' goes on" (Thomas 1983, 92).

Now, we will turn our attention from physics to biology. Dr. Jennifer Doudna at the University of California, Berkeley, exemplifies the individual scientist acting on feelings of personal responsibility by communicating to the nonscientist public about possible implications of her research. Doudna is a developer of the CRISPR gene-editing technology. In an essay published in late 2015, Dr. Doudna describes her personal journey of discovering a technology that is currently revolutionizing biology. Her essay also tells how she came to think about possible social and ethical implications of the technology and took action to help the general public, policy makers, and other scientists grapple with ethical issues raised by CRISPR (Doudna 2015).

Acting as a biochemist doing basic research, Doudna did not often think about the societal implications of her research. But within a year after showing that CRISPR-mediated gene editing in humans and other organisms was feasible, she describes herself "lying awake at night wondering whether I could justifiably stay out of an ethical storm that was brewing around a technology I had helped to create." She responded decisively to her sleepless nights with specific, carefully conceived actions to communicate to diverse stakeholders about ethical issues raised by her research. Within two years, she had helped organize a conference of scientists to discuss and take a position on CRISPR research and use, given over sixty talks for a variety of audiences worldwide, written articles, and appeared before the US Congress and staff members of the White House Office of Science and Technology Policy. Doudna's creativeness as a scientist and her engagement as a humanitarian and world citizen are models and inspiration for scientists across disciplines and career stages.

Many other examples exist of individual scientists acting to inform others about possible misuses of scientific knowledge and/or about how scientific knowledge can be used to correct injustices or relieve risks to humanity's welfare. Some examples of such action by organized groups of scientists are detailed in the later Research Ethics section.

Communication

In early 2015, the outgoing CEO for the American Association for the Advancement of Science (AAAS) commented about the current state of the relationship between science and society: "Science has never been more productive. And yet, the overall climate for science is more difficult than I have ever seen in my scientific career. . . . The relationship between scientists and the rest of the public seems to be slipping backward. Climate change, genetically modified foods, and the teaching of evolution are among topics that now trigger unproductive tension in the fragile relationship between the scientific community and the rest of society. A weakened science-society relationship not only undermines public support for science but also makes it difficult for science to contribute to the solutions of societal problems" (Leshner 2015). One reason for tension between the science community and the nonscientist public is the politicization of high-profile science-related issues with potentially great societal impact like those named by Leshner. The American physicist, who served as executive officer of the US National Academy of Sciences (1994–2011) and was science and technology advisor to the US secretary of state (2011–14), William Colglazier (2016) wrote that "providing objective, high-quality advice that is free of politics and special interests is an important civic responsibility for the science community." He encourages scientists to ramp up their level of encouragement and insistence that government create more and better science-advice systems. Of course, government officials must first wish to solicit high-quality scientific advice in order to create these opportunities, a trait that seems sadly lacking in the highest tier of US government at the present time.

Responsibility for a productive science-government relationship lies with both scientists and policy makers. As for scientists' responsibilities for improving their relationship with government, Colglazier suggests that universities offer courses on science and technology policy and train students for vocations that combine scientific and technological knowledge with an understanding of policy making in both public and private sectors. For improved credibility, scientists and science journalists must practice transparency when their scientific advice incorporates value judgments. Scientific societies could encourage high-quality science communication by offering awards for outstanding science journalism and other forms of science communication. Later, we consider the reciprocal responsibility of government to listen to and heed advice from the science community.

Colglazier believes that science is losing its influence in current policy debates in the UK and the United States because political value judgements trump science. He explains that three varieties of value judgements—distributional, procedural, and evidential—contribute to weaker scientific clout in government:

> Distributional values assess fair outcomes—costs, benefits and risks to individuals, groups, generations, ideologies, things people value and so on. Procedural values assess the fairness of a decision-making process—people are sometimes willing to accept an outcome they disagree with if the process is viewed as fair. Finally, evidential values concern the weight of evidence needed to justify a decision, and the question of "how sure is sure enough"?

These evidential values are the most tangible to a scientist interested in policymaking. But they can be heavily influenced by often conflicting perspectives on distributional and procedural aspects.

The contrasting views of US Republican and Democratic administrations concerning two issues clearly illustrate this phenomenon. Republican administrations have emphasized the scientific uncertainties in evidence on human-induced climate change, and minimized uncertainties in the long-term safety of nuclear-waste disposal. Democratic administrations have done just the opposite. The difference in views of the United States and Europe on genetically modified foods and the "precautionary principle" is another example. (Colglazier 2016)

In the US government, reliance on and respect for scientific information appears to be in decline. Soon after inauguration, President Trump purged all references to climate change from the White House website (Davenport 2017) and deleted references to carbon pollution as a cause of climate change on the EPA website (Kahn 2017). Trump has left the White House Office of Science and Technology Policy virtually vacant (Kang and Shear 2017). Eighteen months into his presidency, Trump had not established a bioethics advisory commission, despite the many issues that demand attention including CRISPR, genome editing and gene drives, robotics, and brain research. With the exception of George H. W. Bush, every US president since 1974 has had a national bioethics commission. Hopefully, both scientists and the Trump administration will take and make opportunities to improve their lines of communication. Consider now communication between scientists and the general public.

Many scientists are and have been exceptional communicators, making scientific findings available to the nonscientific public. Striking examples include the books of Oxford professor and evolutionary biologist Richard Dawkins, the late paleontologist Stephen J. Gould, theoretical cosmologist Stephen Hawking, and the *Cosmos* television productions of astronomers Neil deGrasse Tyson and the late Carl Sagan. Countless lesser-known scientists create similar opportunities for nonscientists to learn about current scientific findings and incorporate them into their worldviews.

Just how receptive is the general public to accepting scientific findings as a key component of individual decision making and public policy making? Will the average citizen enthusiastically embrace new scientific information and use it to guide decision making in her or his personal lifestyle and/or political life? A battery of sociological and political science studies over the past three decades gives complicated and nonabsolute answers to these important questions. Scientists who invest time and attention to communicate about science to the lay public find some of these studies encouraging and others discouraging.

Since about 1990, nonscientist public opinion about science's credibility has become progressively divergent and politically polarized (Gauchat 2012). There is the generalization that conservatives are antiscience evidenced by widespread denial of scientific evidence on climate change, the value of child vaccinations, evolution, and the benefits of embryonic stem cell research. Others claim that liberals are equally skeptical of science as shown by their prevalent opposition to nuclear power as a viable alternative to fossil fuel as well as their doubt over the safety of genetically engineered food organisms.

A so-called intrinsic hypothesis posits that differences between conservative and liberal views on science either reflect real, biologically based, psychological differences between the groups or reveal a strong bias, for whatever reasons, for conservative institutions to more often take a skeptical view of scientific evidence (Mooney 2012). However, recent data suggest that Republicans are not unusually skeptical about policy recommendations based on scientific evidence but rather that Democrats are especially receptive to advice from the scientific community compared to Republicans and Independents (Blank and Shaw 2015).

What most social scientists seem to agree upon is that we all engage in *motivated reasoning*, the processing of new information so that one's conclusions agree with one's preexisting beliefs. For example, if you already strongly oppose government regulating industry in any way, you will likely view scientific information about climate change that argues for stronger greenhouse gas emission standards with greater skepticism than a person not opposed to government regulations would. Likewise, if your religious beliefs do not give full moral status to an individual until the moment of birth, you are more likely to be swayed toward ESC research that helps us learn how it may someday cure diabetes or Parkinson's disease. The opposite conclusion might be reached by someone who assigns full moral status to fertilized eggs.

Motivated reasoning occurs on both sides of most issues. Should this dishearten scientists from communicating their findings to nonscientists? A recent study says, "no." Despite biased processing of scientific information by many persons on diverse topics, a majority of Americans still give credibility to scientific findings and view them as relevant for decision making regardless of whether they self-identify as Democrat, Republican, or Independent (Blank and Shaw 2015). This does not mean that a person gives equal credibility to scientific information on every topic. In fact, deference to scientific findings in policy matters varies greatly by topic and is influenced by ideology. But the results do suggest that science can sometimes provide common ground. This is important at a time when extreme polarization often makes effective communication about relatively objective scientific information difficult.

Presentation style in communicating scientific findings to the public is important. Three political scientists at Stony Brook University point out that factors affecting one's emotions—such as the appearance of the bearer of information, accompanying music, and even the wording used to convey scientific findings or survey questions—influence one's reaction to scientific information. Such factors can have a determinative effect on one's acceptance even before initiating what the receiver perceives to be an open-minded response to new information (Nisbet et al. 2015). Thus, what is believed to be a rational evaluation of scientific information can be more a rationalization for unconscious feelings and/or prior attitudes. Recognizing this phenomenon can help communicators of science accomplish their aim. For example, the Stony Brook authors give evidence that using the term *global warming* to communicate information on climate science is more likely to produce skepticism among conservatives than is the term *climate change*. Ultimately, scientists must persist in communicating their findings to the nonscientist public regardless of the initial reactions of receivers of that information. Awareness that we all engage in biased information processing via motivated reasoning can aid scientists and others by making them more patient and less judgmental of audiences that do not immediately embrace what they have to say.

While communicating research results and their implications to nonscientists, scientists also have an opportunity to explain that science is a "way of knowing." Especially important for nonscientists to appreciate is the tentative nature of scientific findings and the meaning of the word *theory* used in science—not just a guess but a well-established concept with abundant empirical evidence to support it, such as the germ theory for disease or the heliocentric theory for our planetary system.

Necessary risks accompany teaching about the tentative nature of scientific findings and the central role of skepticism in the scientific process. Nonscientists hearing about the revisionist nature of science may bring undue skepticism to the totality of science's findings, especially revelations that challenge their previously held beliefs or preferred life styles. Well-established theories such as evolution via natural processes and climate change due to human activities challenge many laypersons. Science-based recommendations such as childhood vaccinations, decreasing carbon emissions, or attention to exercise and diet for disease prevention may be disregarded due to misunderstanding of scientists' use of the word *tentative* to describe their findings. *Tentative* emphasizes the self-correcting nature of the scientific process, which is a good thing.

Even worse than misunderstanding science's method is when certain groups purposefully misconstrue the scientific process in the eyes of non-scientists or willfully undermine the validity of scientific findings out of real fear or self-interest. For example, Exxon Mobile gains long-term profit if it successfully convinces stakeholders that climate change is not linked to fossil fuel emissions. Similarly, some religious fundamentalist denominations, certain textbook publishers, and some private Christian schools may foment distrust in the scientific theory of biological evolution for ignoble purposes.[1]

Scientists communicating their findings are most effective when they have the trust of citizens representing a broad spectrum of political and religious views. It can sometimes be difficult for scientists to gain and hold such trust when their findings become politicized or acquire philosophical overtones that unfairly, and to the detriment of the public, reflect back at the messenger. Climate change research and ESC research are examples. Scientists must be vigilant for media propensities to attribute partisan or philosophical motives to messengers reporting high-interest, ethically laden scientific findings. Immediately correcting a journalist's misinterpretation during an interview is one way to stem politicization of scientific discoveries. When misinterpretations become published, follow-up letters or interviews help to steer public understanding into the realm of objectivity.

Another way to thwart the media or others from placing research reports into inappropriate philosophical contexts is for scientists to preempt that practice by commenting themselves on the philosophical significance of their work. In fact, evolutionary biologist and author Ernst Mayr wrote: "It should be the aim of every scientist to eventually generalize his views of nature so that they make a contribution to the philosophy of science (Mayr 1997, 36).

Public misunderstanding and media misuse and misinterpretation of science is rampant. Scientists may feel engaged in a losing battle to keep nonscientists adequately and accurately informed about their research, but it is a battle in which scientists must persist and prevail. The alternative is widespread ignorance and even greater misunderstandings about scientific advances that have profound implications for human values and society. Such ignorance and misunderstanding have their own far-reaching, negative consequences for humankind, including the establishment of a small, elite group of persons—scientists and their benefactors—charting the technological future for everyone. This is bad not because scientists and funders are malicious, unsavory, or devoid of commonly held values. It is bad because it is undemocratic.

Research Ethics

A category of research called *dual-use research of concern* is biological research that could reasonably be predicted to provide knowledge, products, or technologies that, if misapplied, could endanger public health and safety, the environment, agricultural plants and animals, or national security (NIH, Office of Science Policy 2017). The phrase *dual use* means that there are not only benefits but also anticipated dangers from the research. The challenge in overseeing dual-use research of concern is to minimize its risks without unduly dampening its benefits or stifling the independence of researchers. Next, we look at three cases of dual-use research of concern and some responses of the scientific community to them.

The 1975 Asilomar Conference on Recombinant DNA exemplifies a community of scientists taking responsibility for the safety and ethical implications of dual-use research of concern. Recall that recombinant DNA refers to laboratory-produced DNA molecules containing regions (base sequences) from two or more sources. For example, the recombinant DNA of a bacterium that produces human insulin has both bacterial and human DNA within its genome. Organized by DNA researcher Paul Berg, who later shared the 1980 Nobel Prize in Chemistry with two other recombinant DNA researchers, the Asilomar Conference of more than one hundred biologists plus some lawyers and physicians developed voluntary research guidelines to help ensure the safety of their work. The guidelines provide practices and mechanisms to help prevent pathogenic microbes from escaping researchers' laboratories. The group also informed the public about their research. The conference illustrates an application of the precautionary principle, which states that when a new technology with unknown effects on human well-being emerges, it is incumbent upon developers of the technology to demonstrate or ensure its safety before proceeding rather than upon others to demonstrate its hazards. The second case involves virus research.

During the better part of 2012, a contentious controversy raged over research done on a subtype of avian influenza virus A named H5N1 (Ross 2013). The issues included gain-of-function research, dual-use research of concern, and assessing the benefits of freedom in scientific communications versus risks to public safety. Gain-of-function research refers to the genetic alteration of a virus or an organism that enhances an existing trait or creates an entirely new trait. Gain-of-function mutations can happen spontaneously in nature or be purposefully created in the laboratory.

The H5N1 virus harbors itself in both wild and domestic bird populations. It is especially lethal to poultry. Transmission from birds to humans is rare but has been documented in about seven hundred cases, mainly among poultry workers in Asia where the mortality rate is about 60 percent. Bird to human transmission is probably due to contact with blood or feces from infected birds. H5N1 infects very deep areas of the respiratory system rather than the upper respiratory tract. Therefore, natural strains of H5N1 do not presently move between individuals via airborne droplets from coughs or sneezes. If H5N1 were to mutate to an airborne form infectious to humans, a pandemic killing tens or even hundreds of millions of people could result. For this reason, billions of dollars have been spent researching H5N1 and stockpiling vaccines against it.

The controversy erupted late in 2011 when two groups, one at the University of Wisconsin and the other in the Netherlands, sought to publish gain-of-function research on H5N1 in the world's two premier science journals, *Nature* and *Science* (Faden and Karron 2012; Roos 2013). Both groups received funding from the NIH to investigate the likelihood of H5N1 mutating to an airborne form transmissible between humans. Both groups also successfully modified the H5N1 genome so that virus transmission among laboratory ferrets became airborne. The transmission between individual laboratory mammals without needing an intermediary host raised the specter of a naturally mutated virus causing a pandemic in humans. Two alarming possibilities of the groups' research are (1) accidental escape of the pathogen from laboratories and (2) details of the research methodology reaching the hands of bioterrorists or other nefarious persons able to replicate the work. The US National Science Advisory Board for Biosecurity (NSABB) recommended withholding methodological details from the papers, essentially a partial censorship of communication between scientists.[2] The US government accepted the recommendations and asked that details of the work be redacted from the manuscripts. Discussion about the pros and cons of withholding or not withholding details of published dual-use research of concern ensued.

Next, the lead authors of the two research articles, along with thirty-seven other influenza researchers, called for a voluntary sixty-day moratorium on using live H5N1 virus particles. They also called for a meeting similar to the 1975 Asilomar Conference to discuss the ethical and safety issues raised by gain-of-function H5N1 research. Within a month the World Health Organization (WHO) convened such a meeting in Geneva. Attending were the lead authors of the controversial manuscripts, representatives from the two journals, bioethicists, and other scientists researching influenza viruses. The Geneva meeting participants reached consensus on two issues:

1. The moratorium on H5N1 research imposed on the newly modified and highly transmissible strains of H5N1 should continue, but should be lifted to allow research on naturally-occurring H5N1.
2. Publication of the complete manuscripts on the modified, airborne strains of H5N1 would have a greater public health benefit than a rushed publication of partially censored manuscripts.

The moratorium was largely adhered to by influenza researchers for a full year after which the organizing group for the research suspension called it off. After the WHO recommendations, the NSABB reversed its earlier recommendation that methodological details be redacted from the papers, and eventually both manuscripts were published, one in *Science* and the other in *Nature* (Herfst et al. 2012; Imai et al. 2012).

What can we learn from the H5N1 controversy regarding responsibilities? First, we must acknowledge that there is no clear-cut formula for dealing with dual-use research of concern. When each case arises, it must be thoughtfully and deliberately analyzed, the perceived near- and long-term benefits and risks weighed, and a wide cross section of stakeholders should be involved in decisions for action. Laudable actions during the H5N1 uproar were that the two lead researchers of the controversial research advocated for a research moratorium, that they called for a meeting of stakeholders to discuss benefits and risks of the research, and that they achieved widespread participation in the WHO's Geneva meeting.

Some people faulted the researchers for performing the work in the first place. But arguments for the public health benefit of their research are convincing. Learning what specific mutations in natural H5N1 strains could lead to a pandemic allows health officials to monitor natural H5N1 genomes for its evolution in that direction, thereby thwarting the threat. In 2014, the US government took the unprecedented step of halting two dozen dual-use research of concern virus research projects for a year while the NSABB developed a new risk assessment process to evaluate gain-of-function, dual-use research (Kaiser 2014). What if NSABB, the scientific community, and/or WHO someday decide that H5N1 or similar dual-use research will cease to be funded, allowed, or published? One likely outcome is that talented young researchers will decline to devote careers to influenza research or related fields. Foretelling the detrimental effects of such a situation is difficult, as is assessment of the risk that a bioterrorist may gain the expertise, resources, and motivation to use published scientific information nefariously. Regular dialogue among scientists, bioethicists, health officials, and the general public along with great care by researchers seem like the most responsible paths forward. The third case involves the genome-editing tool, CRISPR.

In 2015, three groups of scientists met to discuss ethical issues associated with the use of CRISPR technology. Each made somewhat different recommendations. One group called for a voluntary moratorium on CRISPR-mediated editing of the human germ line, while another group acknowledged that it was too early for clinical trials of CRISPR-mediated gene editing in humans and called for international dialogue on the issue. The third group met late in the year after Chinese researchers had already applied CRISPR technology to edit genes in nonviable human embryos. This group was international, including representatives from the US NASEM, the British Royal Society, and the Chinese Academy of Science. Their three-day meeting in Washington, DC, concluded that safety considerations now preclude CRISPR-mediated human germ line editing in viable embryos, but research that could ultimately lead to germline editing should be allowed to proceed due to its potential to prevent human suffering and produce other benefits.

Research Funding and Mission

Writer, cultural critic, and fellow of the American Academy of Arts and Sciences Wendell Berry (2015, 14-15) observes that "the thing most overlooked by scientists . . . is the complicity of science in the Industrial Revolution, which science has served not by supplying the 'scientific' checks of skepticism, doubt, criticism, and correction, but by developing marketable products, from refined fuels to nuclear bombs to computers to poisons to pills." Berry goes on: "Scientists in general, like humanists and artists in general, have accepted the industrialists' habit, or principle, of ignoring the contexts of life, of place, of community, and even of economy."

An example of what Berry is talking about is the Strategic Defense Initiative (SDI) of the 1980s. During that period, several university and industry scientists received research funding from Pres. Ronald Reagan's SDI program. The SDI mission was to develop a reliable national defense system against incoming intercontinental ballistic missiles carrying nuclear warheads. At first look, what could be wrong with that? Why would it not be a good idea to develop a foolproof defense system? Upon deeper analysis, it became clear that such research would upset the principle of mutual assured destruction (MAD) credited with preventing a nuclear holocaust since the end of World War II.

The rationale behind MAD is that if adversary nations (e.g., Russia and the United States) both have enough weaponry to destroy each other, and if neither nation has a leak-proof defense against incoming missiles, then neither will be tempted or foolish enough to launch a first strike. The problem with SDI is that if the program were to succeed in creating a reliable nuclear shield, it would give the United States first-strike capability. It was argued that the moral character of the United States would never permit its launching a nuclear first strike. But that argument holds little sway with Russia or other nuclear adversaries, in light of the nuclear events of 1945. Thus, it was effectively argued that the SDI program, instead of increasing US security, actually threatened it, encouraging a first strike on the United States, despite MAD, in order to deter the United States from developing a first-strike capability. During the 1980s, several university presidents instituted policies to prevent SDI research being performed on their campuses. After Reagan left office, the SDI program dissipated, although research on antimissile defense systems never entirely disappeared, and the flawed rationale for the program shows signs of revival now almost forty years later.

A trap difficult for scientists to avoid, particularly when their research requires an extraordinarily large budget as is common in areas like high-energy physics and biomedicine, is becoming beholden to the source of funding. Scientists receiving support from nonpublic sources must be especially vigilant not to allow their research findings to be misused or suppressed to further the political or philosophical agendas of their benefactors.

Government and corporate entities sometimes ignore, misrepresent, or suppress scientific information for their own ends and often at public expense. Scientists must be alert for such departures from truthfulness. An example of corporate interest overriding science is the well-documented case of seven tobacco companies in the 1950s and 1960s suppressing and undermining results of their own internal research showing that tobacco is harmful and addictive.[3] A more recent example is Exxon-Mobile's suppressing evidence about the harmful effects of burning fossil fuels on global warming gathered by its own corporate scientists. The company withheld this information from stockholders and the public at large for decades while promoting misinformation about climate change (Hall 2015; Gillis and Krauss 2015).

An especially dramatic example of a scientist and a medical doctor using their findings and strong social conscience against incredible opposition from industry is the story of geochemist Clair Patterson and pediatrician Herbert Needleman. Together they discovered a toxic pervasiveness of lead in the environment and in the bodies of inner-city children. They traced the lead's origin to paint, food cans, and leaded gasoline. Through numerous appearances before the US Congress, in courtrooms, and at the EPA, the two were finally able to trump economic interests with scientific information and enable legislation to remove and ban lead from those products in the 1970s.

Finally, increasing privatization of big science by billionaires with pet research projects (Broad 2014) poses a special challenge to scientists and institutions benefiting from wealthy patrons. Temptation to interpret research results in the direction of a wealthy patron's interests may be great, especially if that funding includes extra salary allowances to hire classroom teachers that relieve researchers from preparing lectures, administering tests, and meeting with students. In cases where a patron's funding actually supports a fully staffed research institute, pressure to produce results compatible with the patron's hopes or interests is even greater. Scientists and institution administrators must hold Bronowski's "habit of truth" especially close to their hearts and minds in such situations. Examples of billionaires privatizing American science include Microsoft cofounder Paul Allen, real estate mogul Fred Kavli, financier Leon Black, and CEOs at eBay, Google, and Facebook. Allen and Kavli have established brain research institutes, Black a Melanoma Research Alliance, and the internet CEOs a satellite-borne telescope to peruse the solar system for space rocks that could endanger Earth.

Another concern about large private donations to science is their tendency to go to the wealthiest universities. This further increases the disparity between research opportunities at historically leading universities and other qualified institutions. An examination of sources for research money at the fifty leading universities in science-research spending showed that 30 percent of the schools' research funds come from private donors (Broad 2014).

Finally, there is danger of large private donations to medical research favoring work on diseases prevalent in self-identified Caucasians at the expense of orphan diseases more common in minority populations. Countering these concerns is that funding from private sources can help take up the slack when science funding goes through lean periods due to changes in governmental priorities. Also, as American biologist and conceiver of DNA sequencing machines Leroy Hood points out, science philanthropy "lets you push the frontiers" (Broad 2014) often rejected by public funding agencies. Estimated private funding for science in the United States in the foreseeable future approaches $100 billion. Billionaires are increasing in numbers. They are becoming major players in setting the nation's research priorities, and their priorities may or may not reflect consensual values of the general public.

Teaching

Science educators have a responsibility to see that student scientists learn communication skills appropriate for explaining their work and its societal implications to others, particularly to nonscientists. This is vital for graduate students who represent the next generation of established scientists. Marcia McNutt (2015), editor-in-chief of *Science*, suggests ways for graduate students to gain communication skills and for senior scientists to nurture such skills in their students. She points out that the students who present posters instead of just reading an oral presentation at large scientific meetings gain skill at communicating science one on one.

In 2003, biologist and associate director for science at the White House Office of Science and Technology Policy Jo Handelsman authored a succinct and insightful essay on the need for science graduate programs to deliver a strong foundation in teaching and oral communication. She likens research-focused universities that neglect training future scientists as teachers to hypothetical music schools that train pianists "to play with only their right hand, leaving them on their own to figure out the left hand's responsibility" (Handelsman 2003). Handelsman points out that even scientists without teaching responsibilities must make science compelling to all nonscientists, including policy makers, industrial managers, patent examiners, and broad public audiences.

Scientists should be adept at explaining the nature of the scientific process and also communicating scientific findings. To this end, Handelsman offers suggestions for teaching scientists how to teach, including offering formal instruction in the art of teaching, requiring graduate students to design and deliver lectures for undergraduates and other nonscientist audiences, and giving opportunities for graduate students to mentor under-graduates. A major benefit to the society of scientists who teach well is "a citizenry that not only has an enhanced sense of the power and limits of scientific inquiry but can also profit from the intellectual and experimental foundations of that inquiry. . . . Research universities should raise a generation of future scientists who, like pianists who play with both hands, practice their art with a dynamic complement of skills, to the great benefit of society" (Handelsman 2003). Unfortunately, many scientists responsible for training future generations of scientists received no training to teach while they were graduate students and have not acquired teaching skills on their own. Therefore, meeting the responsibility to adequately train future scientist-teachers may sometimes require that scientists establish collaborations with colleagues in other disciplines where acquiring teaching skills is emphasized.

Simply teaching students how to teach is empty without exemplifying good teaching. Scientists who can teach must demonstrate so not only in the classroom but by talking science to civic groups, Sunday morning classes, boards of education, grade school classes, city councils, Congressional committees, life-long learning groups, other community enrichment gatherings, and to local newspapers via letters to editors. Students ought to be made aware of and invited to participate in their mentors' community teaching activities. Department heads, deans, and vice presidents need to recognize the importance of these outreach activities and reward, or at least refrain from penalizing, faculty members who commit time and attention to them.

In "Why Researchers Should Resolve to Engage," the editors of *Nature* (2017) argue for serious engagement by scientists outside the university community, especially on topics of climate science and genome editing. They stress the need to promote scientists' voices outside the inner circles of policy making and the long-term benefits of promoting academic recognition for researchers who expend time and effort to speak up. Finally, the editors remind us that engagement with the public is most effective when one tries to understand the values of those with different views. It takes a "village" to nurture a scientifically literate and engaged citizenry and government. It is worth the investment, and it may be essential for humanity's welfare in both the near and distant future.

I cannot conclude this section without naming three organizations of scientists that act diligently, courageously, and with great integrity and expertise to communicate science beyond the borders of the scientific community: the American Association for the Advancement of Science (AAAS), the Union of Concerned Scientists (UCS), and the National Academy of Sciences (NAS).

The AAAS is the world's largest science organization. Its mission is to "advance science, engineering, and innovation throughout the world for the benefit of all people" (AAAS 2017). AAAS members are primarily, but not restricted to, scientists and science students. The organization has programs and whole offices devoted to government relations, public engagement, and public science education. A specific education venture, Project 2061, aims to help all Americans become literate in science, mathematics, and technology before the return of Haley's Comet in 2061. AAAS's annual meeting always includes free public lectures by leading scientists on topics of public concern and hands-on science discovery displays for children.

The overall mission of the UCS is to use a science-based approach to build a healthier planet and a safer world. Currently, specific projects address climate change, agricultural practices and our food sources, clean energy, nuclear power, nuclear weapons, and clean vehicles. An outgrowth of the UCS is the Center for Science and Democracy, whose mission is to facilitate the cooperation of scientists and the general public in working to help ensure that government policies are informed by science and evidence (Center for Science and Democracy 2017).

The NAS was established by President Lincoln in 1863 to "whenever called upon by any department of the Government, investigate, examine, experiment, and report upon any subject of science or art, the actual expense of such investigations, examinations, experiments, and reports to be paid from appropriations which may be made for the purpose, but the Academy shall receive no compensation whatever for any services to the Government of the United States" (NAS 2017). Originally, 50 scientists were appointed as founding members of the NAS. Currently the NAS has close to 2,400 members. New members are nominated and elected by existing members of the academy. Membership is for life and is one of the highest honors bestowed upon a US scientist. The NAS is headquartered in Washington, DC, and publishes statements, reports, and more than two hundred books each year on science subjects of importance for the well-being of the nation and human society as a whole. Categories receiving attention from the NAS include science education, energy, environment, climate, foreign policy, national security, and public health. Many NAS reports and other publications are available free online.[4]

Nonscientists' Responsibilities

Nonscientist citizens of the general public have one primary responsibility—strive to be informed about developments in science that may affect personal and societal well-being. Three corollaries to this responsibility are (1) ask questions of scientists and science journalists about their work and reporting, (2) never purposefully misrepresent the findings of science, and (3) question aspirants to policy-making positions about their knowledge of and positions on science-related issues with societal significance. To clarify, the term *nonscientist* includes those scientists who are experts in one area but unversed in other scientific disciplines in which current discoveries have ethical and social implications. Such scientists must recognize themselves as "nonscientists" in those disciplines and take action to inform themselves like any other nonscientist. Thus, a physicist studying high-energy plasmas is a "nonscientist" when it comes to stem cell research, cloning, or gene editing. The physicist in this case has the same responsibility as a banker or a farmer to inform himself or herself about stem cells, cloning, and gene editing. And it works both ways—cell and molecular biologists bear responsibility to inform themselves about developments in fields such as physics, nanotechnology, and AI that have unique ethical and social consequences.

Once one commits to being scientifically literate, the challenge is to choose reliable sources of information. There are many excellent sources of scientific information for laypersons. They range from books and periodicals to television, radio, Internet websites and videos, and documentary films. Appendix 3 details several of these, focusing on information sources for the biotechnologies examined in this book.

Scientists who make a special effort to communicate to the nonscientist public may comprise a culture distinct from science itself. What I mean by this requires reference to English novelist and physical chemist C. P. Snow (1905–80), who wrote an essay about what he called the "two cultures," aimed particularly at literary intellectuals (Snow [1959] 2001). He lamented the gulf of misunderstanding and incomprehension between science and the humanities. In the early 1990s, I had a personal conversation with the late cell biologist and author Lynn Margulis (1938–2011) about Snow's two cultures. She suggested to me that scientists who write about science for nonscientists constitute a third culture. Others suggest that engineers comprise a culture separate from science. One can imagine other "cultures" as well, such as the theologically inclined and the environmentally minded.

Whether communicative scientists comprise a third, fourth, or fifth culture is of no consequence. What is important is that it exists, is growing, and forms an important source of information for nonscientist citizens. If misunderstanding or distrust is a criterion for identifying separate cultures, as Snow described for scientists and literary scholars, Margulis's suggestion may be on the mark. Too often scientists who are nearly exclusively research focused look down upon their scientist colleagues who commit time and energy to make science legible to the general public. University department heads, deans, and vice presidents who value and reward teaching and science communication on par with obtaining large research grants do much to merge these two cultures into one cooperative group for discovering and disseminating information about the world and ourselves.

Policy Makers

Elected representatives in democratic governments tasked to create and enforce laws, rules, regulations, programs, guidelines, or other policies are policy makers. They include school board members, city council members, mayors, state representatives and governors, representatives in national governments, and heads of governments. Policy makers have at least four major obligations when it comes to science. First is to become and remain informed about science issues, particularly those with the greatest impact on their constituents. Second is to seek out and listen to the counsel of scientists. Third is to facilitate education of their constituencies about science-related issues without politicizing the information. And lastly is to form and/or vote on science-related policies with the well-being of all citizens in mind and not just the interests of their own electorates or their reelection prospects.

How Policymakers Can Meet Their Obligations

The first and second responsibilities, staying knowledgeable about scientific matters that impact society and seeking out information and advice of scientists, are closely related. Policy makers have the same opportunities as other nonscientists to educate themselves about scientific issues through reading, attending lectures, watching science education programs on television, at the theater, and on the Internet, and listening to podcasts. In addition to these sources of information, policy makers have unique opportunities to plumb the knowledge and advice of scientists whose work may benefit, endanger, or challenge society in unique ways. In an essay for the journal *Nature*, William Colglazier, major contributor to the worlds of both science and government as described earlier, details four specific ways policy makers can enhance their access to information and advice from scientists (Colglazier 2016): (1) appoint a chief science advisor who reports to the head of state, and an advisory committee comprised of nongovernmental scientists, (2) create civil service positions for scientists and technologists, (3) solicit independent advice from nongovernmental science and technology institutions and then publicize the advice, and (4) offer fellowships for early- and midcareer scientists, engineers, and biomedical professionals to serve in government positions related to science. Colglazier concludes his essay by noting that "in the long run, any political leader who disregards scientific evidence does so at his or her peril, and most politicians know this." As for the third and fourth responsibilities, educating their constituencies about science and subsequently voting on science-related bills with self-interest secondary is the desired path. However, most legislators will need strong encouragement from within and without their constituencies before this becomes the norm. Science-literate constituencies and colleagues already in government can and must provide this encouragement.

Science in the Trump Era

As leaders of a technologically advanced and globally influential nation, US presidents have a unique opportunity to lead the community of nations in the ethical applications of new technologies and in wise responses to scientific discovery. As of this writing, Donald Trump has been president for twenty-eight months. It is fair and responsible for US citizens to give inquiring and thoughtful attention to the relationship between Trump and science. Unfortunately, several of the president's actions and inactions indicate that his relationship with science is not healthy. He appears to lack interest in being informed about pressing scientific issues of the day and on some occasions seems outright hostile to the notion of a scientifically informed government and public.

With the exception of Trump, every president since Franklin D. Roosevelt in the early 1940s has had a science advisor to the president, whose job it is to give advisory opinions and analysis on science and technology issues to the US president. In 1976, the US Congress established the Office of Science and Technology Policy (OSTP) for the purpose of advising the executive office of the president about effects of science and technology on domestic and international policies. The president nominates a person to head the OSTP, and this person becomes the chief science advisor to the president. After nearly one and one-half years into his presidency, Trump had not nominated a person to head the OSTP. Instances during that period where a science advisor's input could have been valuable to Trump include responses and decisions related to Hurricanes Irma, Harvey, and Maria, climate change and the US withdrawal from the Paris Climate Accords, deadly wildfires in the western United States, an unusually lethal flu season, and the nuclear programs of Iran and North Korea. Andrew Rosenberg, director of the Center for Science and Democracy at the Union of Concerned Scientists (UCS) says this about the lack of scientific input into many of Trump's decisions: "They don't want any dissenting voice that gets in the way of a wholly political decision, and if you put science into the mix, it's harder to make a wholly political decision because you have to follow the evidence, and they don't want to do that on any of the issues" (Waldman 2018).

In 2018, Trump finally did nominate a scientist to head the OSTP. Meteorologist and extreme weather expert Kelvin Droegemier was confirmed in the position in August 2018 (Polansky 2018). In addition to climate change, current issues facing Droegemier include the effects of mining companies' mountain top removal on waterways, limits on toxic heavy metal contamination of rivers and streams, regulations limiting asthma-causing ground-level ozone levels and worker exposure to lung disease-causing silica dust, and a plethora of potential human, agricultural, and other field applications of powerful CRISPR-based genome-editing tools. Time will show whether Droegemier can speak truth to power and remain in his position.

Information and professional advice on current bioethical issues have also been lacking in the Trump administration. Since 1974, presidential or congressional commissions and committees to study pressing bioethical topics have provided reports and advice for the US government and public. For example, President Clinton's National Bioethics Advisory Commission reported on stem cell research and human cloning. Pres. George W. Bush's Presidential Council on Bioethics examined human genetic enhancement and assisted reproduction technologies, and Pres. Barack Obama's issued reports and recommendations on synthetic biology and human brain research. President Trump has not established a study group for bioethical topics despite the emergence since 2015 of the most powerful tool humankind has ever had for altering the genetic makeup of virtually any organism on Earth including itself. How we apply CRISPR genome-editing technology during the next few years will have irrevocable ramifications for the biosphere and human nature itself. If ever it were time for expert advice on a bioethical topic, the time is now. And the same can be said for robot, AI, and brain technologies.

Added to this lack of interest in being informed on emerging scientific issues is Trump's active and disturbing suppression of the use and dissemination of existing scientific information. One Orwellian example is Trump's list of words and phrases (e.g., science-based, evidence-based, fetus, diversity) forbidden for use by the Centers for Disease Control and Prevention (CDC) in documents prepared by the agency for its 2018 budget. Another is the administration's alteration or expunging of public, scientific information about climate change on websites for the EPA, DOE, and State Department. And most incredible is an order from the Interior Department prohibiting climate experts at Glacier National Park from meeting with Facebook founder Mark Zuckerberg when he toured the park in 2017 to view melting glaciers.

The UCS published a report detailing Trump's hostility toward science, which it estimated to average "one major attack on science every four days" (Carter et al. 2017). As for the 2019 fiscal year budget for science, Trump proposed deep cuts for most science-related government departments and agencies compared to their 2017 or 2018 budgets (Guglielmi et al. 2018). Then in bipartisan fashion for the second year running, Congress ignored Trump's recommendations and gave science a boost in the budget bill. Trump threatened a veto but eventually signed the FY 2019 Congressional budget bill into law (Koren 2018). The NIH, NSF, DOE, and US geological survey all received significant increases over the previous year. The EPA, which Trump had targeted for a 31 percent decrease, received flat funding.

Other Persons of Influence

Other persons influencing science literacy in society include teachers, science journalists, and clergy. The primary responsibilities of these and others who speak or write about science to nonscientists are accuracy and objectivity. Let us consider teachers first.

Teachers have an absolute obligation to be well versed in the subjects they teach. Unfortunately, this obligation is sometimes unmet due to weak science curricula in university schools of education. How is a prospective K-12 teacher to assess his/her competence to teach a particular science discipline? How is an education student to recognize her or his deficiencies? This is a challenging dilemma. The greatest assets a scholar can have are the ability to recognize that he does not know everything in his discipline and also the ability to identify personal areas of weakness. Assessing one's ignorance requires a fairly high level of competency in the discipline already or a minimum awareness of the existence of those areas where one's personal knowledge is lacking. Education students are unlikely to have this level of competence, and many grade school science teachers lack it as well. Certification exams evaluate a prospective teacher's readiness to teach. The involvement of scientists is a must, and it is incumbent upon those scientist evaluators to hold future K-12 science teachers to a high standard. One way to improve the quality of teachers in public schools is to add a fifth year to science education programs and require that future teachers earn a full major in their area of concentration using a curriculum identical to that of science majors. Federal support for teachers' fifth year of training would be an investment with high dividends.

Both scientists and science journalists reporting on current scientific discoveries are obligated to ensure that the transfer of information occurs with as little misunderstanding and distortion as possible. Scientists must guard against making dramatic overstatements about their research in hopes of attracting the attention of funding sources. Likewise, science journalists must guard against over dramatization of current research findings in order to enhance the size and enthusiasm of their audiences and supervisors. Current examples of overdramatization are media reports that characterize mitochondrial replacement therapy, three-parent baby technology, as being just a step away from "designer babies" and a new eugenics (as discussed in chapter 2).

In chapter 9, I argue that public science education needs revamping to formally expose all citizens to reliable information about biological evolution and ecology, but it is not only science education that needs a makeover. Changes in religious education are also needed. This is especially true for theological and geographical pockets of Christianity where the literal interpretation of scripture still holds sway. Teaching children that biblical scripture supersedes science as a credible source of information about the age of the Earth and the origin of biodiversity undermines both science and the positive influences of religion. Giving so-called equal time to biblical creationism alongside the evidence for evolutionary theory at home or in private religion-based schools harms students' understanding of the nature of science. Insisting that the Bible be treated as a science textbook can actually jeopardize much of the religious teaching a student receives as a child. Demanding that such equal time be observed in public school science classrooms is also inappropriate. I have treated this subject in detail elsewhere in an essay aimed specifically at parents and other teachers of fundamentalist Christian views on the history of Earth and life upon it. An excerpt is relevant here:

> When a child's religious training teaches that evolutionary theory lacks good scientific evidence and rigor, that evolution is promoted as an atheistic scheme to destroy belief in God, that believing in evolution precludes religious faith, that the earth is only a few thousand years old, or that a literal reading of sacred scripture is a reliable account of how the physical universe actually came into being, risk for great and impending disillusionment with religion is set up. For when that child begins reading and thinking for himself, gets out into the world and converses with honest and intelligent persons of different educational backgrounds, or takes a good biology course in high school or at the university, he will discover that what he was taught earlier about evolution and scripture is untrue. Any one of a multitude of discoveries about the world around him could jeopardize all of the religious teachings he received as a child. For example, learning the true meaning of the term scientific theory, how geologists date the earth at 4.6 billion years, about Tiktaalik, feathered dinosaurs, or extinct species of the genus *Homo*, meeting a religious evolutionary biologist, or realizing that the two accounts of creation in the Bible's Book of Genesis are mutually exclusive may create a crisis of faith. The result may be to discard religion as deceptive and unworthy of further attention. But religion can nurture courage and hope, kindness and compassion, cooperation and fellowship, and the ability to envision peace during times of war. To jeopardize these and other virtues of religion by railing dogmatically against evolution does a disservice to the future personal lives of children and society as a whole. Galileo cited Cardinal Baronius (1598) when in 1615 he wrote to his friend, the Grand Duchess

Christina of Tuscany, that scripture teaches "how to go to heaven, not how the heavens go." Galileo was writing about the conflict between his new telescopic discoveries and the Roman Catholic Church's insistence that Aristotle's earth-centered universe was correct because clerics agreed with literal readings of scripture. In 1633 Galileo was arrested by The Holy Office of the Inquisition and convicted of teaching that the earth orbits the sun. Nearly 350 years later, the Church officially forgave Galileo. Perhaps those who teach that evolution has not and does not occur can learn from this episode in the history of Christianity and consider the possibility that scripture teaches us how to live life, not how life came to be. (Bradley 2013b, 236–37)

Where does the responsibility lie for honest and accurate treatment of scientific topics in the context of religious education? With pastors and other teachers of religion in the churches? At seminaries? With parents who insist on honesty and accuracy? I do not have the answers. It is a difficult problem. I do believe that an increased emphasis on evolution and ecology in grades K-12 and higher education as proposed in chapter 9 will be very helpful in this regard. Alert readers will notice that I have yet said little or nothing about four of the six persons I named earlier as greatly admired by me: Pico della Mirandola, Gerald Feinberg, Aldo Leopold, and Shannon Vallor. Their ideas appear in chapter 9.

Chapter Summary

The good life with biotechnology now and far into the future requires informed actions and attitudes now from both scientists and nonscientists. Scientists must commit themselves to truthfulness and to educating the nonscientist public about their research and its societal implications. Truthfulness includes both honesty in the reporting of research results to other scientists and also the informing of the general citizenry about possible risks, as well as the benefits, of their research. Additional responsibilities of scientists include preparing higher education students for engagement with the nonscientist public while practicing a research ethics that includes discernment in assessing motivations and objectives behind sources of research funding.

Nonscientists' primary responsibility is to be informed about scientific advances that have major implications for personal and societal well-being or harm. Examples include therapeutic cloning, germ line engineering of human and nonhuman organisms, development of autonomous and semi-autonomous robots, brain-scanning and manipulation technologies, nanotechnology for health and surveillance, age retardation, and synthetic biology. Obtaining and disseminating accurate information about science is especially important for science journalists and policymakers. Questioning political candidates about their knowledge of science-related issues is the responsibility of voters. The same applies to the relationship between clergy and parishioners.

Policy makers' responsibilities are to seek advice from scientists and be informed about scientific matters, encourage scientific literacy among their constituencies, and make science-related policy with the well-being of all in mind.

Questions for Reflection and Discussion

1. What are three important scientific issues that you feel you ought to become better informed about? Detail how you will go about doing so.
2. Which of the three issues above (Question 1) do you believe is the most important one for your representatives in state and federal governments to be well informed about? Do you know whether they are well informed about this topic? If not, how can you find out? If you learn that these representatives are not well informed on this topic, what will be your next action?
3. If a person in a position of authority, such as a political representative, teacher, or clergyperson, speaks inaccurately about a scientific topic (e.g., evolution, climate change, safety of vaccinations), what do you believe is your best response? Have you ever been in a situation like this? What was it like? What did you do? What would you do differently next time?
4. How do you feel about billionaires funding research that may have profound societal implications?
5. What do you believe is the best way for citizens to respond when elected policy makers show little interest in being informed about scientific issues that greatly impact societal well-being?

Notes

1. The meaning of the word *theory* in the context of science is widely misunderstood and sometimes purposefully misrepresented. It does *not* have the same meaning as it does in common parlance to indicate a guess about the how

or why of something with little or no supporting evidence. Instead, it is the closest thing to "fact" that science offers—a unifying explanation for a natural phenomenon based on validating outcomes of numerous and diverse empirical tests of relevant hypotheses. Thus, the theory of evolution is comparable in scientific status and credibility to the atomic theory for matter, the germ theory for disease, the tectonic plate theory for the distribution of continents over the Earth's surface, and the theory of gravity to explain planetary movement around the sun and apples falling from trees.

2. The NSABB is an advisory body to the government established to inform the science community about research on agents that pose a potential threat to national security and to redact key methodological details from research manuscripts in order to make replication of the work more difficult for persons intending to harm society.

3. The WHO has published an online manual of documents revealing how the tobacco industry obstructed or delayed tobacco control measures and policies: The Tobacco Industry Documents—What They Are, What They Tell Us, and How to Search Them: A Practical Manual. http://www.who.int/tobacco /communications/TI_manual_content.pdf (accessed April 11, 2017).

4. Articles in NAS's journal, *Issues in Science and Technology*, are available at http://www.nasonline.org/publications/issues-in-science-and-technology/, and entire books are available at https://www.nap.edu/content/about-the-national -academies-press (both accessed January 7, 2017).

Sources for Additional Information

Denworth, L. 2008. *Toxic Truth: A Scientist, A Doctor, and the Battle Over Lead.* Boston: Beacon Press.

Elliot, D., and J. E. Stern, eds. 1997. *Research Ethics: A Reader.* Hanover, NH: University Press of New England.

Gregory, J., and S. Miller. 1998. *Science in Public: Communication, Culture, and Credibility.* Cambridge, MA: Perseus Books Group.

Kang, C., and M. D. Shear. 2017. "Trump Leaves Science Jobs Vacant, Troubling Critics." *New York Times*, March 20. https://www.nytimes.com/2017/03/30 /us/politics/science-technology-white-house-trump.html?_r=0 (accessed April 8, 2017).

Lowrance, W. W. 1985. *Modern Science and Human Values.* New York: Oxford University Press.

Presidential Commission for the Study of Bioethical Issues. 2016. *Bioethics for*

Every Generation: Deliberation and Education in Health, Science, and Technology. Washington, DC. https://bioethicsarchive.georgetown.edu/pcsbi/sites/default/files/PCSBI_Bioethics-Deliberation_0.pdf (accessed October 31, 2018).

Sun, L. H., and J. Eilperin. 2017. "CDC Gets List of Forbidden Words: Fetus, Transgender, Diversity." *Washington Post*, December 15. https://www.washingtonpost.com/national/health-science/cdc-gets-list-of-forbidden-words-fetus-transgender-diversity/2017/12/15/f503837a-e1cf-11e7-89e8-edec16379010_story.html?utm_term=.1cff7cc30a21 (accessed May 25, 2018).

Tobias, J. 2017. "A Brief Survey of Trump's Assault on Science." *Pacific Standard*, December 15. https://psmag.com/environment/a-brief-survey-of-trumps-assault-on-science (accessed May 25, 2018).

9
The Urgency of Now

> The earth that directed itself instinctively in its former phases seems
> now to be entering a phase of conscious decision through its human
> expression. This is the ultimate daring venture for the earth, this
> confiding its destiny to human decision, the bestowal upon the human
> community of the power of life and death over its basic life systems. . . .
> Teaching children about the natural world should be treated as one of
> the most important events in their lives.
>
> —Thomas Berry, *The Dream of the Earth*

Now is the time for all humans to learn about life, its antiquity, the inter-dependency of its myriad forms, and to make judgements about where humans are taking themselves and the rest of the living world. In this chapter, I tell why humanity needs to consider long-range goals for itself and the planet, explain why increasing public literacy in bioscience is a crucial prerequisite for selecting long-range goals, describe a proposal for increasing bioscience literacy, and discuss the need for a common ethics to guide humanity in its biotech future. Specifically, we seek answers to the following questions:

1. Why should humanity seek consensus on long-range goals for itself and the planet?
2. Why is public literacy in bioscience important for making wise decisions about developing and using modern biotechnologies?
3. What action is needed to dramatically increase bioscience literacy?
4. What ethics can best guide our decisions about developing and using modern biotechnologies?

Humanity's Need for Long-Range Goals

Humankind urgently needs to initiate a well-organized and sustained effort for identifying long-range goals for itself and our planet. If we do not do this, the inertia of modern technologies—including CRISPR, gene drives, embryo selection, synthetic biology, robotics, AI, nanotechnology, and neuroenhancement—will choose directions for us, some of which we may later regret. Whether we seek a consensus plan for our future or not, these technologies will spawn irrevocable trajectories for *Homo sapiens* and the biosphere.

Consider the meaning of *long-range goal* in the context of technologies discussed in this book. *Long-range* means much further out than a few decades or even a few generations. In 1969, the young physicist and futurist Gerald Feinberg published a short book titled *The Prometheus Project: Mankind's Search for Long-Range Goals*. Observing advances in molecular biology and fledgling computer science, Feinberg foresaw irretrievable, technology-mediated changes to human nature itself coming sooner rather than later. He proposed what he called the Prometheus Project (from the Greek word for foresight) by which humans would reach consensus on long-range goals for their species. Feinberg defined a long-range goal as "some future state of affairs whose realization would require an effort lasting over many generations" (1969, 4). I adopt Feinberg's concept of long-range goal for our discussion here. A useful perspective on our present situation, at the beginning of what may be an unending era of life manipulation via biotechnologies, is gained by viewing ourselves in the framework of life's history on Earth.

We are now 13.8 billion years into the universe's known existence and 3.8 billion years into life's sojourn on planet Earth. Vertebrates appeared about 420 million years ago, the first placental mammals about 160 million years ago, the first primates 5 million years ago, the first member of our genus, *Homo habilis*, 2.8 million years ago, and modern *Homo sapiens* 200,000 years ago. Roughly 10,000 years ago, humans began manipulating the gene pools of plants and animals by simple selection for desired traits. About 400 years ago, Francis Bacon, René Descartes, Galileo Galilei, Isaac Newton, and others devised a way to gain new information about the universe, which we now call the scientific method and which has carried us to our present position as overlords for the planet's life systems. The ideas of Nicolaus Copernicus, Galileo, and Charles Darwin challenged and defeated the medieval belief that humans are at the center and peak of creation, showing rather that we, and all the cosmos, are integral parts of the same nature. We are not separate from nature. Biological evolution driven by natural selection and random genetic drift has driven biological change for billions of years. Only recently have agriculturalists and molecular biologists added crossbreeding, cross-pollination, cell fusion, mutation-inducing chemical and radiation treatments, and genetic engineering and editing technologies to life's repertoire for altering itself.

Despite these technologies, Darwinian evolution still clings precariously to its nearly four-billion-year-long role as sculptor of the germ plasm for most of Earth's life forms. Now though, life is at a crossover point for consciously, rapidly, and dramatically altering itself in human-devised directions. Irrevocable outcomes of re-creating life with these technologies await us regardless of whether our creations are products of consensus-driven decisions.

More than five centuries ago, the young Italian Renaissance philosopher Pico della Mirandola wrote of humankind's special place in nature as choice makers: "Constrained by no limits, in accordance with thine own free will . . . with freedom of choice and with honor, as though the maker and molder of thyself, thou mayest fashion thyself in whatever shape thou shalt prefer. Thou shalt have the power to degenerate into the lower forms of life, which are brutish. Thou shalt have the power, out of thy soul's judgment, to be reborn into the higher forms, which are divine" (Pico della Mirandola [1486] 2005, 287).

In Pico's view, and in mine, it is more human for us to make choices than to allow choice-making opportunities to slip away. When the trajectory of applications of a technology forks and moves down one path, it will often be impossible to retrace our steps and choose a different trajectory. Unplanned outcomes of twenty-first-century biotechnologies may preclude other more desirable outcomes. Claiming that the technologies discussed in this book can produce irrevocable changes to human nature and to the rest of nature as a reason to formulate long-range goals calls for examples of such irrevocable outcomes. Here are some:

Example 1:

Transformation of the earthly environment from its present state to some stable state that is better suited for human habitation. This may include the eradication of certain "undesirable" species of plants and animals and the creation of entirely new ecosystems designed to thrive under human-caused conditions not conducive to the health of natural ecosystems. There could also be genome-edited food crops and animals designed to grow in places previously inhabitable only by natural ecosystems.

Realizing this outcome could preclude:

1. Preservation of the genetic biodiversity contained in natural ecosystems,
2. Maintaining a large and diverse base of plants and fungi as sources
   of new pharmaceuticals,
3. A comprehensive understanding of the 3.8-billion-year-long process
   of biological evolution,
4. Sanctuaries of wilderness—habitats for naturally evolved communities of organisms and sources of rejuvenation for the human psyche.

Example 2:
*Genetically designing specialized human types to optimally perform
particular tasks such as space travel, negotiating with intelligent robots,
and reproduction.* This would produce a calculated, genetic-based stratification of society.

Realization of this outcome would preclude or exacerbate:

1. Autonomy in choosing one's life work or profession,
2. Equalizing life opportunities.

Example 3:
*Making available age-retardation procedures to expand healthy hu-
man life spans to centuries or millennia.* Genetic manipulations, regenerative medicine, and pharmaceuticals may collaborate to greatly
expand human life spans. Google has already spawned the company
Calico, which is funded at over $1 billion and whose mission is to
counter humans' mortality problem.[1] If age-retardation technology is
not available to all, society will become further stratified on the basis
of life span. Germ line engineering to create centuries-long life spans
will solidify the stratification into virtual permanence, likely giving
those with expanded life spans greater economic and political power.

Realization of this outcome would likely preclude/exacerbate:

1. Equalizing life opportunities for all people,
2. Democratization in virtually all sectors of social and political life.

Example 4:

Creation of a race of conscious machines whose abilities surpass human intellect and whose decision making and authority guarantee comfortable, peaceful living for all humans. Such robots could become the enforcers of peace through drastic punishment of humans planning or perpetrating violence, injustice, or other antisocial behavior ranging from a personal to global scale.

This outcome would preclude or exacerbate human autonomy.

These outcomes may sound like science fiction. But remember that we are considering outcomes that will take many generations and likely many centuries to materialize. It is important to note that the rationale for humankind thinking now about its long-range goals is not that selecting a particular goal now will avoid precluding alternative goals later. Rather, the rationale for selecting long-range goals now is to give purpose and consensus to the technological trajectories that humankind embarks upon, realizing that nearly any trajectory selected will preclude certain other choices.

In order to preserve the option of choice over the future of our species and living environment, the global public first needs to educate itself about the need to work toward a consensus on long-range goals. Feinberg offers suggestions about how to go about this. His suggestions were designed for media technologies of the 1970s, pre-Internet and pre–social media, so our current communication technologies ought to facilitate the goal-selecting process. Proposing how to actually solicit, vet, and select long-range goals on a global scale is not my objective here. Rather, I propose three actions to take immediately to help prepare the global community to engage in Feinberg's Prometheus Project:

1. Convene a series of global summits to define and publicize the problem. These would be similar to the meetings of the International Climate Study Group, which was formed by collaboration between the United Nations and the World Meteorological Association. A similar collaboration between the UN and professional societies such as NASEM and the UK's Royal Society could initiate these summits.

2. Overhaul curricula in public education systems to include strong components on ecology and evolution. Knowledge in these areas is essential for an appreciation for the antiquity and interconnected-

ness of all living things. Such knowledge is essential for the wise selection of long-range goals for life on Earth.

3. Promote a global ethics appropriate for engagement with modern
   biotechnologies. Traditional ethical systems based on religious
   dogma or on rules, such as utilitarian and deontological ethics, are
   inadequate for the job at hand. Philosopher of science and technology Shannon Vallor (2016) tackled this problem in a recent
   book, *Technology and the Virtues: A Philosophical Guide to a Future
   Worth Wanting*. She argues strongly for a virtue ethics to guide us in
   the age of biotechnology.

I am not in a position to convene a global summit, but I am willing to
work on that with others more critically positioned. Recent calls to reach a
global commons ethic for justly sharing the planet's land resources among
all of Earth's citizens are ambitious on a scale similar to humanity reaching
consensus on long-range goals (Creutzig 2017; Ostrom 2015; Sreenivasan
1995). The need for these large projects has been described. It is now a matter of education and will to achieve. Felix Creutzig, professor at the Berlin
Technical University, outlines two preliminary steps toward governing land
as a global commons (Creutzig 2017). These are also applicable as steps toward humankind's search for long-range goals: (1) educate the public to the
importance and benefits of the project with help from individual scientists
and the United Nations, and (2) establish a system for international governance to coordinate activities critical to the project.

Regarding step one, I detail in the next section why a dramatic improvement of public literacy in bioscience is needed to prepare for discussions about
long-range goals. I solicit the advice and assistance of readers and readers'
associates in implementing such improvements via engagement with teachers, local school boards, and state boards of education. The UN and national
and international science organizations need to engage in step two.

Why Public Biological Literacy Is Important

Given the power of twenty-first-century biotechnologies to irrevocably alter
Earth's biosphere including human nature itself, certain current and prevailing public attitudes and government policies concern me. Living wisely
as one component of nature requires that humans make good decisions to
protect what we value. In my view, four factors combine to jeopardize good

decision making about the well-being of life on Earth: (1) commodification of nature, (2) human exemptionalism (3) trained incapacities in the disciplines of evolution and ecology, and (4) the sixth extinction now underway on Earth. How do these three attitudinal phenomena, coupled with ongoing catastrophic losses in biodiversity, combine to argue for dramatic changes in K-12 and higher public education curricula? Let us consider each of the four factors, imagine their combined impact on life's future, and then consider how we ought to respond to our current situation.

Commodification of Nature

Commodifying nature is valuing nature mainly, if not solely, for its economic potential. In the celebrated essay *The Land Ethic*, Wisconsin naturalist and conservationist Aldo Leopold wrote that land is far more than the board feet of lumber it can produce, more than its agricultural value or its commercial worth for tourism. To Leopold, "land" is all of nature, not just soil. The "key-log"" that must be removed before an effective "land ethic" can be developed is to "quit thinking about decent land-use as solely an economic problem," wrote Leopold ([1949] 1970, 224).[2] Seven decades later, a logjam of ignorance about evolution and ecology perpetuated by shortsighted, profit-centered attitudes still thwarts the development of a land ethic that recognizes the interdependency of nature's countless components, of which we are just one.

Harvard University biologist E. O. Wilson has devoted his professional life to understanding how communities of organisms evolved and to preserving Earth's biodiversity. In his book, *The Creation*, Wilson identifies habitat loss, invasive species, pollution, human overpopulation, and overharvesting as human-caused factors responsible for the current dramatic decline in Earth's biodiversity.[3] Each cause stems directly from humankind's commodification of nature. The ETC Group, a Canadian action group on erosion, technology, and concentration, cites as an extreme example of commodification-based thinking from the suggestion of one synthetic biologist: genetically redesign tree seeds to grow into houses rather than trees.[4] The synthetic biology literature is replete with examples of life viewed as a commodity to be reconstructed for human purposes. With proper controls, creating new life forms for human benefit is not necessarily a bad thing. But in combination with three other factors discussed next, the commodification of life becomes especially threatening to all of nature, the *land* of Aldo Leopold's concern.

Genomic information about humans and agricultural plants and animals, along with genome-altering technologies, now support a multibillion-dollar international biotechnology industry. Among positive attributes of the biotech industry are its capacity to improve healthcare, its ability to bolster the quantity and quality of the world's food supply, and its potential to help solve the global energy crisis while dampening climate change. But the industry also exacerbates our commodification of nature and creates other ethical pitfalls related to marketing life-related products.

Business ethicist Chris MacDonald (2004) at Saint Mary's University in Halifax, describes three corporate ethics issues related to the biotech industry: product safety, corporate social responsibility, and corporate governance. Although these issues are not peculiar to the biotech industry, the dynamic nature of biotechnology and scientist entrepreneurs relatively unexperienced in the business world pose special challenges for biotech companies striving to be responsible to consumers and society at large.

As with most businesses, the aim of a biotech company is to produce and market a product for a profit to its shareholders. In its enthusiasm to announce a discovery or market a new product that appears safe in the current social context, a company may overlook or ignore foreseeable harms the discovery or product could cause in the future. For example, if a biotech company discovers and announces that a particular human population is especially susceptible to a certain pathogen, does it share responsibility with future terrorists who may use that information to harm the population in question? This and similar scenarios make it difficult to develop criteria to assess the "safety" of a biotech product.

Related to safety is the issue of corporate social responsibility. Biotech companies have responsibilities to both company shareholders and product consumers. But MacDonald points out that in the realm of biotechnology, the range of stakeholders can expand to include every living thing on the planet. So major challenges for socially responsible biotech corporations are to assess the social implications of their products and then to act to mitigate possible social harms arising from use of the products. How might a biotech company assess the potential social harm caused by a product that some view as contributing to the commodification of nature and that others view as contributing to general well-being. Germ line genome editing to strengthen the human immune system or using a gene drive to make *Anopheles* mosquitoes resistant to the malaria parasite are not far-fetched examples of such ambiguity.

Finally, what about corporate governance? At issue here is how information and decision-making authority flows between company shareholders, managers, and boards of directors. Decisions based on deception about a product's safety or made primarily with short-term profit gains in mind can have grave consequences for consumers and the ever-expanding range of stakeholders in biotechnology applications. MacDonald notes that the rapid process from discovery through biotech product development and commercialization can impair careful, responsible planning by governance structures in the biotechnology industry.

Public views about products of the biotech industry such as GMOs, therapeutic cloning and associated stem cell therapies, and synthetic and genome-edited life forms are polarized. Some persons with a naive optimism see humankind moving toward a Nirvana of exceptionally long, disease-free lives of plenty and leisure, whereas others succumb to cynicism and fear a loss of our humanity and an irrevocable destruction of nature. But an informed and vigilant middle ground of common sense also exists. We can embrace science for its potential to relieve human suffering and protect and restore the biosphere's health. Simultaneously, we can ready ourselves to recognize and thwart public attitudes and business activities that threaten to extinguish our awe and respect for life with a relentless tsunami of profit-oriented development and marketing of life products that would further feed an already bloated consumerism.

Human Exemptionalism

Human exemptionalism is the belief that humans can thrive outside the laws of nature. Formalized into a Western sociological paradigm soon after the eighteenth-century Enlightenment, exemptionalist thinking held sway through the Industrial Revolution and well into the mid-twentieth century. Human exemptionalism is now severely critiqued by environmental sociologists and ecologists. Nevertheless, much of the general public and governance structure in the United States and other industrialized nations still live, vote, and behave as though humankind can live independently from the natural world. Symptoms of exemptionalist thinking include lack of interest in or denial of the role human activity plays in the current wave of extinctions and period of climate change. The attitude that mass extinction and loss of forests and other natural habitats do not affect human welfare and the view that preserving biodiversity is less urgent than economic growth, expansion of capitalism, military defense, or developing cures for cancers epitomize human exemptionalism. Unfortunately, it is an easy step from believing that humans can thrive outside the laws of ecological interdependence to believing that human actions and naturally occurring events are processed and responded to differently in the arena of nature. Thus, we may lament the natural demise of climax forests on the slopes of Mount St. Helens after its eruption, but we lack the will, foresight, and belief in science to regulate industrial emissions that produce acid rain, which also destroys mountain forests. Exemptionalist thinking in the context of new life forms created via synthetic biology or CRISPR-mediated genome editing, which are endowed with gene drives and released into nature, is alarming to contemplate.

Trained Incapacities in Evolution and Ecology

The concept *trained incapacity* is credited to Norwegian-American economist and sociologist Thorstein Veblen (1857–1929) who wrote of business transactions being "carried out with an eye single to pecuniary gain" with little regard for their effect on the welfare of society, thereby putting workers, the community, and business people at cross purposes (Veblen [1914] 2017). The implication is that business persons are trained to act with one goal in mind, corporate profit, and that this training produces the incapacity to see broad implications of their activities. The concept is also applied in sociology in the context of the eugenic potential of human genetic engineering (Duster 2003).

The trained incapacity concept is also germane for understanding underlying causes of our environmental crisis. Unawareness of and lack of appreciation for basic principles of biological evolution and ecology make wise environmental stewardship impossible. Evolution tells how the biosphere came to be and ecology describes how it works. Making decisions about generating and releasing new life forms (or eradicating existing ones) in the absence of an appreciation for the biosphere's astounding antiquity and complexity is far worse than problematic. For example, using CRISPR-mediated genome editing and synthetic biology, guided mainly by human desire for profit and entertainment, to supplant biological evolution and replace disappearing components of natural biological communities with designer organisms is unlikely to produce stable ecosystems. Such action is more likely to cause ecosystem collapse. The "wisdom" of the marketplace cannot substitute for the "wisdom" of natural selection. "Wisdom" in the marketplace acts to increase profit, whereas the "wisdom" of natural selection has a 3.5-billion-year-old history of acting to create organismal and ecosystem integrity and stability.

That education in the United States fosters trained incapacities in evolution and ecology is irrefutable. A Gallop poll released on the eve of Darwin's two hundredth birthday in February 2009 reported that only 39 percent of Americans "believe in the theory of evolution" (Newport 2009). Biological evolution is often not discussed in high school biology classes for fear of conflict with religious fundamentalists who favor scriptural accounts of creation over the plethora of evidence for life's common genetic ancestry and the evolution of ecosystems over hundreds of millions of years. The upshot of this was reported in a 2017 Gallup poll that found 38 percent of Americans believe God created humans in their present form about ten thousand years ago, a percentage relatively unchanged since the 1980s (Gallup n.d.).

At most colleges and universities, only some life sciences curricula require courses in evolution and ecology, and nonlife science curricula require none at all. The result is that future voters, legislators, business persons, lawyers, lawmakers, CEOs, agriculturalists, medical professionals, physical scientists, engineers, and even gene editors and synthetic biologists acquire a trained incapacity to think wisely about the future of our species and the rest of nature. Deprived of understanding how the evolutionary origins of biodiversity relate to the structure and functioning of today's ecosystems, they do not comprehend humankind's absolute dependence on nature. Noteworthy science education initiatives by the American Association for the Advancement of Science and the US National Academy of Science have sought for several years to combat students' trained incapacities in evolution and ecology.[5] But a global commitment to bioscience literacy is needed.

The Sixth Extinction

Another factor that gives urgency to increasing public literacy in bioscience is the current astonishingly tragic rate of species extinctions, the sixth extinction.[6] The biosphere is in the throes of a crisis so vast that it is difficult to envisage. Environmental biologists estimate that known and unknown species of plants, animals, and microbes are now disappearing at the rate of several thousand per year.[7] The fossil record shows that the normal rate of extinction, when the biosphere is not in crisis, is only about one species every four years.

During the Big Five episodes of past extinctions, the biosphere lost 50 to 80 percent of its marine organisms and 20 to 70 percent of its terrestrial species. Modern humans had nothing to do with those extinctions since we were not on the scene until about two hundred thousand years ago. Naturally occurring climate change, marine estuary habitat loss due to continental coalescence, and one or more asteroid impacts all probably contributed to the Big Five. On the other hand, consensus among scientists identifies human activities as the major cause of the current sixth extinction. Environmental biologist Edward O. Wilson (2006, 118) describes these activities with the acronym HIPPO, where letter order corresponds to the rank of each activity's destructiveness:[8]

H = habitat loss from diverse factors including deforestation and climate change
I = invasive species that displace native plants and animals
P = pollution which has been accumulating since the Industrial Revolution
P = population (human) excess which exacerbates the other four causes
O = overharvesting of non-domestic food and game plants and/or animals

If HIPPO activities continue unabated, half of Earth's plant and animal species will disappear by 2050. Even if humans halted all five HIPPO activities within the next few decades, estimates based on past major extinctions show that nature's restoration of its lost biodiversity would require about ten million years. Presuming that humankind survives the sixth extinction, it will face major ecological voids during this century. It is doubtful that humans will patiently wait ten million years for nature to fill these voids via the evolutionary process. What, if any, role technologies like synthetic biology, CRISPR-mediated genome editing, and gene drives ought to play in species replacement after the sixth extinction is a question that needs careful thought by both nonscientists and scientists knowledgeable about evolution and ecology.

Preparing for Discussions and Decisions on Humanity's Long-Range Goals

Consider a species-specific trait of *Homo sapiens*—foresight—and our failure so far to apply this trait to forming long-range goals for ourselves and Earth's biosphere. This book describes biotechnologies that humans can use now or in the future to manipulate, redesign, or re-create virtually all living things. Although a few nations and states in the United States restrict the use or production of some biotechnology products like human clones or ESCs, there has been no sustained attempt by humankind to consider long-range goals for the future and what roles twenty-first-century biotechnologies ought to play in moving toward those goals. Lacking a global consensus or worldwide plan for how, when, or whether different life forms ought to be redesigned or newly created, such decisions are left primarily to the marketplace, government agencies, private and military institutions with undisclosed research and development objectives, and independent researchers.

This is not to say that an international, public consensus on long-range goals for humankind would or ought to prevent research or the application of technologies that run counter to those goals. The identification of long-range goals could at least provide guidance for major funding agencies for democratic nations, decision-making boards of private corporations, and policy makers tasked with regulating specific technologies. For example, if a majority of humans, particularly in societies with substantial resources for scientific research, made it known that it places higher value on living a high-quality, healthy life for one hundred years than on the prospect of retarding the aging process to create one-thousand-year lifespans, more research funding would likely be directed at preventative and curative medicine than at radical age retardation. Alternatively, a consensus that we should strive for lifespans of a millennia would necessitate a commitment of vast funding to understand and thwart the aging process. Also, sociological research would need to identify and solve problems inherent in a society whose members live much of their lives as though they are immortal. Our current institutions of education, religion, work, and retirement would require drastic reframing.

In 1969, Gerald Feinberg envisioned his proposed goal-identifying Prometheus Project taking several generations. He worried then that even if we began the process immediately, we would be into the twenty-first century before the project's completion and that it might then already be too late to direct some biotechnologies into pathways consistent with humankind's goals for itself. Now, as we near a quarter of the way through the twenty-first century, humankind still has no set of consensual long-range goals. Feinberg died in 1992, and probably none but a handful of intellectuals have ever heard of him or his writing. Feinberg's idea deserves to be revived, so next we consider how humanity can best prepare to set long-range goals for itself.

My first premise in this section is that it is better to deliberate about our long-range future and the future of life on Earth rather than simply allow technology itself, human hubris, and human whims for profit and entertainment dictate our future relationship with biotechnology. My second premise is that informed deliberation is better than ignorant or misinformed deliberation. Given these premises, all nations, but particularly those developing genetic, robotic, brain, AI, nano- and other transformative technologies, need to make major commitments to improve education in the life sciences. Such commitments ought to include formal, mandated evolution and ecology components of public education curricula.

Students in grades K-12 need hands-on life science experiences during every year of their education. The goal is for students to *feel* their place in nature's grand pageant and not simply memorize information about food pyramids, the carbon cycle, frog innards, DNA, and proteins. Both children and adults must experience nature regularly so they come to view themselves as planetary citizens and feel their kinship with, dependence upon, and responsibilities for all of Earth's inhabitants.

Science education is flagging in the United States due largely to unconfident, ill-prepared science teachers who substitute pedagogical skills and training for knowledge about science, especially biology (Mervis 2015). Aggressive and dramatic changes in our education systems are needed to nurture a knowledge-based appreciation of the natural world. This is evidenced by currently pervasive ways that nature is viewed by much of humankind. Of particular concern are policies and attitudes prevailing in highly industrialized nations where measures of national well-being are linked to the production and sale of material things whose creation expends nonrenewable natural resources and/or compromises the basic elements of air, soil, and water. On the other hand, there are some developments that give hope.

In 2013, the final draft of the Next Generation Science Standards (NGSS) in the United States was released. NGSS for grades K-12 was a collaboration between educators from twenty-six states, the AAAS, the National Research Council, the National Science Teachers Association, and Achieve, a nonprofit organization previously involved in developing math and English standards. By the end of 2016, sixteen states had adopted the standards and several others have indicated interest. The standards for education about evolution and ecology are ambitious and excellent.[9] What seems to be missing in the NGSS for the life sciences though is information on and encouragement for nurturing emotional ties between students and the living world. Enjoyable, regular, and frequent outdoor experiences are crucial to students' education—from collecting and examining pond water and seeing microscopic life to exploring the immensity of the universe on dark nights away from city lights and during overnight camping trips. Although some locations and school districts are better situated to offer regular outdoor experiences than others, all could devise ways to nurture the "spiritual" connection between nature and students. Keeping a journal about the changes of one's feelings about things natural and one's connection to them, writing poetry, and sketching or sculpting are some ways to encourage the development of empathy for and oneness with nature.

College and university undergraduate and graduate curricula also need strengthening in the life sciences. Sadly, many universities that developed strong core curricula in the 1970s and 1980s are now dropping core courses or weakening requirements, under pressure to make students competitive for jobs in business, engineering, and industry. Broad education with a diverse core curriculum that nurtures cross-disciplinary thinking skills is suffering under these pressures. Ideally, every college graduate should have formal studies in evolution and ecology, including outdoor experiences. Cultural historian and ecotheologian Thomas Berry proposed similar mandates three decades ago in his book *The Dream of the Earth* (Berry 1988).[10]

From personal experience on university curriculum committees, I know that suddenly requiring biology courses for every graduate is not realistic. As a beginning, I propose that curricula for all molecular biology majors, business school majors, and medical and law students in the United States require that students take a field ecology course and a course on the basic principles of biological evolution. Future genome designers, policy makers, and persons in positions to influence policy making must have opportunities to develop an appreciation for the antiquity of life and the inextricable interdependence of all living things. Particularly in groups of highly educated, influential people like these, such appreciation is vital for the future health of the biosphere and for the preservation of islands of the natural world created by 3.8 billion years of evolution. When manifestations of the benefits of life-science training become apparent in an entire generation of professionals, pressure will mount to extend the benefits to other curricula and eventually across disciplines.

I am not a politician, and I have no experience at high-level administration of public education. The expertise and connections needed to institute systemic and dramatic changes in public education at state, national, and global levels are not in my background. My hope is that some persons reading this book who do have the wherewithal are moved by the arguments here to take action to overhaul education systems in the directions described.

Worldwide initiatives to improve education in science, technology, engineering, and mathematics (STEM subjects) offer templates for raising the quality and increasing the reach of education in evolution and ecology.[11] In the United Statas, both Republican and Democratic presidents have supported STEM initiatives. A STEM Education Coalition joins over five hundred professional, business, and education organizations, including the US Department of Education and the NSF, to raise legislative government's awareness of the importance of high-quality STEM education. Many universities and corporations have established STEM initiatives of their own that include actions ranging from lesson plans and curriculum improvement to multimedia resource development.

Ethics in the Age of Biotechnology

How are we to behave as creators of new life forms and architects of our own nature? We have never been in such a position. Decisions we make, actions we take, and paths upon which we embark have consequences that are difficult to predict and will affect all future generations. What ethical system(s) should we adopt and teach to our worldwide grandchildren and great-grandchildren to help prepare them to live good lives with technologies of the twenty-first century and beyond?

Philosopher of science and technology Shannon Vallor (2016) tackled this problem in a recent book, *Technology and the Virtues: A Philosophical Guide to a Future Worth Wanting*. She argues convincingly that we do need a particular and common ethics to inform our use of what she terms NIBC technologies: nanotechnology, information technology, biotechnology, and cognitive science. These include the very technologies discussed in this book. Vallor contends that neither rule-based ethical systems such as Kantian deontological ethics and utilitarianism nor religion-based ethical systems are appropriate for helping us live a good life in the age of these life-transforming technologies. In her view, a virtue ethics is the most appropriate guide for our entry into the world of designer genomes, synthetic life forms, autonomous robots, privacy-threatening surveillance technologies, cognitive-enhancing technologies, and related technologies. Let us now recall the major rule-based ethical systems of the West described in chapter 7, deontological, utilitarian, and religion-based ethics. Then we will consider why those rule-based systems are inadequate for the job at hand. Finally, we examine Vallor's opting for a virtue ethics, which is also favored for social robots (as discussed in chapter 7).

The ideas of eighteenth-century German philosopher, Immanuel Kant, still have great influence on how persons in the West treat each other today. One of the greatest moral philosophers of the modern era, Kant formulated a rule called the categorical imperative for judging the morality of actions between humans. Recall from chapter 7 that Kant formulated two versions of his categorical imperative. The first formulation is most relevant to our present discussion: "Act only according to the maxim by which you can at the same time will that it should become a universal law" (Kant [1785] 1959; Rachels 2003, 131). This means that for an action to be morally right, one must be able to logically will that everybody in the world perform the same action. As an example, we saw earlier that lying fails this test because if everybody lied, nobody would believe anybody else, so nothing would be gained by lying.

Another major rule-based ethical system is utilitarianism. Credited mainly to nineteenth-century British philosophers Jeremy Bentham and John Stuart Mill, utilitarian ethics requires that we act to provide the greatest overall happiness or the least harm to sentient beings. Any animal is sentient if it can suffer. So the feelings of nonhuman animals as well as humans must enter into calculations that determine which course of action is right.

Finally, religion-based rules and norms may guide our actions. Because religion-based rules are many and varied, depending upon the religion, they speak foremost to respective believers. This renders religion-based norms problematic as a base for developing a common ethics to guide humanity's actions in the realm of modern bio- and related technologies. But what about Kantian or utilitarian ethics?

A phenomenon Vallor calls *acute technosocial opacity* makes it impossible to use the rules of Kantian or utilitarian ethics to secure the ultimate goal of ethics—a good life. In the case of Kant's categorical imperative to will that an action be universal, Vallor argues that possible outcomes of the NIBC technologies make it impossible to will their use by all persons. For example, one cannot reasonably will that *all* persons at some future time use gene-editing technology and embryo selection to transform their bodies and minds in ways that we cannot imagine. We simply do not have enough information about what that future might look like to be able to will it for the future. Similarly, we cannot will a world full of social robots without knowing the roles of these robots and their varied effects upon the lives of individuals. This lack of information about our distant technological future constitutes technosocial opacity.

Utilitarian ethics is hindered by the same lack of knowledge. One cannot possibly calculate the overall global happiness or harm that will come from developing particular technologies such as brain-computer interfacing, brain circuitry manipulations, dramatic genome alterations to humans and other organisms, the creation of new life forms, the purposeful extinction of certain life forms, or the use of nanotechnology to produce virtually any material item we desire.

What we need, contends Vallor, is a common, flexible ethical framework that cultivates a moral character that will help us cope and even allows us to flourish with the complexities and uncertainties of the rapidly developing technologies we have discussed. She proceeds to develop a set of twelve *technomoral virtues* for living a good life in the twenty-first century and beyond:

1. Honesty
2. Self-Control
3. Humility
4. Justice
5. Courage
6. Empathy
7. Care
8. Civility
9. Perspective
10. Flexibility
11. Magnanimity
12. Technomoral wisdom

Vallor details why each of these twelve virtues is important, if not essential, for living well with modern and emerging technologies, particularly with information technology, robotics and AI technology, and human enhancement biotechnologies. She suggests that our education systems commit to developing the first eleven virtues in ways that make them moral habits. Religious diversity can continue to flourish within social systems that teach and nurture these universal virtues. With very few exceptions, the list is compatible with other ethical systems and even enriches them. One need not abandon previously held moral systems in order to develop habits necessary for a vibrant virtue ethics. The twelfth virtue, technomoral wisdom, is different from the other eleven. Technomoral wisdom "is not a specific excellence or disposition, but a general condition of well-cultivated and integrated moral expertise that expresses successfully—and in an intelligent, informed, and authentic way—each of the other virtues of character that we, individually and collectively, need in order to live well with emerging technologies" (Vallor 2016, 154).

Humankind will ultimately choose its future relationship with biotechnology. It will do so either directly and wisely through engaged dialogue and discernment or indirectly through nonengagement and widespread ignorance about the technologies already profoundly changing life on Earth and our connections to it. To choose our biotechnology future consciously and wisely, we need to be collectively scientifically literate, especially in regard to the nature of life. Secondly, we need a moral compass to help guide both near-term and long-range decisions about our treatment of all life and the uses of biotechnologies.

Achieving global consensus on the kind of future we wish for our descendants and other things living on our small planet will require herculean effort, commitment, and creativity. Current information and social media technology make reaching such a consensus at least feasible. We must act immediately to prepare ourselves to engage in long-range goal discussions. We can do this by becoming better informed about the scientific and ethical issues associated with twenty-first-century biotechnologies. Also critical is to nurture virtues in ourselves, our children, and grandchildren that will prepare humanity to wisely plan how best to live with twenty-first-century biotechnologies. Technology is not prone to standing idle while we ignore it or wring our hands.

Vallor makes the important point that to be antitechnology is to be antihuman. We cannot and should not turn backward and try to be something other than what we have always been—changers and molders of ourselves and our surroundings to accommodate our needs, desires, and artistic and engineering impulses. What we need are knowledge about our place in nature, a dose of humility, and a search for wisdom. As Vallor puts it (2016, 13), "the only meaningful questions are: *which* technologies shall we create, with what knowledge and designs, affording what, shared with whom, for whose benefit, and *to what greater ends.*"

Chapter Summary

Our use of twenty-first-century technologies—including CRISPR genome editing, gene drives, embryo selection, human neuroenhancement, synthetic biology, robotics and AI, and nanotechnology—will produce irrevocable outcomes for humanity and life on our planet. For this reason, humanity would do well to select long-range goals for itself and the rest of nature. Choosing long-range goals wisely requires knowledge about nature, our place in it, and the nature of the technologies being developed. Structuring K-12 and university curricula to educate all students about the evolutionary origins of biodiversity and the ecological interdependency of all life forms and nurturing in them an awe and reverence for nature are ways to prepare for discussions about humanity's long-range goals. Another is to nurture an ethics appropriate for guiding our development and use of modern technologies. A virtue ethics that ultimately results in technomoral wisdom can provide an appropriate ethical framework for living well with the life-altering technologies of the twenty-first century well into the future.

Questions for Reflection and Discussion

1. Do you agree that certain twenty-first-century technologies will create irrevocable outcomes for humanity and/or life on planet Earth? If not, why not? If so, what might be one such outcome not described in this chapter? What other paths for humanity would that outcome preclude?
2. Do you agree that increased public knowledge about evolution and ecology would help citizens make wiser decisions about the use of technologies like CRISPR and synthetic biology? If not, why not? Are there other areas of bioscience literacy that you believe are just as important as these two?
3. Are there good reasons to preserve as much of the naturally evolved biosphere as possible? If so, what are they?
4. How do you feel about the prospect of designing new ecosystems using modern genetic biotechnologies and supplanting struggling natural ecosystems with new ones engineered to thrive under Earth's present conditions?
5. Do you believe that ethical systems derived from religious traditions or from Western thinkers like Jeremy Bentham and John Stuart Mill (utilitarianism) and Immanuel Kant (deontological ethics) are adequate to

guide us in living well with the modern technologies described in this book? Why or why not?

6. Is virtue ethics a good prospect for guiding our living with modern technologies? Why or why not? Would some other ethical system be better for this?

7. Is Feinberg's Prometheus Project feasible to undertake within the foreseeable future? If so, how can you personally help to make it reality? If not, why not?

Notes

1. Please see http://www.lifeextension.com/magazine/2014/4/Google-Wants-To-Extend-Your-Life/Page-01 and http://www.lifeextension.com/magazine/2014/4/Google-Wants-To-Extend-Your-Life/Page-01 (both accessed April 14, 2017).

2. *Key-log* is a term from the lumberjack days of early twentieth-century northern Wisconsin. It refers to the single log among thousands in a flotilla of logs that is jammed, preventing the entire mass from moving downstream to the saw mill.

3. Invasive species, such as predators, disease-causing organisms, and competitors that displace native species, are often a product of commodifying nature as in the long-distance transport of fresh food products to satisfy humans' appetite for out-of-season produce that also provides profit for the transporters and results in the transport of alien species into new territories.

4. Synthetic biologist Drew Endy, Department of Bioengineering at Stanford University, is credited with this statement. In the ETC Group report on Extreme Genetic Engineering (www.etcgroup.org). In Endy's defense, the web page for his laboratory describes his research in the engineering of genetically encoded memory systems as having potentially profound impacts on the study and treatment of diseases.

5. For the efforts of American Association for the Advancement of Science, see http://www.project2061.org/publications/guides/evolution.pdf and http://www.project2061.org/publications/rsl/online/TRADEBKS/REVS/GLOBALEC.HTM (both accessed January 8, 2017); for the US National Academy of Science, see http://www.nationalacademies.org/evolution/Reports.html (accessed January 8, 2017).

6. There are several books with this term in their titles. One of the most accessible of these is by paleoanthropologist Richard Leakey and R. Lewin, *The*

Sixth Extinction: Patterns of Life and the Future of Humankind (New York: Random House1995).

7. Estimates of the number of species that will be lost *annually* in this century range from seventeen thousand to one hundred thousand. Estimates differ due to differences in the estimated total number of species on Earth. This rate of extinction, along with the rate at which species enter trajectories toward premature extinction, is about one hundred times the rate at which evolution brings new species into existence. As the remnants of many ecosystems disappear later in this century, the extinction rate is expected to rise to more than one thousand times the rate at which species are born.

8. The estimated number of species of living things on earth ranges from 3.6 to 112 million. Of these, fewer than 2 million are known to biologists.

9. One may view specific objectives for each grade K-12 via keyword searches on the NGSS website at http://www.nextgenscience.org/search-standards (accessed April 17, 2017).

10. Berry proposed six specific courses for college students to help remedy widespread ignorance about the deep history of the universe and Earth. The first five courses include information about Earth's origin, the antiquity and emergence of life on Earth, principles of ecology, humanity's evolutionary and cultural history, and humanity's recognition of its uniqueness and responsibilities in the unfolding universe that includes life and consciousness (Berry 1988, 99–105). The final course Berry proposes is about the origin and identification of values, which he argues must be discovered within our experience of reality itself.

11. An excellent summary of STEM initiatives is at https://teach.com/what/teachers-know/stem-education/ (accessed June 11, 2017).

Sources for Further Information

Enriquez, J., and S. Gullans. 2015. *Evolving Ourselves: How Unnatural Selection and Nonrandom Mutation Are Changing Life on Earth*. New York: Penguin Group.

Gaskell, G., and M. W. Bauer. 2006. *Genomics and Society: Legal, Ethical and Social Dimensions*. London: Earthscan.

Harris, Sam. 2010. *The Moral Landscape: How Science Can Determine Human Values*. New York: Simon and Schuster.

Jasanoff, S. 2016. *The Ethics of Invention: Technology and the Human Future*. New York: W. W. Norton.

Kelly, Kevin. 2010. *What Technology Wants*. New York: Penguin Books.

Leakey, Richard, and Roger Lewin. 1995. *The Sixth Extinction: Patterns of Life and the Future of Humankind*. New York: Random House.

Naisbitt, J. 1999. *High Tech High Touch: Technology and Our Search for Meaning*. New York: Broadway Books.

Shermer, Michael. 2015. *The Moral Arc: How Science Makes Us Better People*. New York: St. Martin's Press.

Appendix 1

The Central Dogma of Biology, CRISPR, and Gene Drive

The central dogma of biology is a statement about the flow of genetic information inside cells. This is an unusual use of the word *dogma* since science and dogmatic thinking are incompatible. In 1957, at a meeting of the Society for Experimental Biology held at University College in London, Francis Crick proposed a relationship between DNA and proteins that he referred to as "the doctrine of the triad."[1] His proposal later became known as the central dogma of biology and is diagramed like this:

DNA ➔ *RNA* ➔ protein

The schematic shows that genetic information in DNA flows to an intermediary RNA molecule (messenger RNA; mRNA), and the information in mRNA then directs production of protein. The scheme is "central" to all of biology because DNA is the genetic material for all cells on Earth and because proteins are directly or indirectly responsible for all parts of the cells, tissues, and organs in every living thing.[2]

The first half of the central dogma (**DNA ➔ mRNA**) is called *transcription*, and the second half (**mRNA ➔ protein**) is called *translation*. Important to note is that DNA is not transformed into mRNA, and mRNA is not transformed into protein. Rather, the arrows signify a transfer of information so that the genetic information present in DNA is used to make a new molecule, mRNA, which acquires that information. The genetic information in mRNA is then used to construct a protein. At the end of the process, DNA still exists as it did at the outset, unchanged by the process of transcription, and mRNA is unchanged by the process of translation.

To appreciate the power of biotechnologies like embryo selection, genomics, genetic enhancement, and synthetic biology (discussed in chapter 2) and the powerful gene-editing tool called CRISPR (from chapter 4), we need some basic information about the three components of the central dogma: DNA, RNA, and protein. Let us begin with the end product, protein.

Proteins

Proteins are among the largest molecules in a cell. They are long, linear molecules that fold and bend to acquire complicated three-dimensional shapes (fig 1.3). One way to classify a protein is on the basis of its function: structural, catalytic, regulatory, or nutritional. Structural proteins produce structures inside the cell like microtubules and muscle filaments or substances secreted by cells such as hair, silk, and spider web material.[3] Catalytic proteins are also called enzymes. They make it possible for the thousands of biochemical reactions inside a cell to occur at the rapid rates required to maintain life. Regulatory proteins help an organism to adjust the biochemical reactions occurring inside its cells in ways appropriate for the various internal and external conditions confronting the organism. For example, hormones like insulin and prolactin are regulatory proteins that respond to elevated blood glucose levels and a suckling infant, respectively.[4] Other proteins called transcription factors control the rate and timing of gene transcription to produce mRNAs. Nutritional proteins include proteins in mothers' milk and egg yolk proteins that supply nutrients to infants and to developing embryos. The different functions performed by different proteins reflect differences in their three-dimensional shapes.

Individual proteins are made of smaller subunits called amino acids joined together in linear arrays specific for each protein. An average-sized protein contains about three hundred amino acids. The cell uses twenty different amino acids to build its protein chains. The sequence of amino acids in a protein determines its three-dimensional structure (fig. 1.3).[5] A genetic code in mRNAs specifies the sequences of amino acids in a cell's proteins. Let us see how this process of translation works.

Messenger RNA and the Genetic Code

For each protein, there is a corresponding mRNA encoding the amino acid sequence for the protein. Like proteins, mRNAs are long molecules comprised of small subunits linked together. Unlike proteins, which have twenty different kinds of subunits, mRNAs have only four kinds of subunits. These are called *bases* and are designated by the letters A, G, U, and C. Bases are attached at regular intervals to a backbone of alternating sugar (ribose) and phosphate groups depicted simply as a dotted line in figure A1.1. The sequence of bases in a mRNA determines the order of amino acids in the protein encoded by that mRNA.

How can four bases specify the order in which twenty different amino acids join together to form a protein? Each set of three adjacent bases in an mRNA specifies one and only one of the twenty amino acids. These sets of triplet bases are called *codons*, and they constitute the genetic code. Since four different letters (A, G, U, and C) can be arranged into sixty-four (4^3) different triplets (AUG, AAC, CCG, UAC, etc.), there are more than enough triplets to code for each of the twenty amino acids. In fact, some of the "extra" codons give redundancy to the genetic code; that is, a given amino acid is specified by more than one triplet codon. During protein formation, ribosomes move along the mRNA, translating codons as they travel along the mRNA and joining together the specified amino acids into amino acid chains that become multiple copies of the protein encoded by that mRNA. The ribosome itself contains many types of proteins and also some RNA molecules called ribosomal RNAs. One cell contains thousands of ribosomes, subcellular machines whose job is to build proteins. The ribosome builds a protein one amino acid at a time in a sequence corresponding to the sequence of codons in the mRNA (fig. A1.1). Now consider the transcription process; that is, how DNA specifies the formation of particular mRNA molecules.

DNA and Its Transcription to Produce mRNA

DNA is a huge molecule comprised of only five kinds of atoms: carbon, oxygen, hydrogen, nitrogen, and phosphorus; but oh, what a molecule it is! It gives continuity to all past, present, and future generations of cells and is the raw material upon which biological evolution acts. The acronym *DNA* stands for *deoxyribonucleic acid*. Each chromosome in our and other eukaryotic cells contains a single, very long molecule of double-stranded DNA. Within each of the twenty-three human chromosomal DNA molecules reside up to thousands of genes, a gene being a segment of DNA that encodes a protein (fig. 1.4).

Messenger RNA molecules are transcribed from genes. DNA and RNA structures are very similar. Like RNA, DNA is a long chain of bases (A, G, T, and C) attached to a sugar-phosphate backbone. Notice that three of the

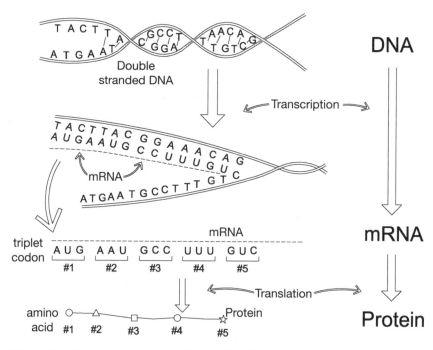

Figure A1.1. The central dogma of biology. Double-stranded DNA opens in the region to be transcribed to produce a messenger RNA (mRNA) complementary to one of the DNA strands. The mRNA moves from the nucleus into the cytoplasm, where it is translated to produce protein. Each set of three bases in the mRNA (#1–5 in the diagram) is a codon specifying a particular amino acid (#1–5) in the protein. (Courtesy of the author)

four bases in DNA are also in RNA (A, G, and C) and that the U in RNA replaces the T in DNA. DNA usually occurs in a double-stranded form consisting of two complementary DNA molecules.

Rules of base pairing determine the complementarity of the two strands of a double-stranded DNA molecule. The C in one strand pairs with the G in the complementary strand, and A and T pair with each other (fig. A1.1). Each pair of complementary bases in a double-stranded molecule is called a base pair, and the length of a DNA molecule is measured in numbers of base pairs. A human egg or sperm cell contains twenty-three different DNA molecules, corresponding to twenty-three chromosomes; together, the twenty-three chromosomes contain more than three billion base pairs of DNA. A typical gene consists of only a few thousand base pairs.

When a gene is transcribed, the complementary strands of DNA in the region of the gene's start point separate to expose the bases. Just one strand serves as a template for transcription to produce an mRNA molecule. Transcription follows the base pairing rules so that wherever C occurs in the template DNA, G is positioned in the corresponding mRNA (and vice versa). Similarly, whenever T occurs in the template DNA, an A is positioned in the corresponding mRNA. If A occurs in the DNA, a U is positioned in the mRNA.

Applying the base-pairing rules, a strand of template DNA reading TACTTACGGAAACAG is transcribed to produce a mRNA molecule reading AUGAAUGCCUUUGUC. From left to right, this small mRNA contains five triplet codons (AUG/AAU/GCC/UUU/GUC) that specify five adjacent amino acids in a protein (fig. A1.1). In a functional RNA, the last codon is a stop codon that does not specify an amino acid but simply tells the ribosome to stop adding amino acids to the protein.[6] A protein three hundred amino acids long requires three hundred triplet codons (nine hundred bases) in its mRNA plus one stop codon at the end.[7]

Discovery of CRISPR in Nature

In the 1990s Ruud Jansen and his colleagues at Utrecht University and other molecular biologists around the world noticed an unusual pattern of base sequences in the DNA of many species of bacteria, from the common gut bacterium *E. coli* to unnamed bacterial species in soil and seawater. This observation led to the discovery of CRISPR function in nature and ultimately to development of the most powerful genetic tool yet devised by humans. CRISPR function in nature is to help bacteria defend themselves against bacteria-killing viruses. How was that discovered and how was the system redesigned as a genetic tool for humans?

The unusual pattern of DNA base sequences found in those many different species of bacteria is this: five segments of DNA, each with the same twenty-nine base sequence and each separated by spacer sequences thirty-three bases long. The five segments of repeated, interspaced DNA comprise CRISPR. Unlike the CRISPR segments, the spacer sequences are different from each other, and many of them correspond to short segments of DNA from viruses that infect and kill bacteria. In bacterial genomes, this pattern of CRISPR and the nonrepeated spacers lie next to genes coding for DNA-cutting enzymes, a class of proteins called nucleases. The nuclease genes lying near CRISPR were christened CRISPR-associated genes, or simply *Cas* genes.[8] Several different *Cas* genes occur in different species of bacteria. This arrangement of CRISPR sequences, spacer sequences, and *Cas* genes is collectively called the CRISPR/Cas system.

When a gene is cut, it becomes nonfunctional. Why should genes for DNA-cutting proteins lie next to CRISPR and the spacer sequences in bacteria? By experimentally testing hypotheses for the natural function of CRISPR/Cas system, researchers finally revealed it to be a bacterial cell's defense against invading viruses. When a virus attacks a bacterium that has the CRISPR/Cas defense system, the bacterium activates its *Cas* genes to make Cas proteins that cut out pieces of DNA from the viral genome. The bacterium then incorporates the viral DNA segments into its own DNA where they become the CRISPR spacer sequences described earlier. These viral DNA segments serve as a molecular memory for the bacterium about the enemy status of the virus. The nearby *Cas* gene is ready to produce a Cas protein to attack the virus should it enter the cell again. Bacterial cells replicate CRISPR's molecular memory of its viral enemies every time they duplicate their DNA in preparation for cell division. Embedded in the DNA of every new generation of bacteria are molecular warnings about the enemy status of maleficent viruses encountered by its ancestors and also the means to incapacitate the viruses. It is like a mother mouse teaching its pups about the danger of cats and supplying each one with a little can of pepper spray to fend off cats whenever they appear.

The CRISPR/Cas system mounts an attack against any incoming virus recognized as containing a segment of DNA in its genome that matches a segment embedded as a spacer sequence among the CRISPR sequences. How this recognition occurs involves base pairing between an RNA molecule copied from the viral spacer DNA and the DNA of the invading virus. RNA copies from the viral spacers (CRISPR RNA) join up with a Cas enzyme to home in on enemy viruses and cut their DNA. CRISPR RNAs recognize foreign viral DNA sequences and guide the hitchhiking Cas proteins to the invading DNA. Once there, the Cas protein attacks the foreign DNA by cutting it and thereby defeating the virus on the bacterium's home turf. Researchers are now discovering that some viruses have evolved means to thwart the bacterial CRISPR defense mechanism, but we need not delve further into this interesting evolutionary arms race here. More relevant is that scientists have redesigned the CRISPR defense system into a powerful genome-editing tool.[9]

Once the workings of bacteria's ingenious CRISPR/Cas defense system became understood in 2007, Jennifer Doudna, a biochemist at the University of California, Berkeley, whose research specialty is RNA structure and function, had a eureka moment. In an interview, she reported thinking, "Oh my gosh, this could be a tool" (Zimmer 2015). Indeed, she was right. Shortly after her startling realization, Doudna and her colleague Emmanuelle Charpentier and others adapted the bacterial CRISPR/Cas defense system into the most powerful gene manipulation tool that humankind now has. With CRISPR technology, humans can now alter at will the genome of virtually any bacterium, plant, or animal including *Homo sapiens* (Jinek et al. 2012; Cong et al. 2013).

CRISPR as a Gene-Editing Tool

Recall that genes are sequences of molecular subunits (A, C, G, and T) called bases. A gene's base sequence specifies the function of the gene. Genes vary in size from less than one hundred to several thousand bases long. Base sequences of genes specify the structure of proteins. The sequence of bases acts as a kind of blueprint for a protein and directs the actions of the protein-synthesizing machinery inside the cell.

Proteins are the "workhorses" of a cell. They comprise specific structures, regulate metabolism, and facilitate biochemical reactions inside cells. Using CRISPR technology, researchers can change the base sequence of a specific gene and thereby alter the function of the protein that the gene codes for. With CRISPR technology, researchers can activate, deactivate, or temporarily silence specific genes, producing profound changes in the physiology, structure, or behavior of cells, tissues, and organs. CRISPR can also correct a disease-causing mutation or replace a defective gene with a normal copy of the gene. For example, changing just a single base in the gene that codes for the blood protein hemoglobin can either cause or cure sickle cell anemia.

To edit a gene or introduce a new gene into a cell using CRISPR technology, scientists introduce two CRISPR components into the cell: a guide RNA and a Cas protein (fig. A1.2). The guide RNA is like an address on an envelope that can guide delivery of the piece of mail to a specific city and street address. In the case of CRISPR, the guide RNA directs delivery of the Cas protein to a specific site in the genome. Of the several Cas proteins available, researchers use one called Cas9 almost exclusively. Cas9 is like a pair of scissors that cuts DNA molecules. A guide RNA is constructed so that Cas9 attaches to it and is carried along with the guide RNA wherever it goes. Finally, the city and street address correspond to a particular chro-

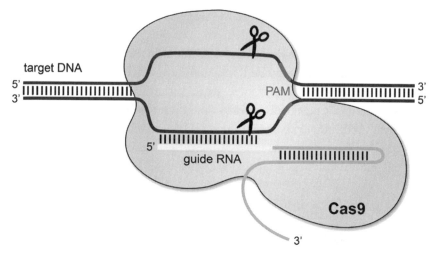

Figure A1.2. CRISPR/Cas9 complex for gene editing. The CRISPR/Cas9 complex consists of a guide RNA custom made by researchers and the Cas9 nuclease, a bacterial protein. The two bind to each other and move together into the nucleus of the cell containing the target DNA to be edited. The guide RNA contains a base sequence complementary to the target sequence in the nuclear DNA. The guide RNA binds to its complementary target sequence in the DNA, positioning Cas9 to make a double-stranded cut in the DNA at the sites marked with scissors in the illustration. The cell has two pathways to repair the DNA breaks. The nonhomologous end joining pathway leads to indels (single-base insertions or deletions) that inactivate the genes in which they occur. The homologous recombination pathway results in addition of DNA base sequences or even entire genes at the repair sites. Researchers provide these added DNA base sequences to the cell along with the CRISPR complex and use them to edit the genome. (Original illustration used with permission from CRISPR Therapeutics, additional labeling by author)

mosome and a specific site within a gene in that chromosome, respectively. For simplicity, we will now call the guide RNA and the Cas9 protein together (i.e., the address and the scissors) a CRISPR complex. Putting all of this together, researchers deliver a CRISPR complex to the interior of a cell whose genome they desire to alter in a specific way.[10] Then what happens?

Near potential target sites for CRISPR/Cas9 gene editing there must be a common three-base sequence called a PAM sequence. Cas9 is very good at finding and binding to PAM sequences.[11] When Cas9 binds to a PAM sequence, the DNA double helix in that region becomes destabilized, allowing the guide RNA to interact with its target DNA sequence. The action of the guide RNA depends upon the base-pairing rules described earlier. If a sequence of bases in the target DNA strand is complementary to a sequence of bases in the guide RNA strand, the two can form a DNA-RNA double-stranded, hybrid molecule. Guide RNAs are designed and synthesized in the laboratory to be complementary to DNA at the spot(s) where gene editing is desired. Since the guide RNA carries the Cas9 nuclease with it, the nuclease becomes positioned correctly to cut double-stranded DNA precisely at the spots in the genome that researchers wish to alter. What happens next is due largely to the presence of DNA repair mechanisms evolved in all cells that repair breaks in DNA.

Cells have two types of repair mechanisms. One type simply joins the broken ends together, but in doing so, an extra base is often inserted or deleted at the repair site. This creates a so-called indel (for *in*sertion and *del*etion) that subverts the function of the gene, inactivating it, which in some cases is exactly what researchers wish to do.[12] The other type of repair mechanism uses a segment of DNA as a template that can add to or precisely change the base sequence at the break point.[13] The cell's repair machinery copies the base sequence information in a specially designed DNA segment provided by researchers along with the CRISPR complex. In this way, the base sequence in a disease-causing gene can be corrected or entirely new genes can be added to a cell's genome.

By introducing multiple and different RNA guide sequences into a cell, researchers can edit several genes simultaneously in the same cell or embryo, a technique called multiplex genome engineering (Cong et al. 2013). In 2016, workers at Yale University used a multiplex CRISPR "toolbox" to simultaneously delete three cancer-associated genes from human cancer cells growing in laboratory cultures. They accomplished the same thing in cancer cells growing as tumors in mice (Cao et al. 2016). In both the laboratory cell cultures and the mouse tumors, the cancer cells used in the research were the human HeLa cells described in chapter 1.

CRISPR-Mediated Gene Drives

Researchers construct a CRISPR-mediated gene drive by inserting a DNA cassette of CRISPR elements directly into the genomes of individual members of a population. DNA sequences in the cassette encode a guide RNA, the Cas9 protein, or some other enzyme that cuts DNA, and whatever genetic alteration (e.g., a genetic correction or addition) researchers wish to impart to the population. The DNA cassette also contains short stretches of DNA complementary to the target organism's genome at the site of the desired genetic alteration to ensure correct placement of the alteration. Reproductively active individuals carrying the CRISPR DNA cassette in their eggs and sperm must be engineered in the laboratory, an act already accomplished with fruit flies (Gantz and Bier 2015) and mosquitoes (Gantz et al. 2015).

When the CRISPR cassette is expressed in a cell, the action of the Cas9 enzyme and its guide RNA, along with the cell's own DNA repair mechanisms, ensure that the CRISPR cassette becomes copied and incorporated into the chromosome homologous to the one already carrying the desired genetic alteration. Recall that homologous chromosomes are members of a pair of chromosomes, one derived from each parent at the time of fertilization. Thus, all non–germ line cells of a sexually reproducing organism normally possess two homologs of each chromosome.

All of this is to say that if researchers insert a CRISPR cassette into one homolog, the cassette becomes copied into the other homolog by the cell's own biochemical metabolism. So when a CRISPR cassette is inserted into the DNA of a fertilized egg, virtually all of the eggs or sperm in the resulting individual possess the cassette. Likewise, when individuals like these mate with non-CRISPR-carrying individuals, virtually all of the progeny and all of the progeny's germ cells carry the CRISPR cassette and also possess whatever genetic alteration that the CRISPR system has mediated. Thus, a human-designed genetic alteration can be spread very rapidly through a wild population. A variety of CRISPR gene drive mediated genetic alterations may be designed including deletion, permanent inactivation, or temporary silencing of a specific gene, correction of an existing mutation in a specific gene, or insertion of one or more new genes into the genome.

Notes

1. Crick's musings in preparation for his 1957 talk are preserved in handwritten notes he made to himself in October 1956. These are available to view on the National Institutes of Health Profiles in Science site at http://profiles.nlm.nih.gov/SC/B/B/F/T/_/scbbft.pdf (accessed July 19, 2017). Years later, in his 1990 book, *What Mad Pursuit: A Personal View of Scientific Discovery* (New York: Basic Books), Crick explains that he came to use the word *dogma* simply to describe "a grand hypothesis that, however plausible, had little direct experimental support" (109) at the time.

2. Many viruses use RNA, not DNA, as their genetic material. But viruses are not cells nor are they considered living since their propagation depends upon being inside a living cell.

3. Microtubules are very small hollow tubes comprised of proteins called tubulins. Hundreds of microtubules reside inside every eukaryotic cell. They help maintain cell shape, direct the flow of objects inside the cell, and aid in separating duplicate sets of chromosomes during cell replication.

4. Insulin facilitates the conversion of glucose to the energy storage molecule glycogen in the liver. Prolactin released from the pituitary gland in response to stimulation of the nipple by a suckling infant facilitates the production of breast milk.

5. Although the genetic code specifies twenty different amino acids, all twenty of them are not necessarily used in every protein.

6. There are three stop codons: UAG, UGA, UAA.

7. Actually, pre-mRNAs in eukaryotic cells are about five times as long as their corresponding mRNAs because many of the bases in transcribed genes do not code for amino acids. These noncoding regions of genes are called *introns*, while the coding regions are called *exons*. The transcribed introns in pre-mRNA must be discarded and the exons joined together to produce a functional mRNA.

8. By convention, the names of genes are italicized, and the names of the corresponding proteins that the genes encode are not italicized. Thus, *Cas* genes code for Cas proteins.

9. Actually, bacteria have evolved more than one CRISPR defense mechanism, but one of these, the so-called type II system, is the most studied, the best understood, and the one being redesigned and used for genome editing.

10. Components of the CRISPR complex can be delivered to a cell in different ways. For example, a segment of DNA coding for the guide RNA and the Cas protein may be conveyed to the cell via a viral vector. Once inside the cell, the DNA encoding the guide RNA and Cas protein becomes expressed to produce the guide RNA and a messenger RNA coding for the Cas protein. The cell then translates the Cas messenger RNA to produce the Cas protein, which finds its way to the guide RNA to form the CRISPR complex. Alternatively, researchers may inject a cell with copies of the guide RNA and copies of Cas protein messenger RNA molecules. Then the cell needs only to translate the Cas messenger RNA to produce copies of Cas proteins, which attach to the guide RNAs to produce CRISPR complexes.

11. A PAM sequence is NGG, where N is any of the four bases and G is guanine (its complementary base in the other DNA strand being C), moving in the 5' to 3' direction on one of the two strands of double-stranded DNA.

12. This type of DNA repair is called nonhomologous end joining.

13. This type of DNA repair is called homology-directed DNA repair.

Appendix 2

Tools for Neuroscience and Clinical Neurology

DBS

DBS consists of electrodes implanted deeply in the brain and connected to a subcutaneous power source that emits adjustable electrical impulses to specific brain regions. Sometimes called a brain pacemaker, DBS is used to treat movement, mood, and pain disorders, particularly in patients disabled by symptoms nonresponsive to traditional medical management strategies. Adverse psychological or behavioral side effects may accompany DBS in some patients, but these are usually reversible by adjusting the strength of the electrical impulses or the position of the electrodes. As Okun describes it, "a couple of millimeters in the brain is like the difference between Florida and California. . . . There's an abnormal conversation going on between different regions of the brain" and those conversations are interrupted by DBS (quoted in Noonan 2014). Since the brain has no sensory nerve receptors, implantation of DBS microelectrodes may be done with the patient wide awake. This is a boon for brain research because the patient's cognitive, motor, and emotional states can be monitored during surgery, thereby providing information about the location and function of specific neuronal circuits. For example, one patient undergoing DBS surgery for obsessive-compulsive disorder became ecstatically happy when an electrode "tickled" a brain region associated with pleasure and reward (Noonan 2014). It follows that complex ethical issues arise at the prospect of altering mood with elective surgery, and these are discussed later in the appendix.

DBS was first used in the 1960s to relieve chronic pain in patients whose pain no longer responded to medication. Parkinson's disease symptoms have been treated by DBS since the mid-1990s, and since then, other treated conditions include essential tremor, obsessive-compulsive disorder, dystonia, epilepsy, Tourette syndrome, and major depression.[1] Current research aims to extend DBS treatment to posttraumatic stress disorder, memory loss due to Alzheimer's disease, borderline personality disorder, general anxiety disorder, traumatic brain injury, anorexia, and addiction (Lozano 2013; Noonan 2014). Placement of the implanted electrodes depends upon the condition to be treated, but in most cases they are positioned in clusters of nerve cells called nuclei lying deep within the brain stem. For example, one region commonly stimulated in patients with Parkinson's disease is the subthalamic nucleus. Important to note is that DBS may relieve symptoms of certain brain diseases or disorders, but it does not cure them. Little is known about exactly how DBS exerts its dramatic effects beyond the ability of the electrical impulses to stimulate or dampen electrical activity in neural circuits close to the electrodes. According to neurologist Michael Okun, DBS action in the brain includes stimulating the growth of new brain cells, altering blood flow, and causing electrical wave patterns called neurological oscillations.[2]

MEG and TMS

MEG and TMS are noninvasive neurotechnologies used on humans. The former measures electrical activity in the brain, while the latter stimulates targeted, localized areas of the brain and can affect behavior (Schnitzler and Hirschmann 2012). Used in concert, MEG guides TMS in manipulating the electrical activity of specific neural circuits to aid research and therapy. For example, work is underway to develop this combination of technologies for noninvasive, nonpharmaceutical treatments to correct the abnormal neural activity associated with Parkinson's disease, epilepsy, schizophrenia, and other neurologically based disorders.

To better grasp how MEG works, recall from high school physics class that even a very small electrical current creates a magnetic field around the wire(s) carrying the current. Now just substitute *nerves* (bundled neuronal axons) for *wires*. MEG can detect the combined electrical activity of several thousand neurons by measuring the minute magnetic fields created by that activity. Along with its noninvasiveness, MEG has the advantage of being precise within about one-quarter of an inch in identifying the source of the signal it is detecting. In research, MEG and TMS used together can locate neural circuitry underlying myriad behaviors and mental states. MEG can also help to improve the precision of DBS. By using MEG to monitor magnetic fields resulting from DBS's effects on neural circuitry activity, the location and strength of DBS treatment can be optimized.

fMRI and Diffusion MRI

fMRI relies on the fact that active nerve cells require more oxygen than inactive ones. When neurons suddenly become active, that region of the brain initially experiences a decreased level of oxygen due to its use by the neurons. Quickly though, increased oxygenated blood flow overcompensates for the loss of oxygen, and the neuronally active region is richer in oxygen than less active regions. During an fMRI scan, the subject is exposed to a magnetic field about fifty thousand times stronger than the Earth's field. This powerful force aligns the nuclei of hydrogen atoms in water (H_2O). The magnetic field from aligned atomic nuclei differs from that of randomly arranged atoms and is also influenced by the level of oxygenation of hemoglobin, the oxygen carrying protein in blood. Differences in magnetic signals coming from atomic nuclei in the blood therefore reflect regionalized differences in the blood's state of oxygenation and, by association, the level of neuronal activity.

Diffusion MRI is a noninvasive imaging technique based on the fact that the diffusion pattern of water molecules in a defined space is affected by objects occupying that space. In the case of the brain, the diffusion of water molecules in a space occupied by nerve fibers (bundles of axons) ensheathed with myelin (white matter) is hindered in a direction perpendicular to the fibers compared to its diffusion in a direction parallel to the fibers. MRI instrumentation detects these different water diffusion patterns, and computer analysis of the diffusion patterns reveals the shape, location, and routes of the nerve fibers within the living brain (Hagmann et al. 2006).

PET

PET imaging produces a three-dimensional picture of metabolic activity in the target tissue or organ of a living organism. For PET imaging, the subject takes a dose of a radioactive element, called a tracer, either intravenously, by inhalation, or by ingestion. The tracer emits particles called positrons. Researchers then follow the tracer in the body by measuring positron emissions. In the case of brain PET scans, the tracer is often radioactive oxygen, which attaches itself to hemoglobin, the oxygen-carrying blood protein. Blood containing tracer particles flows through the brain. Those regions with higher levels of the tracer correspond to regions with elevated blood flow. These in turn correspond to regions of higher neuronal activity.

PET depends upon detecting the emission of positrons from radioactive tracers introduced into a subject's body. A positron is the antimatter equivalent of a negatively charged electron. How does the emission of positrons result in a three-dimensional image? Discovered in 1932, positrons are the antimatter equivalent of negatively charged electrons. When a positron is emitted from its tracer, it travels a short distance (less than 1 mm) in the tissue before it encounters an electron. The collision between the matter (electron) and antimatter (positron) annihilates both resulting in the creation of light particles (gamma photons) that are detected as tiny bursts of light by a scanning device called a scintillator. The scintillator rapidly scans very thin layers (sections) of tissue, and computer technology integrates the bursts of light from each section into a coherent three-dimensional image. In fact, the word *tomography* refers to imaging using multiple, thin sections and some type of penetrating ray. In the case of PET, the penetrating rays are gamma photons. By using different tracers that associate with particular biologically active molecules, researchers and clinicians can use PET scans to locate the site(s) and relative levels of specific neurotransmitters (such as dopamine and serotonin) in the brain and to diagnose Alzheimer's disease and other brain disorders, cancer, and heart malfunctions (Demitri 2007; Dugdale 2012; Gersten 2013).

Functional Near Infrared Spectroscopy (fNIRS)

As in fMRI, fNIRS uses real-time, localized changes in oxygenated blood flow in the human brain as a proxy for cerebral neuronal activity. Advantages of fNIRS over other imaging methods are its noninvasiveness, the portability of the necessary equipment, and the opportunity to take simultaneous measurements from multiple sites in the cerebral cortex. Processing information from fNIRS data results in real-time maps of brain activity from unrestrained, nonsedated subjects in normal environments (Ferrari and Quaresima 2012). fNIRS will be invaluable for BRAIN Initiative researchers seeking to identify and map neuronal circuits controlling both normal and pathological aspects of human brain function.

fNIRS relies on the principle of neurovascular coupling, the fact that neuronal activity requires cellular, chemical energy, the production of which requires oxygen carried by the blood-borne protein, hemoglobin. Since hemoglobin absorbs near infrared light more strongly than surrounding bone and nonblood tissues, the absorption of near infrared light is an indicator of the amount of blood in the region of the brain being examined and therefore an indicator of neuronal activity (Ferrari and Quaresima 2012).[3] An interesting and important question remaining unanswered is exactly how the brain controls regionalized dilation of blood capillaries to obtain more oxygenated blood for areas of neuronal activity. The fact that this does happen though makes fNIRS possible.

Optogenetics

In 2010, optogenetics was named "Method of the Year" by the research journal *Nature Methods* and was listed among "Breakthroughs of the Decade" in the top-tier scientific journal *Science*. Why all the accolades? Optogenetics allows researchers to turn on or turn off the activity of specific neurons simply by shining pinpointed light beams of particular wave lengths at them. The technique requires input from several disciplines including microbiology, ecology, biochemistry, genetic engineering, and optical engineering. So far, optogenetics is used only in nonhuman animals such as mice and insects, but results from these experiments are dramatic and point the way to an eventual understanding of how human neural circuitry gone awry results in psychiatric disorders such as depression, obsessive-compulsive behavior, and schizophrenia and in conditions like Parkinson's and autism spectrum disorder. Optogenetics allows researchers to manipulate neuronal circuitry in living animals by activating and deactivating specific neurons with high temporal and spatial precision. Future treatment of brain disorders with beams of light is more than just wishful imagining.

Optogenetics makes use of certain genes from bacterial and algal microbes that code for cell membrane proteins called opsins. Opsins are light sensitive and serve as channels for ions (charged elements) to flow into or out of the cell. Bacteria and algae use opsins to capture energy from light for cellular processes. In optogenetics, neuroscientists use opsins and the genes for opsins to control the electrical activity of nerve cells.

Using genetic engineering technology, researchers began inserting opsin genes into the neurons of mammals and other experimental animals in 2004. Accurate incorporation of opsin genes into an animal's genome results in the genes becoming active and producing opsin proteins, which insert themselves into cell membranes. By shining light of certain wavelengths at particular spots in the brain, experimenters can control the opening and closing of opsin channels in the neurons at that spot. The sudden flow of ions across nerve cell membranes can stimulate cells to propagate electrical signals to neurons with which they are in contact. These in turn send signals on to thousands of other neurons which via a domino effect may activate an entire neuronal circuit and dramatically affect behavior.

Alternatively, light beams may inactivate neurons, causing them to stop propagating electrical signals to other neurons. By using highly focused fiber optics, extremely precise light activation of opsin-engineered neurons virtually any place in the brain is possible.

Current optogenetic research includes searching unique ecological niches for microbes containing opsins with novel properties, genetically engineering already discovered opsin genes to produce opsins with new properties such as their sensitivity to specific wavelengths of light, further fine-tuning methods to deliver opsin genes just to specific neuron types at certain locations in the brain, and improving the precision and reach of opsin-activating light beams.

Optochemistry

Optochemistry is similar to optogenetics in that light is used to control neuronal activity. However, an advantage of optochemistry over optogenetics is that the former bypasses a labor-intensive genetic engineering step required for the latter. This is a new technology for modulating the level of neurotransmitters at their site of action, the synapse. Neurotransmitters are molecules that move between neurons at synapses and thereby put nerve cells in communication with each other. Neurotransmitters prevalent in popular media include serotonin and dopamine. The effects of drugs such as Prozac and Ritalin on neurotransmitter activity are described in the companion volume to this book (Bradley 2013a, 232–37).

Optochemistry differs from optogenetics in that no genetic engineering is required to introduce opsin genes into neurons. Instead, researchers introduce neurotransmitter molecules attached to a light-sensitive molecule called a "cage" into the animal, often orally. When the cage molecules reach the brain, a light pulse of an appropriate wavelength is aimed at the area of interest. The light induces a shape change in the cage that releases the neurotransmitter molecules. The neurotransmitters in turn bind to receptors or channel proteins in the neuronal cell membrane resulting in channel opening and a flow of ions across the membrane. The ion flow propagates an electrical signal that is then passed on to many other neurons and from them on to others. In this way researchers can activate a neuronal circuit at will with a simple light pulse.

Minibrains

To create minibrains, researchers first coax unspecialized ESC or iPS cells to enter the pathway toward specialization into brain cells by culturing them in a relatively simple neural induction medium (Xia and Zhang 2009). Introducing these cells into a gelatin-like matrix of protein for three-dimensional support, adding a nutrient solution, and then gently spinning the small globules of cell-containing gelatin to facilitate nutrient absorption results in pea-sized minibrains within twenty to thirty days (Lancaster et al. 2013). Other workers use a similar matrix called Matrigel produced by cultured, mouse cartilage tumor cells (Bagley et al. 2017). The developing minibrains recapitulate several features of human embryonic brain development and produce at least seven discrete brain regions including a cerebral cortex region subdivided into functional areas analogous to those in a normal human brain. Neurons in the developing minibrains spontaneously acquire electrical activity, a defining characteristic of neurons in functioning human brains.

Calcium Imaging

Free (not bound to other chemical entities) calcium levels rise inside neurons when they propagate or receive signals to or from other neurons respectively. Therefore, measuring calcium levels inside neurons gives information about neuronal activity. Calcium imaging is a technique for visually monitoring the free calcium levels in single cells or in large groups of cells.

Calcium inside cells exists as free calcium (Ca^{++}) and as bound calcium. The latter forms when Ca^{++} attaches to some other chemical entity such as a protein. Free calcium is an important regulator of cellular activity in all cell types. Calcium imaging visually and indirectly monitors free calcium levels in either single cells or in large groups of cells. Neurons with higher levels of free calcium are more actively receiving/sending electrochemi-

cal signals from/to other cells. Using calcium imaging and special microscopes, neuroscientists can literally see which neurons are active in specific
regions of the brain. They can even monitor calcium levels in different regions within a single neuron.

Visualization of calcium is accomplished by introducing into the cells
of interest certain molecules called calcium indicators that fluoresce when
calcium binds to them. The higher the level of free calcium in a cell, the
more calcium indicators are bound to calcium and the more intense is the
observable fluorescence. Researchers use two general categories of fluorescent molecules for calcium imaging: chemical fluorescent calcium indicators and protein-based genetically encoded calcium indicators (Grienberger
and Konnerth 2012). Chemical indicators are introduced directly into the
target cells by microinjection or via pores or channels in the cell membrane. Genes for genetically encoded protein indicators are usually carried
into target cells by viruses whose genetic constitution is altered to contain
the gene for the indicator or using an electrical pulse to induce transient
pores in the cell membrane. If genes for indicators are injected into a live
egg cell, all cells of the animal developing from that egg contain genes for
the indicator. So far, these technologies are used just in nonhuman animals
including mice, rats, and insects. A major advantage of calcium imaging is
its ability to visualize active neurons in living, behaving animals. Researchers can examine functioning neuronal circuitry under real life conditions in
animals subjected to various experimental conditions and in animals modeling human disease states such as Parkinson's and Alzheimer's disease or
autism. Ethical considerations currently preclude using these technologies
in living humans, but using them in nonhuman primates and in the human
minibrains described above and in chapter 2 will give insights into the workings of human neuronal circuits.

Voltage Imaging

Voltage imaging measures and visualizes the electrical state of individual
and/or populations of nerve cells. It does this indirectly by introducing
voltage-sensitive dyes into cells and monitoring the state of the intracellu-
lar dye molecules using specialized microscopes. Voltage imaging research
focuses mainly on nonmammalian organisms such as crabs, squids, worms,
and salamanders, but some work extends to mice and rats. For example,
Ferezou and her coworkers (2009) used a flexible fiber optic bundle attached

to the heads of mice for real-time monitoring of neuronal activity in the neocortex of awake, behaving animals moving about freely within the experimental area. Limitations of imaging with voltage-sensitive dyes include cell damage from light and heat created by the dyes and difficulty introducing dyes deeper than a few millimeters from the brain surface. As with calcium imaging, ethical considerations preclude using current voltage imaging techniques on living human beings, but information gained from other mammals and human minibrains again will give insights into human brain circuitry.

To better understand how voltage imaging works, some information about the biology of neurons is helpful. A neuron's electrical state is described by its membrane potential or membrane voltage. Just as a car battery has positively and negatively charged terminals, the two sides of a nerve cell's membrane are charged differently. Membrane potential arises from this difference in charge across the membrane and reflects the activity of the neuron. In a neuron's resting state, the inside surface (facing the inside of the cell) of its cell membrane is more negatively charged than the outside surface.

During nerve cell activation by incoming chemical signals from other neurons, ion channels open, allowing positively charged particles (Na^+ ions) to rush into the cell at a very small, localized spot. This changes the membrane potential at that spot. If the incoming signals are strong enough, the inner side of the cell membrane become so positively charged compared to the outer side that an action potential is initiated. Action potentials are transient electrical signals that move rapidly (1–100 meters per second) from their site of origin down nerve cell axons to synapses, where they stimulate the release of chemical signals (neurotransmitters) to other neurons.

Propagation of an action potential along an axon reflects a rapidly moving wave of Na^+ ions rushing into the cell. In the wake of an action potential, molecular pumps in the membrane move Na^+ back out of the cell, returning the cell to its resting state in preparation for initiation of another action potential. Since neural circuits consist of thousands of nerve cells in electrical and chemical communication with each other via action potentials and neurotransmitters, visualizing membrane potentials within specific populations of neurons help researchers identify neuronal circuits, environmental factors that activate or inactivate them, and behavior controlled by them. Visualizing membrane potentials by voltage imaging frequently employs voltage-sensitive dyes, molecules whose absorption or emission of light, distribution across the membrane, or orientation within the membrane changes with changes in the membrane potential (Peterka et al. 2011). Since researchers must introduce the dyes into living tissue, they must be minimally disruptive to normal cell function. They must also remain closely associated with the cell membrane, where the membrane potential is localized rather than diffusing throughout the cell as do calcium indicators.

Similar to calcium imaging, voltage imaging may combine with genetic engineering technology, employing genetically encoded, voltage-sensitive membrane proteins that either fluoresce themselves or facilitate the optical activities of dyes associated with them. Genetic engineering offers the opportunity to target genetic indicators of voltage to specific types of neurons, thereby avoiding extraneous imaging signals from neurons not being investigated. Essential equipment for voltage imaging includes specialized microscopes, other imaging devices, and computer software to detect and analyze the small changes in light absorption or emission that reflect neuronal activity. Advantages of voltage imaging over calcium imaging include visualizing changes in membrane potential at specific locations on a single cell, measuring a much greater range of membrane potentials, and mapping temporal neuronal activity with precision below one millisecond.

Creating Detailed Anatomical Maps of the Brain: CLARITY and Brainbow

In an interview about mapping the billions of brain cells and trillions of connections between them, German neuroscientist Moritz Helmstaedter explains the magnitude of the problem like this: "Mapping the human brain is as difficult as creating a 3D map of a city 10,000 times the size of . . . New York, and locating every inhabitant in every building, street, stairwell, lift and subway. . . . The mapping would then have to unearth all lines of communication—personal interactions as well as those by phone, post and e-mail—between all inhabitants of these 10,000 cities" (Marx 2013).[4] Even this graphic imagery of brain structure and activity fails to capture all of the complexity that brain mappers face, for the living brain is a dynamic structure. Connections between cells are constantly changing—forming, disappearing, strengthening, weakening—minute by minute, even second by second as we live and experience our lives.

How do neuroscientists approach the problem of obtaining a static picture of the cell-to-cell connections in a human brain? They do not begin with the human brain. Instead they develop research strategies and refine laboratory technologies by working on smaller, less complex brains of worms, flies, fish, and mice. Brain mapping involves two major projects— (1) obtaining images of all of the cells in the brain and their connections, and (2) analyzing those images to create a comprehensive three-dimensional brain map.

Brain imaging requires collaborations between scientists, the business world, and the public. Collaborative groups are producing new ways of preparing brain tissue to make its component cells visible, developing new types of microscopes with which to view the tissue, and participating in crowd-sourced analysis of the enormous amounts of data collected from thousands of microscopic images.[5]

A recent methodological breakthrough in three-dimensional imaging of cellular aspects of the brain and other organs called CLARITY renders entire organs optically transparent while preserving cell-cell relationships and even the internal protein-based structure of cells (K. Chung et al. 2013). CLARITY is especially good for examining brain tissue. In fact, the researchers who developed the technique used it to relate cellular structure to function in an entire mouse brain by preserving and revealing nerve cell circuitry including the locations and abundance of specific neurotransmitters, the molecules at synapses, that relay signals between neurons. The researchers were even able to visualize different patterns of gene activity among cells. CLARITY is a powerful imaging method that will allow health researchers to study damaged or diseased brain regions while maintaining a three-dimensional, whole-brain perspective; however, one limitation of CLARITY is its requirement for postmortem brains.

Researchers using CLARITY first infuse the whole brain with a clear chemical cocktail that sets into a firm, Jell-O-like substance, which secures the positions of proteins and other molecules like DNA and RNA within each cell. Next, they wash the tissue free of its fatty, oily substances to make the brain transparent. Finally, by treating the brain with fluorescently tagged antibodies that attach to particular proteins of interest, the exact locations of those proteins throughout the brain are revealed using microscopes and computer software designed to map the fluorescence in three dimensions with high precision.

Brainbow is another powerful imaging method (Livet et al. 2007; Cai et al. 2013) that reveals three-dimensional relationships between individual cells inside whole, intact brains. Brainbow employs modern genetic engineering techniques to mark brain cells in worms, flies, fish, mice, and other animals with one of about one hundred different color hues. The results are reminiscent of a real rainbow coursing through the brains (and/or other organs) of experimental animals. Since each cell emits just one color, an individual nerve cell with its axons and dendrites can be accurately tracked within the whole brain. Brainbow enthusiast and neurobiologist Alex Shier of Harvard University compares nerve cell networks to a bowl of pasta: "Imagine you have a bowl of spaghetti—there is no way to follow each [piece of] spaghetti from end to end. But if each [piece of] spaghetti had a different color, you could. That's what Brainbow does for the nervous system" (Ghose 2009).

The Brainbow technique begins by genetically engineering individual cells to possess multiple cassettes of genes coding for variously colored fluorescent proteins. Genes for specific colors are then randomly activated by specific, externally applied, chemical signals in the living animal. The result is a brain with a "brainbow" comprised of about one hundred different colors. Mapping the cellular connections within an entire brain requires a painstaking effort of individually, microscopically examining and analyzing tiny pieces of tissue and then assembling the results into a picture of the entire organ.

For Brainbow data analysis, Helmstaedter and coworkers (2013) took an especially creative approach of a crowd-sourced manual annotation. More than 224 trained volunteers used special computer software to personally examine and manually trace the outlines of tens of thousands of stained cross sections through brain neurons, including their long, thin axonal and dendritic projections, digitally imaged from electron microscopic examination of ultrathin (25 nm)[6] slices of brain tissue.[7] The volunteers together logged more than twenty thousand hours of work for the project. Mistakes were identified and minimized by having four to six volunteers trace the boundaries of each given cell through the thousands of slices containing that cell and comparing their tracings. The end result was a computer-generated, three-dimensional reconstruction of 950 nerve cells and their mutual contacts in a specific region of the mouse retina. The use of crowd-sourcing allowed the researchers to accomplish what no small group of researchers could reasonably achieve by themselves and with accuracy unmatched by any automated process.

Other examples of crowd-sourcing include the National Audubon Society's Christmas Bird Count across North America, National Geographic's Genographic Project using modern DNA analysis and voluntary DNA donors to probe humankind's origin and early migrations, the SETI@home project to search for radio signal signs of extraterrestrial intelligence, and a UCS project to enlist idle time on personal computers to run climate change models to better understand human activity's contribution to climate change.[8] Future Brainbow studies in animal models for human neurological and psychological disorders such as Parkinson's disease, posttraumatic stress syndrome, schizophrenia, and depression should help identify specific neuronal circuits associated with these and other disorders.

Notes

1. Dystonia is a disorder that causes uncontrollable twisting of the body, complications of which can cause shortened life spans in young patients. DBS has returned the twisted trunk and limbs of treated children to normal or near normal function within weeks (Lozano 2013).

2. Michael Okun and neurosurgeon Kelly Foote are codirectors of the University of Florida Center for Movement Disorders and Neurorestoration. Their close collaboration in DBS therapy and research at the center was described by interviewer and science journalist David Noonan (2014).

3. Near infrared light (electromagnetic radiation) has wavelengths (750–1400 nanometers) just above that of visible light (400–700 nanometers). fNIRS employs multiple wavelengths of near infrared light on either side of 810 nanometers, the wavelength at which oxygenated hemoglobin and deoxygenated hemoglobin absorb the same amount of electromagnetic radiation. Analyzing the light reflected (unabsorbed) from hemoglobin at these different wavelengths allows calculation of the relative amounts of oxygenated and deoxygenated hemoglobin in the brain region being examined.

4. The actual words quoted are those of technical editor Vivien Marx, the interviewer and author of the article reporting on what Helmstaedter said during the interview for *Nature* magazine.

5. Crowd sourcing is a process whereby analytic or other services are obtained from a large group of volunteers or part-time workers, particularly from an online community. It is used to divide a large amount of tedious work between many individuals, the small contribution of each participant combining with others' contributions to gain a greater result.

6. Nm = nanometer. One nm is one billionth of a meter, and the width of a human hair is 80,000 to 100,000 nm.

7. In preparation for electron microscopy, the cellular constituents of small pieces of brain tissue were first fixed in place with chemical preservatives and stained so that the boundaries of the cells and their parts would be visible in the electron microscope. Then the water was gradually removed from the tissue and eventually replaced with an Epoxy-like resin resulting in small blocks of a hard plastic-like substance containing the embedded tissue pieces. Next, ultrathin sections of the embedded tissue were sliced from the blocks using an extremely small and sharp diamond-edged knife. Finally, each of the tens of thousands of slices were examined by electron microscopy, and digital images of each were produced for use by the crowd-sourced volunteers.

8. Information about this crowd-sourcing study of climate change is at http://www.ucsusa.org/global_warming/what_you_can_do/climate-change-citizen-science.html#.VQSjsY62rhk (accessed March 14, 2015).

Appendix 3

Sources of Scientific Information for Nonscientists

Cell biology, genomics, brain science, computer science, nanoscience, and the biotechnologies emerging from these basic sciences advance at a breakneck pace. I spend time every day trying to keep up; but what about you students, young parents, nonscientist professionals, and other busy citizens whose days already brim with essential activities claiming time and attention? How can you stay informed about developments in twenty-first-century biotechnologies? There is some bad news but mostly good news. The bad news is that virtually nobody can stay absolutely current when it comes to biotechnology. Reading a book like this provides a foundation for keeping informed. But scientific details in any newly published book about biotechnology are bound to be three or more years out of date, compared to scientists' current research since the route between scientific discoveries at laboratory benches and books is long and winding. Now for the good news: there are diverse and handy sources for current news about biotechnology and its implications for individuals and society. Here are some:

American Association for the Advancement of Science (AAAS): http:// www.aaas.org/news/. This page contains links to current news about scientific discoveries that affect us all, including many on biotechnology and its associated ethical and societal issues.

Ethical, Legal, and Social Implications (ELSI of the Human Genome Project [HGP]): http://www.ornl.gov/sci/techresources/Human_Genome/elsi /elsi.shtml. This section of the government's HGP site offers an extensive suite of pages and links to articles and other websites with current information on the ethical implications of diverse biotechnologies.

Genetics and Public Policy Center, Johns Hopkins University: https:// jscholarship.library.jhu.edu/handle/1774.2/843. This link contains current and yearly archives of reports and news releases on ethical and legislative aspects of genetic biotechnologies, including assisted reproduction, pre-natal genetic testing, genetic therapy trials, stem cells, direct-to-consumer genetic testing, and personalized medicine.

National Human Genome Research Institute: https://www.genome.gov/. This site provides free information, including transcripts of lectures, about the human genome project and its applications. Click on Health, Education, Issues, or Newsroom at the top of the home page for articles written for the general public.

National Public Radio's *Science Friday*: http://www.sciencefriday.com/. This weekly, afternoon call-in talk show with scientists, hosted by science journalist Ira Flatow and underwritten by the National Science Foundation consists of two one-hour programs about nature, science, and technology. Past programs are archived as podcasts.

National Science Foundation (NSF): http://www.nsf.gov/discoveries/. The NSF gives links to short, nontechnical descriptions of NSF–funded research in all scientific disciplines on the Discoveries page.

NOVA/PBS: http://www.pbs.org/wgbh/nova/. NOVA programs air on public television stations and revolve on the premise that the world of science is fun. Programs on special topics air regularly on local PBS stations. Program transcripts and audio and video recordings of material related to program topics are freely available. From this site, one can hear experts discuss scientific and ethical aspects of cloning, stem cell research, genetic engineering, age retardation, neuroscience, DNA and human evolution and diversity, human genome projects, gene therapy, nanotechnology, and synthetic biology.

Periodicals, including *Discover Magazine, Scientific American, Scientist: Magazine of the Life Sciences, Science,* and *Nature,* publish nontechnical articles and news items about current developments in biotechnology and its ethical aspects. These all have their own websites with links to free articles, news releases, podcasts, and blogs. Hard copies are also at most libraries and many bookstores.

Podcasts are a convenient and enjoyable way to learn about modern biotechnologies and their applications. Using the website https://www.podsearch.com/ one can easily find podcasts on specific science and technology topics.

Radiolab: (http://www.radiolab.org/). Radiolab describes itself as "a show about curiosity. Where sound illuminates ideas, and the boundaries blur between science, philosophy, and human experience." It airs on many radio stations nationwide and presents engaging programs about ideas and the products of ideas, scientific ideas, and discoveries. For example, on February 24, 2017, this program on CRISPR technology features one of the technology's discoverers, Jennifer Doudna. She and other CRISPR researchers explain how CRISPR has been and can be used. This and many other episodes of Radiolab are available online as podcasts.

State Academies of Science: https://www.academiesofscience.org/naas -affiliated-academies?start=20. Academies of Science are rich sources of scientific information, events, programs, and publications for the general nonscientist public. You need not be a citizen of a particular state to benefit from that state's Academy of Science. Accessible material from State Academies of Science covers topics ranging from physics and astronomy to biotechnology, marine science, climate change, evolution, genetics, and many others.

References

AAAS (American Association for the Advancement of Science). 2017. "Mission and History." http://www.aaas.org/about/mission-and-history (accessed April 8, 2017).

ADF&G (Alaska Department of Fish and Game). 2015. "2015 Alaska Preliminary Commercial Salmon Harvest and Exvessel Values." October 16. http://www.adfg.alaska.gov/index.cfm?adfg=pressreleases.pr10162015 (accessed March 7, 2017).

Adler, J. 2015. "Mind Meld." *Smithsonian*, May, 45–51.

Aguirre, A. J., R. M. Meyers, B. A. Weir, et al. 2016. "Genomic Copy Number Dictates a Gene-Independent Cell Response to CRISPR-Cas9 Targeting." *Cancer Discovery*, Early content. http://cancerdiscovery.aacrjournals.org/content/early/2016/06/03/2159–8290.CD-16–0154.abstract (accessed June 28, 2016).

Alivisatos, A. P., M. Chun, G. M. Church, et al. 2013. "The Brain Activity Map." *Science* 339: 1284–85.

Al Jazeera. 2013. "Scientists Achieve First Human-to-Human 'Mind Meld.'" August 28. http://america.aljazeera.com/articles/2013/8/28/scientists-achievefirsteverhumantohumanmindmeld.html (accessed August 11, 2015).

Allen, C., and W. Wallach. 2012. "Moral Machines: Contradiction in Terms or Abdication of Human Responsibility?" In: *Robot Ethics: The Ethical and Social Implications of Robotics*. edited by P. Lin, K. Abney, and G.A. Bekey, 55–68. Cambridge: Massachusetts Institute of Technology Press.

American Anthropological Association. 1998. "American Anthropological Association Statement on 'Race.'" May 17. http://www.americananthro.org/ConnectWithAAA/Content.aspx?ItemNumber=2583 (accessed June 15, 2016).

Arria, A. M., and R. L. Dupont. 2010. "Nonmedical Prescription Stimulant Use among College Students: Why We Need to Do Something and What to Do." *Journal of Addictive Diseases* 29 (4): 417–26. http://www.ncbi.nlm.nih.gov/pmc/articles/PMC2951617/ (accessed September 22, 2015).

Asimov, I. 1950. *I, Robot.* Garden City, NY: Doubleday.

———. 1976. "The Bicentennial Man." In: *The Bicentennial Man and Other Stories,* 135–72. Garden City, NY: Doubleday.

———. 1986. Foreword to *The Three Pound Universe,* by J. Hooper and D. Teresi. New York: Macmillan Publishing.

Associated Press. 2016. "Congress Passes GMO Food Labeling Bill." NBC News, Health, July 14. http://www.nbcnews.com/health/health-news/congress-passes -gmo-food-labeling-bill-n609571 (accessed March 5, 2017).

Atherton, K. D. 2017. "The Pentagon's New Drone Swarm Heralds a Future of Autonomous War Machines. *Popular Science.* January 10. http://www.popsci .com/pentagon-drone-swarm-autonomous-war-machines (accessed May 21, 2017).

Bagley, J. A., D. Reuman, S. Bian, et al. 2017. "Fused Dorsal-Ventral Cerebral Organoids Model Complex Interactions between Diverse Brain Regions." *Nature Methods* 14: 743–51.

Ball, Philip. 2004. "What Is Life? Can We Make It?" *Prospect Magazine,* August, 101.

Baltimore, D., P. Berg, M. Botchan, et al. 2015. "A Prudent Path Forward for Genomic Engineering and Germline Gene Modification." *Science* 348: 36–38.

Bara, P., and T. Zidenga. 2003. "GMOs, God, and the Prince of Wales." *AgBioWorld,* August 5. http://www.agbioworld.org/biotech-info/religion/tawanda.html (accessed November 12, 2018).

Barker, K. 2014. "Mission Matters: DARPA's Inclusion in the BRAIN Initiative Is Downright Creepy." Scientists as Citizens. http://scientistsascitizens .org/2014/10/06/mission-matters-darpas-inclusion-in-the-brain-initiative -is-downright-creepy/ (accessed July 19, 2015).

Bassuk, A. G., A. Zheng, Y. Li, et al. 2016. "Precision Medicine: Genetic Repair of Retinitis Pigmentosa in Patient-Derived Stem Cells." *Nature,* Scientific Reports 6, article number 19969. http://www.nature.com/articles/srep19969 (accessed June 27, 2016).

Baum, R. 2003. "Point-Counterpoint: Nanotechnology, Drexler and Smalley Make the Case for and against 'Molecular Assemblers.'" *Chemical and Engineering News* 81 (48): 37–42. http://pubs.acs.org/cen/coverstory/8148/8148cou nterpoint.html (accessed March 10, 2017).

Beitiks, E. 2016. "5 Reasons Why We Need People with Disabilities in the CRISPR Debates." *Biopolitical Times,* September 8. http://www.biopoliticaltimes.org /article.php?id=9661 (accessed October 14, 2016).

Bekey, G. A. 2005. *Autonomous Robots: From Inspiration to Implementation and Control.* Cambridge: Massachusetts Institute of Technology Press.

———. 2012. "Current Trends in Robotics: Technology and Ethics." In: *Robot Ethics: The Ethical and Social Implications of Robotics,* edited by P. Lin,

K. Abney, and G.A. Bekey, 17–34. Cambridge: Massachusetts Institute of Technology Press.

Benjamin, Walter. 2009. *The Work of Art in the Age of Mechanical Reproduction*. New York: Classic Books America.

Benninghoff, A., and W. Hessler. 2008. "Nanoparticles Damage Brain Cells." *Environmental Health News*, November 17. http://www.environmentalhealthnews .org/ehs/newscience/nanoparticles-damage-brain-cells/ (accessed March 11, 2017).

Berreby, D. 2011. "Environmental Impact." *Scientist* 3: 40–44.

Berry, T. 1988. *The Dream of the Earth*. Berkeley, CA: Counterpoint.

Berry, W. 2015. "The Melancholy of Anatomy." *Harper's Magazine*, February, 11–15.

Blank, J. M., and D. Shaw. 2015. "Does Partisanship Shape Attitudes toward Science and Public Policy? The Case for Ideology and Religion." *ANNALS of the American Association for Political and Social Science* 658: 18–35.

Bohannon, J. 2015. "Fears of an AI pioneer." *Science* 349: 252.

Boyden, E. S., F. Zhang, E. Bamberg, G. Nagel, and K. Deisseroth. 2005. "Millisecond-Timescale, Genetically Targeted Optical Control of Neural Activity". *Nature Neuroscience* 8 (9): 1263–68. http://www.nature.com/neuro /journal/v8/n9/full/nn1525.html (accessed January 24, 2015).

Bradley, J. T. 2005. "The Moral Status of the Human Embryo: A Biologist's View." *Bulletin of Human Behavior and Development* 1 (1): 1–7.

———. 2013a. *Brutes or Angels: Human Possibility in the Age of Biotechnology*. Tuscaloosa: University of Alabama Press.

———. 2013b. "Darwin's Great Idea and Why It Matters." In: *Charles Darwin: A Celebration of His Life and Legacy*, edited by J. Bradley and J. Lamar, 217–40. Tuscaloosa: University of Alabama Press.

———. 2013c. "Robots Becoming Persons: How Will We Know?" *Journal of the Alabama Academy of Science* 84 (2): 59.

BRAIN Working Group. 2014. *BRAIN 2025: A Scientific Vision*. Report to the Advisory Committee to the Director, National Institutes of Health. https:// www.braininitiative.nih.gov/2025/?AspxAutoDetectCookieSupport=1 (accessed October 10, 2014).

"Brains in Action." 2014. *Scientist*, February 1. http://www.the-scientist.com /?articles.view/articleNo/38987/title/Brains-in-Action/ (accessed January 13, 2014).

Broad, W. J. 2014. "Billionaires with Big Ideas Are Privatizing American Science." *New York Times*, March 15. https://www.nytimes.com/2014/03/16 /science/billionaires-with-big-ideas-are-privatizing-american-science.html (accessed April 11, 2017).

Bronowski, J. 1965. *Science and Human Values.* New York: Harper and Row Publishers.

———. 1973. *The Ascent of Man.* Boston: Little, Brown.

Brown, J. S., and P. Duguid. 2000. "A Response to Bill Joy and the Doom-and-Gloom Technofuturists." *Standard Media International*, April 13. http://engl102-f12-lombardy.wikispaces.umb.edu/file/view/A+Response+to+Bill+Joy+and+the+Doom+and+Gloom+Technofuturists+by+John+Seely+Brown+and+Paul+Duguid.pdf (accessed March 11, 2017).

Brown, T. R., and J. B. McCormick. 2011. "New Directions in Neuroscience Policy." In Illes and Sahakian 2011, 675–700.

Buchman, D. Z., J. Illes, and P. B. Reiner. 2010. "The Paradox of Addiction Neuroscience.", *Neuroethics* 4: 65–77. http://www.researchgate.net/publication/226241050_The_Paradox_of_Addiction_Neuroscience (accessed November 7, 2018).

Buniak, L., M. Darragh, and J. Giordano. 2014. "A Four-Part Working Bibliography of Neuroethics: Part 1: Overview and Reviews—Defining and Describing the Field and Its Practices." *Philosophy, Ethics, and Humanities in Medicine* 9: 9. http://www.peh-med.com/content/9/1/9 (accessed August 9, 2015).

Bureau of Labor Statistics. 2013. Occupational Employment Projections to 2022. December. https://www.bls.gov/opub/mlr/2013/article/occupational-employment-projections-to-2022.htm (accessed May 30, 2017).

Burkitt, L. 2009. "Neuromarketing Companies Use Neuroscience for Consumer Insights." *Forbes* online, October 29. https://www.forbes.com/forbes/2009/1116/marketing-hyundai-neurofocus-brain-waves-battle-for-the-brain.html#796e18f617bb (accessed November 7, 2018).

Cai, D., K. B. Cohen, T. Luo, et al. 2013. "Improved Tools for the Brainbow Toolbox." *Nature Methods* 10: 540–46. http://www.nature.com/nmeth/journal/v10/n6/pdf/nmeth.2450.pdf (accessed March 1, 2015).

Callaway, E. 2013. "Deal Done over HeLa Cell Line." *Nature* 500: 132–33.

———. 2016. "UK Scientists Gain License to Edit Genes in Human Embryos." *Nature* 530: 18. http://www.nature.com/news/uk-scientists-gain-licence-to-edit-genes-in-human-embryos-1.19270 (accessed June 23, 2016).

———. 2018. "EU Law Deals Blow to CRISPR Crops." *Nature* 560: 16.

Cao, J., L. Wu, S-M. Zhang, et al. 2016. "An Easy and Efficient Inducible CRISPR/Cas9 Platform with Improved Specificity for Multiple Gene Targeting." *Nucleic Acids Research*, July 25. http://nar.oxfordjournals.org/content/early/2016/07/25/nar.gkw660.full (accessed July 30, 2016).

Čapek, K. (1921) 2004. *Rossum's Universal Robots.* Translated by Claudia Novack. New York: Penguin Group.

Carter, J., G. Goldman, G. Reed, et al. 2017. "Sidelining Science Since Day

One: How the Trump Administration Has Harmed Public Health and Safety in Its First Six Months." Center for Science and Democracy at the Union of Concerned Scientists, July. https://www.ucsusa.org/sites/default/files/attach/2017/07/sidelining-science-report-ucs-7–20–2017.pdf (accessed May 25, 2018).

CBCnews. 2007. "Technology and Science Tech Bytes, July 16. Your View: How Would You Define a Robot?" http://www.cbc.ca/technology/technology-blog/2007/07/your_view_how_would_you_define.html (accessed November 5, 2015).

Center for Science and Democracy. 2017. "What's at Stake." http://www.ucsusa.org/our-work/center-science-and-democracy#.WOk087i1vcs (accessed April 8, 2017).

CFAWR. 2010. "Animal Cruelty Syndrome." http://cfawr.org/animal-abuse.php (accessed March 18, 2017).

Chang, L., and D. Y. Tsao. 2017. "The Code for Facial Identity in the Primate Brain." *Cell* 169: 1013–28. doi:10.1016/ j.cell.2017.05.011, 2017 (accessed June 6, 2017).

Charles, Prince of Wales. 2000. "A Speech by HRH the Prince of Wales Titled a Reflection on the 2000 Reith Lectures, BBC Radio 4." Prince of Wales and Duchess of Cornwall, Royal Family website, May 17. https://www.princeofwales.gov.uk/speech/speech-hrh-prince-wales-titled-reflection-2000-reith-lectures-bbc-radio-4 (accessed December 3, 2018).

Chinnery, P. F. 2014. "Mitochondrial Disorders Overview." *GeneReviews [Internet]*, August 14. https://www.ncbi.nlm.nih.gov/books/NBK1224/(accessed November 7, 2018).

Cho, A. 2014. "The Accidental Roboticist." *Science* 346: 192–94.

Chung, K., J. Wallace, S-Y Kim, et al. 2013. "Structural and Molecular Interrogation of Intact Biological Systems." *Nature* 497: 332–37.

Chung, Y. G., J. H. Eum, J. E. Lee, et al. 2014. "Human Somatic Cell Nuclear Transfer Using Adult Cells." *Cell Stem Cell* 14: 777–80.

Clark, A., and D. Chalmers. 1998. "The Extended Mind." *Analysis* 58: 7–19. http://postcog.ucd.ie/files/TheExtendedMind.pdf (accessed January 27, 2015).

Colglazier, B. 2016. "Encourage Governments to Heed Scientific Advice." *Nature* 537: 587. http://www.nature.com/news/encourage-governments-to-heed-scientific-advice-1.20695 (accessed October 30, 2016).

Collins, Francis. 2006. *The Language of God: A Scientist Presents Evidence for Belief.* New York: Simon and Schuster.

Collins, F. 2013. "The Brain: Now You See It, Soon You Won't." NIH Director's Blog, April 11. http://directorsblog.nih.gov/2013/04/11/the-brain-now-you-see-it-soon-you-wont/ (accessed July 23, 2017).

Cong, L., F. A. Ran, D. Cox, et al. 2013. "Multiplex Genome Engineering Using

CRISPR/Cas Systems." *Science*: 339: 819–23. http://science.sciencemag.org/content/339/6121/819 (accessed July 18, 2016).

Conley, J. M., R. Cook-Deegan, and G. Lázaro-Munoz. 2014. "Myriad after Myriad: The Proprietary Data Dilemma." *North Carolina Journal of Law and Technology* 15 (4): 597–637. http://www.ncbi.nlm.nih.gov/pmc/articles/PMC4275833/ (accessed May 31, 2016).

Cosgrove, K. P., C. M. Mazure, and J. K. Staley 2007. "Evolving Knowledge of Sex Differences in Brain Structure, Function and Chemistry." *Biological Psychiatry* 62: 847–55.

Cossins, D. 2015. "Crossed Wires." *Scientist*. January 16. http://www.the-scientist.com/?articles.view/articleNo/41919/title/Crossed-Wires/ (accessed September 20, 2015).

Creutzig, F. 2017. "Govern Land as a Global Commons." *Nature* 546: 28–29.

Crichton, M. 2002. *Prey*. New York: HarperCollins Publishers.

Cross, R. 2017. "CRISPR's Breakthrough Problem." *Chemical and Engineering News* 95 (7): 28–33.

Cyranoski, D. 2017. "Trials of Embryonic Stem Cells to Launch in China." *Nature* 546: 15–16.

Cyranoski, D. and S. Reardon. 2015. "Embryo Editing Sparks Epic Debate." *Nature* 520: 593–95. http://www.nature.com/news/embryo-editing-sparks-epic-debate-1.17421 (accessed February 21, 2017).

Damasio, A. 1994. *Descartes' Error: Emotion, Reason and the Human Brain*. London: Picador.

Darnovsky, M. 2013. "A Slippery Slope to Human Germline Modification." *Nature* 499: 127. http://www.nature.com/news/a-slippery-slope-to-human-germline-modification-1.13358 (accessed February 26, 2017).

Darwin, Charles. 1871. *The Descent of Man, and Selection in Relation to Sex*. Vol. 1. London: John Murray. http://darwin-online.org.uk/content/frameset?itemID=F937.1&viewtype=image&pageseq=1 (accessed July 23, 2017).

Davenport, C. 2017. "With Trump in Charge, Climate Change References Purged from Website." *New York Times*, January 20. https://www.nytimes.com/2017/01/20/us/politics/trump-white-house-website.html?_r=0 (accessed April 11, 2017).

Demitri, M. 2007. "Types of Brain Imaging Techniques." *Psych Central*. http://psychcentral.com/lib/types-of-brain-imaging-techniques/0001057 (accessed July 23, 2017).

Dennis, B. 2015. "FDA Must Develop Plan to Label Genetically Engineered Salmon, Congress Says." *Washington Post*, December 17. https://www.washingtonpost.com/news/to-your-health/wp/2015/12/17/congress-to-fda-no-genetically-engineered-salmon-in-supermarkets-unless-it-is-labeled/?utm_term=.1c49d25df08c (accessed March 7, 2017).

Desai, M., I. Kahn, U. Knowblich, et al. 2011. "Mapping Brain Networks in Awake Mice Using Combined Optical Neural Control and fMRI." *Journal of Neurophysiology* 105(3): 1393–405. http://jn.physiology.org/content/105/3/1393.short (accessed January 24, 2015).

Descartes, René. (1641) 1984. *Meditations on First Philosophy*. In *The Philosophical Writings of René Descartes*. Translated by J. Cottingham, R. Stoothoff, and D. Murdoch, 1–62. Vol. 2. Cambridge: Cambridge University Press.

Devlin, H. 2007. "What Is Functional Magnetic Resonance Imaging (fMRI)?" *Psych Central*. http://psychcentral.com/lib/what-is-functional-magnetic-resonance-imaging-fmri/0001056 (accessed July 23, 2017).

Didier, C., W. Duan, J-P. Dupuy, et al. 2015. *Science* 349: 1064–65.

Doudna, J. 2015. "My Whirlwind Year with CRISPR." *Nature* 528: 469–71. http://www.nature.com/news/genome-editing-revolution-my-whirlwind-year-with-crispr-1.19063 (accessed July 21, 2016).

Dreger, A.D. 2000. "Metaphors of Morality in the Human Genome Project." In *Controlling Our Destinies: Historical, Philosophical, Ethical, and Theological Perspectives on the Human Genome Project*, edited by P. R. Sloan, 155–84. Notre Dame, IN: University of Notre Dame Press.

Drexler, E. 1986. *Engines of Creation*. New York: Anchor Books.

Dugdale, D. C., III. 2012. "PET Scan." MedlinePlus, US National Library of Medicine, National Institutes of Health. http://www.nlm.nih.gov/medlineplus/ency/article/003827.htm (accessed July 23, 2017).

Duster, T. 2003. "The Hidden Eugenic Potential of Germ-Line Interventions." In *Designing Our Descendants—The Promises and Perils of Genetic Modifications*, edited by A. R. Chapman and M. S. Frankel, 156–78. Baltimore: Johns Hopkins University Press.

Einstein, Albert. Letter to President Roosevelt. 1939. Atomic Archive. http://www.atomicarchive.com/Docs/Begin/Einstein.shtml (accessed December 26, 2016).

Esteller, M. 2011. "Epigenetic Changes in Cancer." *Scientist* 3: 34–39.

Esvelt, K. M. 2017. "Unnatural Responsibilities: Synthetic Biology Offers Unusual Rewards and Risks." *Scientific American* 316(4): 50–51.

ETC Group. 2007 "Extreme Genetic Engineering: An Introduction to Synthetic Biology." http://www.etcgroup.org/content/extreme-genetic-engineering-introduction-synthetic-biology (accessed February 2009).

European Commission. 2015. "In Defense of Europe: Defense Integration as a Response to Europe's Strategic Moment." European Political Strategy Centre Strategic Notes 4 (June 15). https://ec.europa.eu/epsc/sites/epsc/files/strategic_note_issue_4_en.pdf (accessed October 25, 2018).

Everding, G. 2016. "Genetically Modified Golden Rice Falls Short on Lifesaving Promises." The Source: Science and Technology, June 2. https://source

.wustl.edu/2016/06/genetically-modified-golden-rice-falls-short-lifesaving
-promises/ (accessed March 1, 2017).

Ewen, S. W., and A. Pusztai 1999. "Effect of Diets Containing Genetically Modified Potatoes Expressing *Galanthus nivalis* Lectin on Rat Small Intestine."
Lancet 354: 1353–54.

Faden, R. R., and R. Al Karron. 2012. "The Obligation to Prevent the Next Dual-
Use Controversy." *Science* 335: 802–4.

Farah, Martha J. 2011. "Neuroscience and Neuroethics in the 21st Century." In
Illes and Sahakian 2011, 761–81.

Farahany, N. A., H. T. Greely, S. Hyman, et al. 2018. "The Ethics of Experimenting with Human Brain Tissue." *Nature* 556: 429–32.

Feinberg, Gerald. 1969. *The Prometheus Project: Mankind's Search for Long-
Range Goals*. Garden City, NY: Doubleday and Company.

Ferezou, I., F. Matyas, and C. C. H. Petersen. 2009. "Imaging the Brain in Action:
Real-Time Voltage-Sensitive Dye Imaging of Sensorimotor Cortex of Awake
Behaving Mice." In *In Vivo Optical Imaging of Brain Function*, edited by Ron
D. Frostig, chapter 6. 2nd ed. Boca Raton, FL: Taylor and Francis. http://
www.ncbi.nlm.nih.gov/books/NBK20229/ (accessed November 7, 2018).

Ferrari, M., and V. Quaresima. 2012. "A Brief Review on the History of Human
Functional Near-Infrared Spectroscopy (fNIRS) Development and Fields of
Application." *Neuroimage* 63: 921–35. https://www.ncbi.nlm.nih.gov/pubmed
/22510258 (accessed July 23, 2017).

Ferreira, M. M. 2016. "CRISPR Moves from Butchery to Surgery." *Genetic
Engineering and Biotechnology News* 36 (9). http://www.genengnews.com
/gen-articles/crispr-moves-from-butchery-to-surgery/5759/?kwrd=CRISPR
%20moves%20from%20butchery (accessed June 22, 2016).

Feynman, R. P. 1959. "Plenty of Room at the Bottom." Talk given to American Physical Society in Pasadena, CA. http://web.pa.msu.edu/people/yang
/RFeynman_plentySpace.pdf (accessed August 31, 2018).

Fields, R. Douglas. 2013. "Map the Other Brain." *Nature* 501: 25–27.

Finn, E. S., X. Shen, D. Scheinost, et al. 2015. "Functional Connectome Finger-
printing: Identifying Individuals Using Patterns of Brain Connectivity." *Nature Neuroscience* 18: 1664–71. doi:10.1038/nn.4135.

Fischbach, R., and J. Mindes. 2011. "Why Neuroethicists Are Needed." In Illes
and Sahakian 2011, 343–76.

Fox, G. E, L. J. Magrum, W. E. Balch, et al. 1977. "Classification of Methano-
genic Bacteria by 16S Ribosomal RNA Characterization." *Proceedings of the
National Academy of Sciences USA* 74: 4537–41.

Frégnac, Y., and G. Laurent. 2014. "Neuroscience: Where Is the Brain in the
Human Brain Project?" *Nature* 513: 27–29.

Freitas, R. A., Jr. 1999. *Nanomedicine*. Vol. 1 Austin, TX: Landes Bioscience.

Gaj, T., C. A. Gersbach, and C. F. Barbass III. 2013. "ZFN, TALEN and CRISPR/ Cas-Based Methods for Genome Engineering." *Trends in Biotechnology* 31: 397–405. http://www.ncbi.nlm.nih.gov/pmc/articles/PMC3694601/ (accessed July 24, 2016).

Gallup. N.d. "Evolution, Creationism, Intelligent Design." https://news.gallup .com/poll/21814/evolution-creationism-intelligent-design.aspx (accessed November 1, 2018).

Gantz, V.M. and E. Bier. 2015. The mutagenic chain reaction: a method for converting heterozygous to homozygous mutations. *Science* 348: 442-44.

Gantz, V., N. Jasinskiene, O. Tatarenkova, et al. 2015. "Highly Efficient Cas9- Mediated Gene Drive for Population Modification of the Malaria Vector Mosquito *Anopheles stephensi.*" *Proceedings of the National Academy of Sciences* 112 (49): E6736–43. http://www.pnas.org/content/112/49/E6736.abstract (pdf of full article available here, accessed July 17, 2016).

Garreau, J. 2005. *Radical Evolution: The Promise and Peril of Enhancing Our Minds, Our Bodies—And What It Means to Be Human*. New York: Doubleday.

Gauchat, G. 2012. "Politicization of Science in the Public Sphere: A Study of Public Trust in the United States. 1974 to 2010." *American Sociological Review* 77: 167–87.

Gaudin, Sharon. 2007. "'I Do, Robot.' Technical Advances Lead to Very Close Relationships with Robots." Computerworld, October 31. http://www.compu terworld.com/article/2539809/computer-hardware/researcher-humans-will -love-marry-robots-by-2050.html (accessed March 6, 2016).

GEN News Highlights. 2015. "CRISPR Clears Major Obstacle Impeding Its Therapeutic Use." *Genetic Engineering and Biotechnology News*, April 2. http://www .genengnews.com/gen-news-highlights/crispr-clears-major-obstacle-impeding -its-therapeutic-use/81251106/ (accessed June 27, 2016).

———. 2017. "CRISPR Variant Produces Tuberculosis-Resistant Cows." *Genetic Engineering and Biotechnology News*, February 1. http://www.genengnews .com/gen-news-highlights/crispr-variant-produces-tuberculosis-resistant-cows /81253807 (accessed March 1, 2017).

Gersten, T. 2013. "Brain PET Scan." MedlinePlus, US National Library of Medicine, National Institutes of Health. http://www.nlm.nih.gov/medlineplus/ency /article/007341.htm (accessed July 23, 2017).

Ghose, T. 2009. "Over the Brainbow." *Scientist*, June 1. https://www.the-scientist .com/uncategorized/over-the-brainbow-44110 (accessed November 3, 2018).

Gibney, E. 2015. "Injectable Brain Implant Spies on Individual Neurons." *Nature* 522: 137–38.

Gillis, J., and C. Krauss. 2015. "Exxon Mobil Investigated for Possible Climate

Change Lies by New York Attorney General." *New York Times*, November 5. https://www.nytimes.com/2015/11/06/science/exxon-mobil-under-investigation-in-new-york-over-climate-statements.html (accessed April 11, 2017).

Goetz, J., S. Kiesler, and A. Powers. 2003. "Matching Robot Appearance and Behavior to Tasks to Improve Human-Robot Cooperation." *Proceedings of the Twelfth IEEE International Workshop on Robot and Human Interactive Communication*. October 31–November 2, Proceedings, Milbrae, CA, 55–60. https://pdfs.semanticscholar.org/3c24/5d6c0758a213e05d71f53753dc5e853ad44f.pdf (accessed November 7, 2018).

Gold, Jordan. 2015. "Is It Ethical to Have Sex with a Robot?" Konbini. http://www.konbini.com/en/lifestyle/japan-sex-robot-dolls/ (accessed February 26, 2016).

Gopnik, A. 2017. "Making AI More Human." *Scientific American* 316 (6): 60–65.

Greely, H. T., K. M. Ramos, and C. Grady. 2016. "Neuroethics in the Age of Brain Projects." *Neuron* 92: 637–41.

Green, R. M. 2014. "The Need for a Neuroscience ELSI Program." *Hastings Center Report* 44 (4). http://onlinelibrary.wiley.com/doi/10.1002/hast.333/pdf (accessed September 11, 2015).

Greene, J. D., R. B. Somerville, L. E. Nystrom, et al. 2001. "An fMRI Investigation of Emotional Engagement in Moral Judgment." *Science* 293: 2105–8.

Grens, K. 2015. "Optogenetics Advances in Monkeys." *Scientist*, October 5. http://www.the-scientist.com/?articles.view/articleNo/44156/title/Optogenetics-Advances-in-Monkeys/ (accessed October 5, 2015).

———. 2017. First Clinic-Ready Stem Cell Repository. *Scientist*, April 6. http://www.the-scientist.com/?articles.view/articleNo/49155/title/First-Clinic-Ready-Stem-Cell-Repository/&utm_campaign=NEWSLETTER_TS_The-Scientist-Daily_2016&utm_source=hs_email&utm_medium=email&utm_content=50176304&_hsenc=p2ANqtz—PJTwzFNyO7OFfyF_MBB-TLZ_WSztvmhZvaR-LNVil0pfQzWE3P4dGzV8Pl8YS-rlWwEbMnFIe5WOf-a2yhGlufJ3j8w&_hsmi=50176304 (accessed April 9, 2017).

Grienberger, C., and A. Konnerth. 2012. "Imaging Calcium in Neurons." *Neuron* 73: 862–85.

Grosch, E. G., and R. M. Hazen. 2015. "Microbes, Mineral Evolution, and the Rise of Microcontinents—Origin and Coevolution of Life with Early Earth." *Astrobiology* 15: 922. DOI: 10.1089/ast.2015.1302 (accessed June 15, 2016).

Guglielmi, G., A. Maxmen, L. Morello, et al. 2018. "Trump Budget Gives Last-Minute Reprieve to Science Funding." *Nature* 554: 284–85. https://www.nature.com/articles/d41586-018-01811-x (accessed November 8, 2018).

Hagmann, P., L. Jonasson, P. Maeder, et al. 2006. "Understanding Diffusion MR Imaging Techniques: From Scalar Diffusion-Weighted Imaging to Diffusion

Tensor Imaging and Beyond." *RadioGraphics* 26, suppl. 1. http://pubs.rsna .org/doi/full/10.1148/rg.26si065510 (accessed April 28, 2015).

Haidt, J. 2001. "The Emotional Dog and Its Rational Tail: A Social Intuitionist Approach to Moral Judgment." *Psychological Review* 108: 814–34.

Hall, L., K. Topinka, J. Huffman, L. Davis, and A. Good. 2000. "Pollen Flow between Herbicide-Resistant *Brassica napus* Is the Cause of Multiple Resistant *B. napus* Volunteers." *Weed Science* 48: 688–94.

Hall, S. 2015. "Exxon Knew about Climate Change Almost 40 Years Ago." *Scientific American*, October 26. https://www.scientificamerican.com/article /exxon-knew-about-climate-change-almost-40-years-ago/ (accessed April 11, 2017).

Hammond, A., R. Galizi, K. Kyrou, et al. 2016. "A CRISPR-Cas9 Gene Drive System Female Reproduction in the Malaria Mosquito Vector *Anopheles gambiae*." *Nature Biotechnology* 34: 78–83. http://www.ncbi.nlm.nih.gov/pmc /articles/PMC4913862/ (accessed July 21, 2016).

Hamzelou, J. 2016. "Exclusive: World's First Baby Born with New '3 parent' Technique." *New Scientist/Daily News*, September 27. https://www.newscientist .com/article/2107219-exclusive-worlds-first-baby-born-with-new-3-parent -technique/ (accessed April 12, 2017).

Handelsman, J. 2003. "Teaching Scientists to Teach." *Howard Hughes Medical Institute (HHMI) Bulletin*, June. https://scholardevelopment.okstate.edu/sites /default/files/Mentoring%20Workshop%20Series%20-%20CEAT%20-%20 Session%201%20-%20Reading_Part2.pdf (accessed January 5, 2017).

Harris, J. 2015. "Germline Manipulation and Our Future Worlds." *American Journal of Bioethics* 15: 30–34.

Harris, S., J. T. Kaplan, A. Curiel, et al. 2009. "The Neural Correlates of Religious and Nonreligious Belief." *PLOS One*, October 1. http://journals.plos .org/plosone/article?id=10.1371/journal.pone.0007272 (accessed September 16, 2015).

Haynes, J-D. 2011. "Brain Reading: Decoding Mental States from Brain Activity in Humans." In Illes and Sahakian 2011, 3–14.

Healy, M. 2015. "International Gene-Editing Conference Declines to Ban Eventual Use in Humans." *Los Angeles Times*, Science Now, December 3. http:// www.latimes.com/science/sciencenow/la-sci-sn-gene-editing-ban-20151203 -story.html (accessed June 23, 2016).

Helmstaedter, M., K. L. Briggman, S. C. Turaga, et al. 2013. "Connectomic Reconstruction of the Inner Plexiform Layer in the Mouse Retina." *Nature* 500: 168–74.

Herfst, S., Eefje J. A. Schrauwen, M. Linster, et al. 2012. "Airborne Transmission of Influenza A/H5N1 Virus between Ferrets." *Science* 336: 1534–41.

Honda News Releases. 2014. "All-New ASIMO Takes a Stroll around New York." November 11. http://hondanews.eu/eu/en/corporate/media/pressreleases /4132/all-new-asimo-takes-a-stroll-around-new-york (accessed July 23, 2017).

Hooper, Judith, and D. Teresi.1986. *The Three Pound Universe*. New York: Macmillan.

Horgan, J. 2013. "Why You Should Care about Pentagon Funding of Obama's BRAIN Initiative." *Scientific American* blogs. May22. http://blogs.scientificam erican.com/cross-check/why-you-should-care-about-pentagon-funding-of -obamas-brain-initiative/ (accessed July 19, 2015).

"How Big Is Science?" 2015. Editorial. *Scientific American* 313: 40–41.

Hug, L. A., B. J. Baker, K. Anantharaman, et al. 2016. "A New View of the Tree of Life." *Nature Microbiology* 1, article number 16048. doi:10.1038/nmicrobiol .2016.48 and available at http://www.nature.com/articles/nmicrobiol201648 (accessed May 18, 2016).

Hull, D. L. 1997. "The Ideal Species Concept—and Why We Can't Get It." In *Species: The Units of Biodiversity*, edited by M. F. Claridge, A. H. Dawah, and M. R. Wilson, 357–80. London: Chapman and Hall.

Human Brain Project. 2014. Mediation Report. http://www.fz-juelich.de /SharedDocs/Downloads/PORTAL/EN/pressedownloads/2015/15–03–19hbp -recommendations.pdf?__blob=publicationFile (accessed July 30, 2015).

Iacoboni, M., J. Freedman, J. Kaplan, et al. 2007. "This Is Your Brain on Politics." *New York Times*, November 11. https://www.nytimes.com/2007/11/11 /opinion/11freedman.html (accessed October 25, 2018).

Illes, J., and E. Racine. 2005. "Imaging or Imagining? A Neuroethics Challenge Informed by Genetics." *American Journal of Bioethics* 5: 5–18.

Illes, Judy, and Barbara J. Sahakian, eds. 2011. *The Oxford Handbook of Neuroethics*. Oxford: Oxford University Press

Imai, M., T. Watanabe, M. Hatta, et al. 2012. "Experimental Adaptation of an Influenza H5 HA Confers Respiratory Droplet Transmission to a Reassortant H5 HA/H1N1 Virus in Ferrets." *Nature* 486: 420–28.

Insel, Thomas R., Story C. Landis, and Francis S. Collins. 2013. "The NIH BRAIN Initiative." *Science* 340: 687–88.

International Human Genome Sequencing Consortium. 2004. "Finishing the Euchromatic Sequence of the Human Genome." *Nature* 431: 931–45.

Jaffe, G. 2017. "The ABCs of BMO Disclosure in the United States." Center for Science in the Public Interest. https://cspinet.org/news/abcs-gmo-disclosure -united-states-20170925 (accessed May 3, 2018).

Jahr, F. 2015. "How Humans Evolved Supersize Brains." *Quanta Magazine*, November 10. https://www.quantamagazine.org/20151110-evolution-of-big -brains/ (accessed November 27, 2015).

James, M., and J. L. Ortiz. 2018. "Jury Orders Monsanto to Pay $289 million to Cancer Patient in Roundup Lawsuit." *USA Today*, August 10. https://www.usatoday.com/story/news/2018/08/10/jury-orders-monsanto-pay-289-million-cancer-patient-roundup-lawsuit/962297002/ (accessed November 12, 2018).

Jasanoff, S. 2016. *The Ethics of Invention: Technology and the Human Future*. New York: W. W. Norton.

Jinek, M., K. Chylinski, I. Fonfara, et al. 2012. "A Programmable Dual-RNA-Guided DNA Endonuclease in Adaptive Bacterial Immunity." *Science* 337: 816–21.

Johns Hopkins Medicine Health Library. N.d. "Electroencephalogram (EEG)." Johns Hopkins University, Johns Hopkins Hospital, and Johns Hopkins Health System. http://www.hopkinsmedicine.org/healthlibrary/test_procedures/neurological/electroencephalogram_eeg_92,P07655/ (accessed July 23, 2017).

Jones, P. A., and R. Martienssen. 2005. "A Blueprint for a Human Epigenome Project: The AACR Human Epigenome Workshop." *Cancer Research* 65 (24): 11241–46.

Joy, B. 2000. "Why the Future Doesn't Need Us." *Wired*, 8.04, April. https://www.wired.com/2000/04/joy-2/ (accessed March 10, 2017).

Kahn, B. 2017. "Trump's EPA Is Removing Climate Change Information from Website." Inside Climate News, February 3. https://insideclimatenews.org/news/03022017/epa-donald-trump-climate-change-science-scott-pruitt (accessed April 11, 2017).

Kaiser, J. 2014. "U.S. Halts Two Dozen Risky Virus Studies." *Science* 346: 404.

———. 2016. "First Proposed Human Test of CRISPR Passes Initial Safety Review." http://www.sciencemag.org/news/2016/06/human-crispr-trial-proposed (accessed June 30, 2016).

———. 2017. "A Yellow Light for Embryo Editing." *Science* 355: 675. http://www.sciencemag.org/news/2017/02/us-panel-gives-yellow-light-human-embryo-editing (accessed February 21, 2017).

Kandel, E. R., J. H. Schwartz, and T. M. Jessell. 2000. *Principles of Neural Science*. 4th ed. New York: McGraw-Hill.

Kang, C., and M. D. Shear 2017. "Trump Leaves Science Jobs Vacant, Troubling Critics." *New York Times*, March 30. https://www.nytimes.com/2017/03/30/us/politics/science-technology-white-house-trump.html (accessed April 11, 2017).

Kang, X., W. He, Y. Huang, et al. 2016. "Introducing Precise Genetic Modifications into Human 3PN Embryos by CRISPR/Cas9-Mediated Gene-Editing." *Journal of Assisted Reproduction and Genetics* 33: 581–88.

Kant, I. (1785) 1959. *Foundations of the Metaphysics of Morals*. Translated by Lewis White Beck. Indianapolis, IN: Bobbs-Merrill.

Kapner, D. A. 2003. "Recreational Use of Ritalin on College Campuses." *Info-*

facts Resources, Higher Education Center for Alcohol and Other Drug Prevention, June. http://w2.mtsu.edu/healthpro/documents/rec_use_of_ritalin.pdf (accessed September 22, 2015).

"The Kill Switch." 2015. Editorial. *Nature* 521: 260.

Koch, C., and R. C. Reid. 2012. "Observatories of the Mind." *Nature* 483: 397–98.

Koren, M. 2018. "Congress Ignores Trump's Priorities for Science." *Atlantic*, March 23. https://www.theatlantic.com/science/archive/2018/03/trump-science-budget/556229/ (accessed June 2, 2018).

Kriebel, D., J. Tickner, P. Epstein, et al. 2001. "The Precautionary Principle in Environmental Science." *Environmental Health Perspectives* 109 (9): 871–76. https://www.ncbi.nlm.nih.gov/pmc/articles/PMC1240435/ (accessed November 8, 2018).

Kurzweil, R. 1999. *The Age of Spiritual Machines*. New York: Viking Press.

Kuzma, J. 2016. "A Missed Opportunity for U.S. Biotechnology Regulation." *Science* 353: 1211–13.

Lancaster, M. A., M. Renner, C. A. Martin, et al. 2013. "Cerebral Organoids Model Human Brain Development and Microcephaly." *Nature* 501: 373–79.

Lanphier, E., F. Urnov, S. H. Haecker, et al. 2015. "Don't Edit the Human Germ Line." *Nature* 519: 410–11. http://www.nature.com/news/don-t-edit-the-human-germ-line-1.17111 (accessed July 21, 2016).

Leopold, A. (1949) 1970. "The Land Ethic." In *A Sand County Almanac and Sketches Here and There*, 201–26. New York: Oxford University Press.

Leshner, Alan I. 2015. "Editorial: Passing the CEO Baton." *Science* 347: 587.

Levy, David. 2007. *Love and Sex with Robots: The Evolution of Human-Robot Relationships*. New York: HarperCollins Publishers.

———. 2012. "The Ethics of Robot Prostitutes." In *Robot Ethics: The Ethical and Social Implications of Robotics*, edited by P. Lin, K. Abney, and G. A. Bekey, 223–31. Cambridge: Massachusetts Institute of Technology Press.

Levy, Neil. 2007. *Neuroethics: Challenges for the 21st Century*. Cambridge: Cambridge University Press.

———. 2011. "Neuroethics and the Extended Mind." In Illes and Sahakian 2011, 285–94.

Lewis, T. 2016. "With CRISPR, Modeling Disease in Mini Organs." *Scientist*, May 6. http://www.the-scientist.com/?articles.view/articleNo/46047/title/With-CRISPR—Modeling-Disease-in-Mini-Organs/ (accessed June 28, 2016).

Liang, P., Y. Xu, X. Zhang, et al. 2015. "CRISPR/Cas9-Mediated Gene-Editing in Human Tripronuclear Zygotes." *Protein and Cell* 6: 363–72. http://link.springer.com/article/10.1007/s13238–015–0153–5 (accessed July 12, 2016).

Liepelt, R., and J. Brooks. 2017. "Understanding Body Ownership and Agency." *Scientist*, May 1. http://www.the-scientist.com/?articles.view/articleNo/49268/title/Understanding-Body-Ownership-and-Agency/&utm_campaign

=NEWSLETTER_TS_The-Scientist-Daily_2016&utm_source=hs_email&
utm_medium=email&utm_content=51981627&_hsenc=p2ANqtz-8p5Lu3q
pUeyxwJgQlEjI9X1JFFqKogCR9huAwlshb3KpOaI5EyNVUJI3BKmf6U2G
TCluLIkZdaeryZHtmUE-FXOixt8A&_hsmi=51981627 (accessed May 30,
2017).

Liptak, A. 2013. "Justices, 9–0, Bar Patenting Human Genes." *New York Times*,
June 13. http://www.nytimes.com/2013/06/14/us/supreme-court-rules-human
-genes-may-not-be-patented.html (accessed April 12, 2017).

Liu, J., T-M Fu, and Z. Cheng, 2015. "Syringe-Injectable Electronics." *Nature
Nanotechnology* 10: 629–36. http://www.nature.com/nnano/journal/v10/n7
/full/nnano.2015.115.html (accessed July 23, 2017).

Livet, J., T. A. Weissman, H. Kang, et al. 2007. "Transgenic Strategies for Com-
binatorial Expression of Fluorescent Proteins in the Nervous System." *Nature* 450: 56–62. http://www.nature.com/nature/journal/v450/n7166/pdf
/nature06293.pdf (accessed July 23, 2017).

Long, J. H., Jr. 2012. *Darwin's Devices: What Evolving Robots Can Teach Us about
the History of Life and the Future of Technology*. New York: Basic Books.

Lozano, A. 2013. "Tuning the Brain." *Scientist*, October 28. http://www.the-scien
tist.com/?articles.view/articleNo/38047/title/Tuning-the-Brain/ (accessed
July 23, 2017).

Lunshof, J. 2014 "Regulate Gene-Editing in Wild Animals." *Nature* 521: 127.
http://www.nature.com/news/regulate-gene-editing-in-wild-animals-1.17523
(accessed July 25, 2016).

Lurquin, Paul F. 2002. *High Tech Harvest*, Boulder, CO: Westview Press.

MacDonald, C. 2004. "Business Ethics 101 for the Biotech Industry." *Biodrugs*
18: 71–77.

MacDorman, K. F., R. D. Green, C.-C. Ho, and C. Koch. 2009. "Too Real for
Comfort: Uncanny Responses to Computer Generated Faces." *Computers in
Human Behavior* 25: 695–710. http://www.macdorman.com/kfm/writings
/pubs/MacDorman2009TooRealForComfort.pdf (accessed January 5, 2016).

Marchione, Marilynn. "First Gene-Edited Babies Claimed in China." Associated
Press, November 26, 2018. https://www.apnews.com/4997bb7aa36c45449b48
8e19ac83e86d (accessed November 30, 2018).

Marcus, G. 2017. "Am I Human?" *Scientific American* 316 (3): 58–63.

Mariani, J., G. Coppola, P. Zhang, et al. 2015. "FOXG1-Dependent Dysregula-
tion of GABA/Glutamate Neuron Differentiation in Autism Spectrum Disorders." *Cell* 162: 375–90.

Markoff, J. 2013. "Connecting the Neural Dots." *New York Times*, February
25. http://www.nytimes.com/2013/02/26/science/proposed-brain-mapping
-project-faces-significant-hurdles.html?pagewanted=all&_r=0 (accessed October 7, 2012).

———. 2015. *Machines of Loving Grace: The Quest for Common Ground between Humans and Robots*. New York: HarperCollins Publishers.

Marshall, E. 2013. "Supreme Court Rules Out Patents on 'Natural' Genes." *Science* 340: 1387–88.

Marx, V. 2013. "Brain Mapping in High Resolution." *Nature* 503: 147–52.

Matthews, D. J. H., P. V. Rabins, and B. D. Greenberg. 2011. "Deep Brain Stimulation for Treatment-Resistant Neuropsychiatric Disorders." In Illes and Sahakian 2011, 441–53.

Mayr, E. 1997. *This Is Biology*. Cambridge, MA: Belknap Press of Harvard University Press.

McCarthy, M. M. 2015. "Sex Differences in the Brain." *Scientist* 29 (10). http://www.the-scientist.com/?articles.view/articleNo/44096/title/Sex-Differences-in-the-Brain/ (accessed January 27, 2017).

McKee, J. K. 2000. *The Riddled Chain: Chance, Coincidence, and Chaos in Human Evolution*. London: Rutgers University Press.

McNutt, Marcia. 2015. "It Starts with a Poster." *Science* 347: 1047.

Medical Discoveries. 2014. "Electroencephalogram (EEG)." http://www.discoveriesinmedicine.com/Com-En/Electroencephalogram-EEG.html (accessed December 6, 2014).

Mervis, J. 2015. "Why Many U.S. Biology Teachers Are 'Wishy-Washy.'" *Science* 347: 1054.

Miller, Kenneth. 1999. *Finding Darwin's God: A Scientist's Search for Common Ground between God and Evolution*. New York: HarperCollins Publishers.

Mooney, C. 2012. *The Republican Brain: The Science of Why They Deny Science and Reality*. Hoboken, NJ: Wiley Publishing.

Moreno, J. D. 2012. *Mind Wars*. New York: Bellvue Literary Press.

Mori, M. 1970. "The Uncanny Valley." Translated by K. F. Macdorman and T. Minato. *Energy* 7(4): 33–35. http://www.androidscience.com/theuncannyvalley/proceedings2005/uncannyvalley.html (accessed January 6, 2016).

Nagel, G., M. Brauner, J. F. Liewald, et al. 2005. "Light Activation of Channelrhodopsin-2 in Excitable Cells of *Caenorhabditis elegans* Triggers Rapid Behavioral Responses." *Current Biology* 15 (24): 2279–84. doi: 10.1016/j.cub.2005.11.032 (accessed July 23, 2017).

Nakamura, T., A. R. Gehrke, J. Lemberg, et al. 2016. "Digits and Fin Rays Share Common Developmental Histories." *Nature* 537: 225–28.

Nakata, H. 2009. "Mind over Matter: Brain Waves Control Asimo." *Japan Times*, April 1. http://www.japantimes.co.jp/news/2009/04/01/business/mind-over-matter-brain-waves-control-asimo/#.VISpoGeLUX4 (accessed December 7, 2014).

NAS (National Academy of Sciences). 2017. An Act to Incorporate the National Academy of Sciences. http://www.nasonline.org/about-nas/leadership/governing-documents/act-of-incorporation.html (accessed January 7, 2017).

NASEM (National Academies of Science, Engineering, and Medicine). 2016. "Gene Drives on the Horizon: Advancing Science, Navigating Uncertainty, and Aligning Research with Public Values." Washington, DC: National Academies Press. https://www.nap.edu/catalog/23405/gene-drives-on-the-horizon -advancing-science-navigating-uncertainty-and (accessed February 28, 2017).

National Academies of Science. 2004. "*Safety of Genetically Engineered Foods: Approaches to Assessing Unintended Health Effects.*" Washington, DC, National Academies Press. https://www.ncbi.nlm.nih.gov/books/NBK215773/ (accessed March 6, 2017).

National Conference of State Legislators. 2018. "Autonomous Vehicles/Self-Driving Vehicles Enacted Legislation." http://www.ncsl.org/research/transportation /autonomous-vehicles-self-driving-vehicles-enacted-legislation.aspx (accessed October 29, 2018).

National Public Radio. 2018. "Clinic Claims Success in Making Babies with 3 Parents' DNA." Morning Edition, June 6. https://www.npr.org/sections /health-shots/2018/06/06/615909572/inside-the-ukrainian-clinic-making -3-parent-babies-for-women-who-are-infertile (accessed June 8, 2018).

Natterer, R. D., K. Yang, W. Paul et al. 2017. "Reading and Writing Single-Atom Magnets." *Nature* 543: 226–28. http://www.nature.com/nature/journal/v543 /n7644/full/nature21371.html (accessed March 12, 2017).

Nemade, R. 2005. "Human 'Epigenome' Project." HUM-MOLGEN Genetic News. http://hum-molgen.org/NewsGen/12-2005/000025.html (accessed October 19, 2018).

Newport, Frank. 2009. "On Darwin's Birthday, Only 4 in 10 Believe in Evolution." Gallup Polls.

NIH (National Institutes of Health). N.d. The Brain Initiative. https://www .braininitiative.nih.gov/pdf/BRAIN_brochure_508C.pdf (accessed October 25, 2018).

———. 1994. Report of the Human Embryo Research Panel, September. http:// bioethics.georgetown.edu/pcbe/reports/past_commissions/human_embryo _vol_1.pdf (accessed January 16, 2012).

———. 2015. National Human Genome Research Institute. "Privacy in Genomics." https://www.genome.gov/27561246/privacy-in-genomics/ (accessed May 29, 2016).

———. Office of Science Policy. 2017. Dual Use Research of Concern. https:// osp.od.nih.gov/biotechnology/dual-use-research-of-concern/ (accessed October 30, 2018).

Nisbet, E. C., K. E. Cooper, and R. K. Garrett. 2015. "The Partisan Brain: How Dissonant Science Messages Lead Conservatives and Liberals to (Dis)Trust Science." *ANNALS of the American Association for Political and Social Science* 658: 36–66.

Noonan, D. 2014. "Inside the Science of an Amazing New Surgery Called

Deep Brain Stimulation." *Smithsonian*, May, 38-46. https://faculty.mtsac.edu/cbriggs/Inside%20the%20Science%20of%20an%20Amazing%20New%20Surgery%20Called%20Deep%20Brain%20Stimulation%202014.pdf (accessed November 2, 2018).

Nordling, L. 2017. "Putting Genomes to Work in Africa." *Nature* 544: 20–22.

Normile, D. 2014. "In Our Own Image." *Science* 346: 188–89.

———. 2018. "First-of-Its-Kind Clinical Trial Will Use Reprogrammed Adult Stem Cells to Treat Parkinson's." *Science News*, July 30. https://www.sciencemag.org/news/2018/07/first-its-kind-clinical-trial-will-use-reprogrammed-adult-stem-cells-treat-parkinson-s (accessed November 10, 2018).

Obama, Barack. 2013. Remarks by the President on the BRAIN Initiative and American Innovation. White House, Office of the Press Secretary, April 2. https://www.whitehouse.gov/the-press-office/2013/04/02/remarks-president-brain-initiative-and-american-innovation (accessed March 24, 2015).

Obama White House Archives. N.d. "Modernizing the Regulatory System for Biotechnology Products: Final Version of the 2017 Update to the Coordinated Framework for the Regulation of Biotechnology." https://obamawhitehouse.archives.gov/sites/default/files/microsites/ostp/2017_coordinated_framework_update.pdf (accessed November 7, 2018).

Offord, C. 2017. "The Rise of the Pangenome." *Scientist* 30 (12): 30–36.

O'Keefe, M., S, Perrault, J. Halpern, et al. 2015. "'Editing' Genes: A Case Study about How Language Matters in Bioethics." *American Journal of Bioethics* 15 (12): 3–10.

Ostrom, E. 2015. *Governing the Commons*. Cambridge: Cambridge University Press.

Oye, K. A, K. Esvelt, E. Appleton, et al. 2014. "Regulating Gene Drives." *Science* 345: 626–28.

Papapetrou, E. P. 2016. "Induced Pluripotent Stem Cells, Past and Future." *Science* 353: 991–93.

Parens, E., J. Johnston, and J. Moses. 2008. "Do We Need 'Synthetic Bioethics'?" *Science* 321: 1449.

Parfitt, G., ed. 1998. *Ben Johnson: The Complete Poems*. New York: Penguin Books.

Pence, G. E., 2002. *Designer Food*. Lanham, MD: Rowman and Littlefield Publishers.

Pennisi, E. 2017. "Pocket-Sized Sequencers Start to Pay Off Big." *Science* 356: 572–73.

Peterka, D. S., H. Takahashi, and R. Yuste. 2011. "Imaging Voltage in Neurons." *Neuron* 69: 9–21.

Pew Research Center. 2013. "Public's View on Human Evolution." http://www.pewforum.org/2013/12/30/publics-views-on-human-evolution (accessed June 17, 2016).

Pick, Adam. 2008. "Pig Valve Transplants for Patients Needing Heart Valve Replacement Surgery." HeartValveSurgery. http://www.heart-valve-surgery .com/heart-surgery-blog/2008/03/24/pig-valve-transplants-tell-me-more/ (accessed May 3, 2018).

Pico della Mirandola. [1486] 2005. "Oration on the Dignity of Man." In *The Human Odyssey—Readings from Original Sources*. Edited by J. T. Bradley and Human Odyssey Faculty. Translated by E. L. Forbes, 286–87. Boston: Pearson Custom Publishing.

Pinstrup-Anderson, P., and E. Schiøler, 2001. *Seeds of Contention*, Baltimore: Johns Hopkins University Press.

Polansky, A. 2018. "After a 19-Month Vacancy, White House Taps an Academic Meteorologist to Head the OSTP." Climate Science and Policy Watch, August 2. http://www.climatesciencewatch.org/2018/08/02/after-19-month-vacancy -white-house-taps-an-academic-meteorologist-to-head-the-ostp/?gclid=EAIaI QobChMIwKKWgPSw3gIVRZNpCh1sPQ1yEAAYASAAEgJ-e_D_BwE (accessed October 31, 2018).

Poo, M., J. Du, N. Y. Ip, et al. 2016. "China Brain Project: Neuroscience, Brain Diseases, and Brain-Inspired Computing." *Neuron* 92: 591–96.

Porteus, M. H. 2009. "Zinc Fingers on Target." *Nature* 459: 337–38.

Precision Medicine Initiative. 2015. All of Us Research Program. https://syndi cation.nih.gov/multimedia/pmi/infographics/pmi-infographic.pdf (accessed April 14, 2017).

Presidential Commission. 2014. *Gray Matters: Integrative Approaches for Neuroscience, Ethics, and Society*. Vol. 1. Washington, DC: Presidential Commission for the Study of Bioethical Issues.

———. 2015. *Gray Matters: Topics at the Intersection of Neuroscience, Ethics, and Society*. Vol. 2. Washington, DC: Presidential Commission for the Study of Bioethical Issues.

Rachels, J. 2003. *The Elements of Moral Philosophy*. 4th ed. New York: McGraw-Hill.

Raffensperger, C., and J. Tickner, eds. 1999. *Protecting Public Health and the Environment: Implementing the Precautionary Principle*. Washington, DC: Island Press.

Raine, A., J. R. Meloy, S. Bihrle, et al. 1998. "Reduced Prefrontal and Increased Subcortical Brain Functioning Assessed Using Positron Emission Tomography in Predatory and Affective Murderers." *Behavioral Sciences and the Law* 16: 319–32.

Rauscher, R. J., III. 2005. "It Is Time for a Human Epigenome Project." *Cancer Research* 65 (24): 11229. http://cancerres.aacrjournals.org/content/65/24 /11229.full (accessed April 23, 2011).

Reardon, S. 2015a. "Global Summit Records Divergent Views on Human Gene Editing." *Nature* 528: 173.

———. 2015b. "The Military-Bioscience Complex." *Nature* 522: 142–44.

———. 2016. "First CRISPR Clinical Trial Gets Green Light from US Panel." *Nature*, June 22. http://www.nature.com/news/first-crispr-clinical-trial-gets -green-light-from-us-panel-1.20137 (accessed April 11, 2017).

Reiner, P. B. 2011. "The Rise of Neuroessentialism." In Illes and Sahakian 2011, 161–75.

Richardson, Kathleen. 2015a. "The Asymmetrical "'Relationship': Parallels between Prostitution and the Development of Sex Robots." Research Position Paper. Published on the ACM Digital Library as a special issue of the ACM SIGCAS newsletter. *SIGCAS Computers and Society* 45 (3): 290–93. http://campaignagainstsexrobots.org/the-asymmetrical-relationship-parallels -between-prostitution-and-the-development-of-sex-robots/ (accessed March 6, 2016).

———. 2015b. "Sex Robots and the Sex Trade." http://asiapacific.anu.edu.au /newmandala/2015/10/28/sex-robots-and-the-sex-trade/ (accessed March 6, 2016).

Robertson, J.A. 1994. *Children of Choice: Freedom and the New Reproductive Technologies.* Princeton, NJ: Princeton University Press.

Robot Vacuum Review. 2015. 2015 Best. http://robot-vacuum-review.toptenreviews .com/ (accessed November 9, 2015).

Rose, N. 2016. "Reading the Human Brain: How the Mind Became Legible." *Body and Society* 22: 140–77. http://journals.sagepub.com/doi/pdf/10.1177 /1357034X15623363 (accessed January 25, 2017).

Rosenberg, M., and J. Markoff. 2016. "The Pentagon's 'Terminator Conundrum': Robots That Could Kill on Their Own." *New York Times*, October 26. http://www.nytimes.com/2016/10/26/us/pentagon-artificial-intelligence -terminator.html?emc=edit_th_20161026&nl=todaysheadlines&nlid=37665992 &_r=0 (accessed October 26, 2016).

Rosenberg, N. A., J. K. Pritchard, J. L. Weber, et al. 2002. "Genetic Structure of Human Populations." *Science* 298: 2381–85. DOI: 10.1126/science.1078311 (accessed June 15, 2016).

Roskies, A. 2002. "Neuroethics for the New Millennium." *Neuron* 35: 21–23. http://www.dartmouth.edu/~adinar/Adinas_homepage/CV_files/Neuroethics .pdf (accessed July 23, 2017).

———. 2007. "Neuroethics beyond Genethics. Despite the Overlap between the Ethics of Neuroscience and Genetics, There Are Important Areas Where the Two Diverge." *EMBO Reports* 8: S52-S56.

Ross, R. 2013. "Scientists Seek Ethics Review of H5N1 Gain-of-Function Research." Center for Infectious Disease Research and Policy, March 29. http:// www.cidrap.umn.edu/news-perspective/2013/03/scientists-seek-ethics-review -h5n1-gain-function-research (accessed January 1, 2017).

Safire, W. 2002. "Our New Promethean Gift." Introduction for DANA Foundation Conference, Neuroethics: Mapping the Field, May 13–14, San Francisco. http://dana.org/Cerebrum/2002/Neuroethics__Mapping_the_Field/ (accessed August 9, 2015).

Sander, J. D., and J. K. Joung. 2014. "CRISPR-Cas Systems for Genome-Editing, Regulation and Targeting." *Nature Biotechnology* 32: 347–55. http://www.ncbi.nlm.nih.gov/pmc/articles/PMC4022601/ (accessed July 18, 2016).

Schmidtke, K. A., J. E. Magnotti, A. A. Wright, and J. S. Katz. 2013. "The Evolution of Comparative Psychology." In *Charles Darwin: A Celebration of his Life and Legacy*, edited by J. Bradley and J. Lamar. 141–55. Montgomery, AL: NewSouth Books.

Schnitzler, A., and J. Hirschmann. 2012. "Magnetoencephalography and Neuromodulation." In *New Frontiers in Brain and Spine Stimulation*, edited by Clement Hamani and Elena Moro, 121–36. Vol. 107 of *Emerging Horizons in Neuromodulation: International Review of Neurobiology*. Amsterdam: Elsevier.

Searle, J. R. 1980. "Minds, Brains, and Programs." *Behavioral and Brain Sciences* 3(3): 417–57.

Sedgwick, Adam. (1845) 1890. Letter of Adam Sedgwick to Charles Lyell, April 9. In *The Life and Letters of the Rev. Adam Sedgwick*, 84. Vol. 2. Cambridge: Cambridge University Press.

Séralini, G. E., C. E. Mesnage, R. Gress, et al. 2012. "Long Term Toxity of a Roundup Herbicide and a Roundup-Tolerant Genetically Modified Maize." *Food and Chemical Toxicology* 50: 4221–31.

Service, R. F. 2002. "Biology Offers Nanotechs a Helping Hand." *Science* 298: 2322–23.

Sessoli, R. 2017. "Nanoscience: Single-Atom Data Storage." *Nature* 543: 189–90. http://www.nature.com/nature/journal/v543/n7644/full/543189a.html (accessed March 12, 2017).

Shanks, P. 2013. "An ELSI Program for the BRAIN Initiative?" *Biopolitical Times*, August 29. http://www.biopoliticaltimes.org/article.php?id=7120 (accessed September 11, 2015).

———. 2017. "Fear vs. Hope: Gene-Editing—Terrible Turning point?" *Deccan Chronicle*, January 1. http://www.deccanchronicle.com/sunday-chronicle/headliners/010117/2016-fear-vs-hope-gene-editing-terrible-turning-point.html (accessed February 26, 2017).

Shen, H. 2014. "First Monkeys with Customized Mutations Born." *Nature*, January 14. doi:10.1038/nature.2014.14611 (accessed February 20, 2017).

———. 2015. "Neuron Encyclopedia Fires Up to Reveal Brain Secrets." *Nature* 520: 13–14. http://www.nature.com/news/neuron-encyclopaedia-fires-up-to-reveal-brain-secrets-1.17232 (accessed April 27, 2015).

Shubin, N. 2008. *Your Inner Fish: A Journey into the 3.5-Billion-Year History of the Human Body*. New York: Pantheon Books.

Sibbald, B. 2001. "Death but One Unintended Consequence of Gene-Therapy Trial." *Canadian Medical Association Journal* 164: 1612. http://www.ncbi.nlm.nih.gov/pmc/articles/PMC81135/ (accessed June 26, 2016).

Singer, N. 2010. "Making Ads That Whisper to the Brain." *New York Times*, Business Day, November 13. http://www.nytimes.com/2010/11/14/business/14stream.html?_r=0 (accessed September 13, 2015).

Singer, P. 2005. "Ethics and Intuitions." *Journal of Ethics* 9: 331–52.

Smith, A., D. Parker, T. D. Satterthwaite, et al. 2014. "Sex Differences in the Structural Connectome of the Human Brain." *Proceedings of the National Academy of Sciences*. 111: 823–26.

Smith, Homer W. 1959. *From Fish to Philosopher*. Summit, NJ: CIBA Pharmaceutical Products. Biodiversity Heritage Library. https://archive.org/details/fromfishtophilos00smit (accessed January 13, 2015).

Snow, C. P. (1959) 2001. *The Two Cultures*. London: Cambridge University Press.

SoftBank Robotics Corp. N.d. "Robots: Who Is Pepper?" https://www.ald.softbankrobotics.com/en/cool-robots/pepper (accessed June 10, 2017).

Song, H., Z. Zou, J. Kou, et al. 2015. "Love-Related Changes in the Brain: A Resting-State Functional Magnetic Resonance Imaging Study." *Frontiers in Human Neuroscience* 9, article 71. http://journal.frontiersin.org/article/10.3389/fnhum.2015.00071/full (accessed September 16, 2015).

Sparrow, R. 2004. "The Turing Triage Test." *Ethics and Information Technology* 6 (4): 203–13.

———. 2012. "Can Machines Be People? Reflections on the Turing Triage Test." In *Robot Ethics: The Ethical and Social Implications of Robotics*, edited by P. Lin, K. Abney, and G. A. Bekey, 301–15. Cambridge: Massachusetts Institute of Technology Press.

Specter, M. 2016. "DNA Revolution." *National Geographic*, August, 30–59. Published online as "How the DNA Revolution Is Changing Us." http://www.nationalgeographic.com/magazine/2016/08/dna-crispr-gene-editing-science-ethics/ (accessed July 31, 2016).

Sreenivasan, G. 1995. *The Limits of Lockean Rights in Property*. Oxford, UK: Oxford University Press.

Staropoli, N. 2016. "Gene Edited Hornless Cow Improve Animal Welfare but Regulatory Fate Unclear." Genetic Literacy Project. https://www.geneticliteracyproject.org/2016/05/11/gene-edited-hornless-cow-improve-animal-welfare-regulatory-fate-unclear/ (accessed March 1, 2017).

Stein, Z., B. della Chiesa, C. Hinton, and K. W. Fischer. 2011. "Ethical Issues in Educational Neuroscience: Raising Children in a Brave New World." In Illes and Sahakian 2011, 803–22.

Stock, G. 2002. *Redesigning Humans: Our Inevitable Genetic Future*. New York: Houghton Mifflin.

Stringer, C., and J. Galaway-Witham. 2017. "On the Origin of Our Species." *Nature* 546: 212–3.

Synlogic News Release. 2018. "Synlogic Presents Clinical and Preclinical Data from Synthetic Biotic™ Medicine Programs for Treatment of Inborn Errors of Metabolism at Annual Meeting of The Society for Inherited Metabolic Disorders." https://investor.synlogictx.com/news-releases/news-release-details/synlogic-presents-clinical-and-preclinical-data-synthetic (accessed October 19, 2018).

Szent-Györgyi, Albert. 1972. *The Living State*. New York: Academic Press.

Szilard, L. 1945. A Petition to the President of the United States. https://www.trumanlibrary.org/whistlestop/study_collections/bomb/large/documents/pdfs/79.pdf (accessed June 16, 2018).

Tachibana, M., P. Amato, M. Sparman, et al. 2013. "Human Embryonic Stem Cells Derived by Somatic Cell Nuclear Transfer." *Cell* 153: 1228–38.

Taylor, J. P., R. H. Brown Jr., and D. W. Cleveland. 2016. "Decoding ALS: From Genes to Mechanism." *Nature* 539: 197–206. http://www.nature.com/nature/journal/v539/n7628/full/nature20413.html (accessed June 4, 2017).

Theil, S. 2015. "Trouble in Mind." *Scientific American* 313: 36–39, 42.

Thomas, L. 1983. "Science and Science." In: *Late Night Thoughts While Listening to Mahler's Ninth Symphony*. New York: Viking Press.

Thomas, N., and K. F. Atkinson. 2005. "Prof Develops Cancer Nanobomb." *UDaily* (University of Delaware), October 13. http://www1.udel.edu/PR/UDaily/2005/mar/nanobomb101305.html (accessed March 10, 2017).

Tobe, F. 2015. "Rethink Robotics Finally Get Their Mojo with Sawyer." Robohub, September 28. http://robohub.org/rethink-robotics-finally-get-their-mojo-with-sawyer/ (accessed November 8, 2015).

Tovino, S. A. 2011. "Women's Neuroethics." In Illes and Sahakian 2011, 701–14.

Trafton, A. 2011. "Illuminating the Brain." MIT News on Campus and around the World, January 28. http://newsoffice.mit.edu/2011/illuminating-brain-0128 (accessed January 23, 2015).

Travis, J. 2015. "Making the Cut: CRISPR Genome-Editing Shows Its Power." *Science* 350: 1456–57.

Trimper, J., P. R. Wolpe, and K. S. Rommelfanger. 2014. "When 'I' Becomes 'We': Ethical Implications of Emerging Brain-to-Brain Interfacing Technologies." *Frontiers in Neuroengineering* 7 (Article 4): 1–4. http://www.ncbi.nlm.nih.gov/pmc/articles/PMC3921579/ (accessed October 2, 2015).

Turing, A. 1950. "Computing Machinery and Intelligence." *Mind* 59:433–460.

UCLA Center for Mental Health in Schools. 2014. Information Resource: Arguments about Whether Overdiagnosis of ADHD Is a Significant Problem.

http://smhp.psych.ucla.edu/pdfdocs/overdiag.pdf (accessed September 28, 2015).

Underwood, E. 2015. "Brain Implant Trials Raise Ethical Concerns." *Science* 348: 1186–87.

United Nations. 1989. Convention on the Rights of the Child. New York. http://www.ohchr.org/en/professionalinterest/pages/crc.aspx (accessed September 28, 2015).

US Department of Health and Human Services. 2009. "Attachment: Recommendations Regarding Research Involving Individuals with Impaired Decision-Making." https://www.hhs.gov/ohrp/sachrp-committee/recommendations/2009-july-15-letter-attachment/index.html (accessed November 5, 2018).

———. 2014. Brain Initiative: Integrated Approaches to Understanding Circuit Function in the Nervous System (UO1). Part 2 Section 1, Objectives, November 5. https://grants.nih.gov/grants/guide/rfa-files/rfa-ns-15-005.html (accessed November 13, 2018).

US Supreme Court. 2013. Opinion of the Court No. 12–398. Association for Molecular Pathology, et al. v. Myriad Genetics, Inc. et al. http://www.supremecourt.gov/search.aspx?Search=Myriad+Genetics&type=Site (accessed May 31, 2016).

US White House. 2016. Modernizing the Regulatory System for Biotechnology Products: An Update to the Coordinated Framework for the Regulation of Biotechnology (Draft). https://www.whitehouse.gov/sites/default/files/micrositesostp/biotech_coordinated_framework.pdf (accessed October 11, 2016).

———. N.d. Office of Science and Technology Policy. https://www.whitehouse.gov/administration/eop/ostp (accessed October 11, 2016).

Vallor, S. 2016. *Technology and the Virtues: A Philosophical Guide to a Future Worth Wanting.* New York: Oxford University Press.

Van Eenennaam, A. L. 2013. "GMOs in Animal Agriculture: Time to Consider Both Costs and Benefits in Regulatory Evaluations." *Journal of Animal Science and Biotechnology* 4: 37. https://jasbsci.biomedcentral.com/articles/10.1186/2049-1891-4-37 (accessed March 9, 2017).

Veblen, Thorstein. (1914) 2017. *The Instinct of Workmanship and the State of the Arts.* New York: Routledge.

Velasquez-Manoff, M. 2017. "The Upside of Bad Genes." *New York Times*, Sunday Review, June 17. https://www.nytimes.com/2017/06/17/opinion/sunday/crispr-upside-of-bad-genes.html (accessed June 20, 2017).

Venere, E. 2003. "Purdue Engineers: Metal Nano-Bumps Could Improve Artificial Body Parts." *Purdue News*, November 3.

Wade, N. 2017. "New Prospects for Growing Human Replacement Organs in

Animals." *New York Times*, Science, January 26. https://www.nytimes.com /2017/01/26/science/chimera-stemcells-organs.html (accessed March 1, 2017).

Waldholz, M. 2017. "Transformers: 2017 Future of Medicine." *Scientific American* 316(4): 46–53.

Waldman, S. 2018. "Will Trump Name a Scientist? A Poli-Sci Grad Runs the Show." *E and E News*, February 14. https://www.eenews.net/stories/1060073811 (accessed October 31, 2018).

Wallach, W., and C. Allen. 2009. *Moral Machines: Teaching Robots Right from Wrong*. New York. Oxford University Press.

Wang, J., Y. Liu, F. Jiao, et al. 2008. "Time-Dependent Translocation and Potential Impairment on Central Nervous System by Intranasally Instilled TiO_2 Nanoparticles." *Toxicology* 254 (1–2): 82–90.

Whitby, Blay. 2012. "Do You Want a Robot Lover? The Ethics of Caring Technologies." In *Robot Ethics: The Ethical and Social Implications of Robotics*, edited by P. Lin, K. Abney, and G. A. Bekey, 233–48. Cambridge: Massachusetts Institute of Technology Press.

White, J. G., E. Southgate, J. N. Thomson, and S. Brenner. 1986. "The Structure of the Nervous System of the Nematode Caenorhabditis Elegans." *Philosophical Transactions of the Royal Society B* 314 (1165): 1–340.

"Why Researchers Should Resolve to Engage in 2017." 2017. Editorial. *Nature* 541: 5. http://www.nature.com/news/why-researchers-should-resolve-to -engage-in-2017-1.21236 (accessed January 12, 2017).

Wilson, E. O. 1984. *Biophilia*. Cambridge, MA: Harvard University Press.

———. 2006. *The Creation: An Appeal to Save Life on Earth*. New York: W. W. Norton.

———. 2014. *The Meaning of Human Existence*. New York: W. W. Norton.

Winkler, A. M., P. Kochunov, J. Blangero, et al. 2010. "Cortical Thickness or Grey Matter Volume? The Importance of Selecting the Phenotype for Imaging Genetics Studies." *Neuroimage* 53 (3): 1135–46. http://www.ncbi.nlm .nih.gov/pmc/articles/PMC2891595/ (accessed January 19, 2015).

Woopen, C. 2012. "Ethical Aspects of Neuromodulation." In *New Frontiers in Brain and Spine Stimulation*, edited by Clement Hamani and Elena Moro, 315–32. Vol. 107 of *Emerging Horizons in Neuromodulation: International Review of Neurobiology*. Amsterdam: Elsevier.

Xia, X., and S.-C. Zhang. 2009. "Differentiation of Neuroepithelia from Human Embryonic Stem Cells." *Methods in Molecular Biology* 549: 51–58.

Yamada, M., B Johannesson, I Sagi, et al. 2014. "Human Oocytes Reprogram Adult Somatic Nuclei of a Type 1 Diabetic to Diploid Pluripotent Stem Cells." *Nature* 510: 533–36.

Yang, Y., L. Wang, P. Bell, et al. 2016. "A Dual AAV System Enables the Cas9-

Mediated Correction of a Metabolic Liver Disease in Newborn Mice. *Nature Biotechnology* 34: 334–38. http://www.ncbi.nlm.nih.gov/pubmed/26829317 (accessed June 27, 2016).

Zimmer, C. 2015. "Breakthrough DNA Editor Born of Bacteria." *Quanta Magazine*, February 6. https://d2r55xnwy6nx47.cloudfront.net/uploads/2015/02/crispr-natural-history-in-bacteria-20150206.pdf (accessed October 22, 2018).

Zuckerman, S. 1982. *Nuclear Illusion and Reality*. New York: Viking Press.

Index

Page numbers in italics refer to figures and tables.